Lecture Notes in Mathematics 1846

Editors:
J.-M. Morel, Cachan
F. Takens, Groningen
B. Teissier, Paris

Habib Ammari
Hyeonbae Kang

Reconstruction of Small Inhomogeneities from Boundary Measurements

Authors

Habib Ammari

Centre de Mathématiques Appliquées
École Polytechnique
UMR CNRS - 7641
Palaiseau 91128
France
e-mail: ammari@cmapx.polytechnique.fr

Hyeonbae Kang
School of Mathematical Sciences
Seoul National University
Seoul 151-747
Korea
e-mail: hkang@math.snu.ac.kr

Library of Congress Control Number: 2004108535

Mathematics Subject Classification (2000): 35R30, 35B30

ISSN 0075-8434
ISBN 3-540-22483-1 Springer Berlin Heidelberg New York
DOI: 10.1007/b98245

Springer is a part of Springer Science + Business Media

http://www.springeronline.com

Typesetting: Camera-ready TEX output by the authors

Printed on acid-free paper
41/3141/du - 5 4 3 2 1 0

Preface

Electrical impedance tomography (EIT) seeks to recover the electrical conductivity distribution inside a body from measurements of current flows and voltages on its surface. The vast and growing literature reflects many possible applications of EIT techniques, *e.g.*, for medical diagnosis or nondestructive evaluation of materials.

Since the underlying inverse problem is nonlinear and severely ill-posed, general purpose EIT reconstruction techniques are likely to fail. Therefore it is generally advisable to incorporate a-priori knowledge about the unknown conductivity. One such type of knowledge could be that the body consists of a smooth background containing a number of unknown small inclusions with a significantly different conductivity. This situation arises for example in breast cancer imaging or mine detection. In this case EIT seeks to recover the unknown inclusions. Due to the smallness of the inclusions the associated voltage potentials measured on the surface of the body are very close to the potentials corresponding to the medium without inclusion. So unless one knows exactly what patterns to look for, noise will largely dominate the information contained in the measured data. Furthermore, in applications it is often not necessary to reconstruct the precise values of the conductivity or geometry of the inclusions. The information of real interest is their positions and size.

The main purpose of this book is to describe fresh and promising techniques for the reconstruction of small inclusions from boundary measurements in a readable and informative form. These techniques rely on accurate asymptotic expansions of the boundary perturbations due to the presence of the inclusions. The general approach we will take to derive these asymptotic expansions is based on layer potential techniques. This allows us to handle inclusions with rough boundaries. In the course of deriving our asymptotic expansions, we introduce new concepts of generalized polarization tensors (GPT's). GPT's contain significant information on the inclusion which will be investigated. We then apply the asymptotic expansions for designing efficient direct reconstruction algorithms to detect the location, size, and/or orientation of the unknown inclusions.

This book would not have been possible without the collaborations and the conversations with a number of outstanding colleagues. We have not only profited from generous sharing of their ideas, insights and enthusiasm, but also from their friendship, support and encouragement. We feel specially indebted to Gen Nakamura, Jin Keun Seo, Gunther Uhlmann and Michael Vogelius. This book is dedicated to our friendship with them.

We also would like to thank our colleagues Mark Asch, Yves Capdeboscq, and Darko Volkov for reading parts of the preliminary version of the manuscript and our students Ekaterina Iakovleva, Eunjoo Kim, Mikyoung Lim, Kaouther Louati, Sofiane Soussi, Karim Touibi, and Habib Zribi for providing us with numerical examples or/and carefully reading our manuscript.

During the preparation of this book we were supported by ACI Jeunes Chercheurs (0693) from the French Ministry of Education and Scientific Research and Korea Science and Engineering Foundation through the grant R02-2003-000-10012-0.

Paris and Seoul,   *Habib Ammari*
May 2004   *Hyeonbae Kang*

Contents

Part II Detection of Small Elastic Inclusions

Part III Detection of Small Electromagnetic Inclusions

1

Introduction

Electrical Impedance Tomography (EIT) is designed to produce, efficiently and accurately, images of the conductivity distribution inside a body from measurements of current flows and voltages on the body's surface. Due to several merits of EIT such as safety, low cost, real time monitoring, EIT has received considerable attention for the last two decades; see for instance the review papers [255, 256, 78, 58] and the extensive list of references therein. The vast and growing literature reflects many possible applications of EIT, *e.g.*, for medical diagnosis or nondestructive evaluation of materials. In medical applications, EIT could potentially be used for monitoring lung problems, noninvasive monitoring of heart function and blood flow, screening for breast and prostate cancer, and improving electrocardiograms and electroencephalograms [78, 37, 82, 83, 163, 164, 143].

However, insensitivity of boundary measurements to any change of inner-body conductivity values has hampered EIT from providing accurate static conductivity images [4]. In practice captured current-to-voltage pairs must be limited by the number of electrodes attached on the surface of the body that confine the resolution of the image [153, 101]. We can definitely increase the resolution of the conductivity image by increasing the number of electrodes. However, it should be noticed that beyond a certain level, increasing the number of electrodes may not help in producing a better image for the inner-region of the body if we take into account the inevitable noise in measurements and the inherent insensitivity mentioned before.

In its most general form EIT is severely ill-posed and nonlinear. These major and fundamental difficulties can be understood by means of a mean value type theorem in elliptic partial differential equations. The value of the voltage potential at each point inside the region can be expressed as a weighted average of its neighborhood potential where the weight is determined by the conductivity distribution. In this weighted averaging way, the conductivity distribution is conveyed to the boundary potential. Therefore, the boundary data is entangled in the global structure of the conductivity distribution in a highly nonlinear way. This is the main obstacle to finding non-iterative reconstruction algorithms with limited data. If, however, in advance we have additional structural information about the medium, then we may be able to determine specific features about the conductivity distribution with a good resolution. One such type of knowledge could be that the body consists of a smooth background containing a number of unknown small inclusions with a significantly different conductivity. This situation arises for example in breast and prostate cancer imaging [37, 82, 83, 163, 164, 26, 243] or mine detection. In this case EIT seeks to recover the unknown inclusions. Due to the smallness of the inclusions the associated voltage potentials measured on the surface of the body are very close to the potentials corresponding to the medium without inclusions. So unless one knows exactly what patterns to look for, noise will largely dominate the information contained in the measured data. Furthermore, in applications it is often not necessary to reconstruct the precise values of the conductivity or the geometry of the inclusions. The information of real interest is their positions and size.

Taking advantage of the smallness of the inclusions, many promising reconstruction techniques have been designed since the pioneering works by Friedman and Vogelius [120, 121, 123]. It turns out that the method of asymptotic expansions of small volume inclusions provides a useful framework to reconstruct the location and geometric features of the inclusions in a stable way, even for moderately noisy data [73, 53, 28, 64, 191, 39, 261].

In this book we have made an attempt to describe these new and promising techniques for the reconstruction of small inclusions from boundary measurements in a readable and informative form. As we said, these techniques rely on accurate asymptotic expansions of the boundary perturbations due to the presence of the inclusions. The general approach we will take in this book is as follows. Based on layer potential techniques and decomposition formulae like the one due to Kang and Seo in [168] for the conductivity problem, we first derive complete asymptotic expansions. This allows us to handle inclusions with rough boundaries and those with extreme conductivities. In the course of deriving our asymptotic expansions, we introduce new concepts of generalized polarization tensors (GPT's). These concepts generalize those of classical Pólya–Szegö polarization tensors which have been extensively studied in the literature by many authors for various purposes [72, 24, 73, 212, 104, 105, 200, 198, 123, 180, 94, 186, 231, 232, 241, 95]. The GPT's appear naturally in higher-order asymptotics of the steady-state voltage potentials under the perturbation of conductor by conductivity inclusions of small diameter. GPT's contain significant information on the inclusion that will be investigated. We then apply the asymptotic expansions for designing efficient direct reconstruction algorithms to detect the location, size, and/or orientation of the unknown inclusions.

The book is intended to be self-contained. However, a certain familiarity with layer potential techniques is required. The book is divided into three parts that can be read independently.

Part I consists of four chapters dealing with the conductivity problem. It is organized as follows. In Chap. 2, we introduce the main tools for studying the conductivity problem and collect some notation and preliminary results regarding layer potentials. In Chap. 3, we introduce the GPT's associated with a Lipschitz bounded domain B and a conductivity $0 < k \neq 1 < +\infty$. We prove that the knowledge of all the GPT's uniquely determines the domain B and the conductivity k. We also provide important symmetric properties and positivity of the GPT's and derive isoperimetric inequalities satisfied by the tensor elements of the GPT's. These relations can be used to find bounds on the weighted volume. In Chap. 4, we provide a rigorous derivation of high-order terms in the asymptotic expansion of the output voltage potentials. The proofs of our asymptotic expansions are radically different from the ones in [123, 73, 259]. What makes the proofs particularly original and elegant is that the rigorous derivation of high-order terms follows almost immediately. In Chap. 5, we apply our accurate asymptotic formula for the purpose of identifying the location and certain properties of the shape of the conductivity inclusions. By improving the algorithm of Kwon, Seo, and Yoon [191] we first design two real-time algorithms with good resolution and accuracy. We then describe the variational algorithm introduced in [28] and review the interesting approach proposed by Brühl, Hanke, and Vogelius [64]. Their method is in the spirit of the linear sampling method of Colton and Kirsch [89].

In Part II we develop a method to detect the size and the location of an inclusion in a homogeneous elastic body in a mathematically rigorous way. Inclusions of small size are believed to be the starting point of crack development in elastic bodies. In Chap. 5, we review some basic facts on the layer potentials of the Lamé system. In Chap. 6, we give in a way analogous to GPT's, mathematical definitions of elastic moment tensors (EMT's) and show symmetry and positive-definiteness of the first-order EMT. The first-order EMT was introduced by Maz'ya and Nazarov [202]. In Chap. 7, we find a complete asymptotic formula of solutions of the linear elastic system in terms of the size of the inclusion. The method of derivation is parallel to that in Part I apart from some technical difficulties due to the fact that we are dealing with a system, not a single equation, and the equations inside and outside the inclusion are different. Based on this asymptotic expansion we derive in Chap. 8 formulae to find the location and the order of magnitude of the elastic inclusion. The formulae are explicit and can be easily implemented numerically.

The problem we consider in Part III is to detect unknown dielectric inclusions by means of a finite number of voltage-to-current pairs measured on the boundary. We consider solutions to the Helmholtz equation in two and three dimensions. We begin by proving in Chap. 9 existence and uniqueness of a solution to the Helmholtz equation. The proof, due to Vogelius and Volkov [259], uses the theory of collectively compact operators. Based on layer potential techniques and two new decomposition formulae of the solution to the Helmholtz equation, established in Chap. 10, we provide in Chap. 11 for such solutions a rigorous systematic derivation of complete asymptotic expansions of perturbations resulting from the presence of diametrically small inclusions with constitutive parameters different from those of the background medium. The leading-order term in these asymptotic formulae has been derived by Vogelius and Volkov in [259]. We then develop in Chap. 12 two effective algorithms for reconstructing small dielectric inclusions from boundary measurements at a fixed frequency. The first algorithm, like the variational method in Chap. 5, reduces the reconstruction problem of the small inclusions to the calculation of an inverse Fourier transform. The second one is the MUSIC (standing for MUltiple-Signal-Classification) algorithm. We explain how it applies to imaging of small dielectric inclusions. Another algorithm based on projections on three planes was proposed and successfully tested by Volkov in [261]. Results similar to those presented in this part have been obtained in the context of the full (time-harmonic) Maxwell equations in [31].

Finally, it is important to note that some of the techniques described in this book can be applied to problems in many fields other than inverse boundary value problems. In this connection we would particularly like to mention the mathematical theory of composite materials [84, 206, 186, 211, 24] and topological shape optimization [222, 196, 126, 132, 239].

Part I

Detection of Small Conductivity Inclusions

Let Ω be a bounded domain in $\mathbb{R}^d$, $d \geq 2$, with a connected Lipschitz boundary $\partial\Omega$. Let ν denote the unit outward normal to $\partial\Omega$. Suppose that Ω contains a finite number m of small inclusions D_s, $s = 1, \ldots, m$, each of the form $D_s = \epsilon B_s + z_s$, where B_s, $s = 1, \ldots, m$, is a bounded Lipschitz domain in $\mathbb{R}^d$ containing the origin. We assume that the domains D_s, $s = 1, \ldots, m$, are separated from each other and from the boundary. More precisely, we assume that there exists a constant $c_0 > 0$ such that

$$|z_s - z_{s'}| \geq 2c_0 > 0 \quad \forall\, s \neq s' \quad \text{and} \quad \text{dist}(z_s, \partial\Omega) \geq 2c_0 > 0 \quad \forall\, s\,,$$

that ϵ, the common order of magnitude of the diameters of the inclusions, is sufficiently small and that these inclusions are disjoint. We also assume that the background is homogeneous with conductivity 1 and the inclusion D_s has conductivity k_s, $0 < k_s \neq 1 < +\infty$, for $1 \leq s \leq m$.

Let u denote the steady-state voltage potential in the presence of the conductivity inclusions $\bigcup_{s=1}^m D_s$, i.e., the solution in $W^{1,2}(\Omega)$ to

$$\begin{cases} \nabla \cdot \left(\chi\left(\Omega \setminus \bigcup_{s=1}^m \overline{D_s} \right) + \sum_{s=1}^m k_s \chi(D_s) \right) \nabla u = 0 \quad \text{in } \Omega\,, \\ \dfrac{\partial u}{\partial \nu}\bigg|_{\partial\Omega} = g\,. \end{cases}$$

Let U denote the "background" potential, that is, the solution to

$$\begin{cases} \Delta U = 0 \quad \text{in } \Omega\,, \\ \dfrac{\partial U}{\partial \nu}\bigg|_{\partial\Omega} = g\,. \end{cases}$$

The function g represents the applied boundary current; it belongs to $L^2(\partial\Omega)$ and has mean value zero. The potentials, u and U, are normalized by

$$\int_{\partial\Omega} u\,d\sigma = \int_{\partial\Omega} U\,d\sigma = 0\;.$$

The problem we consider in this part is to determine unknown inclusions D_s, $s = 1,\ldots,m$, by means of one or a finite number of current-to-voltage pairs $(g, u|_{\partial\Omega})$ measured on $\partial\Omega$.

This problem is called the inverse conductivity problem with one or finite boundary measurements (or Electrical Impedance Tomography) in contrast with the many measurements problem (or Calderón's problem) where an infinite number of boundary measurements are used. In many applied situations, it is the potential u that is prescribed and the current g that is measured on $\partial\Omega$. This makes some difference (not significant theoretically and computationally) in the case of finite boundary measurements but makes almost no difference in the case of many boundary measurements, since actually it is the set of Cauchy data $(g, u|_{\partial\Omega})$ that is given.

For the many measurements problem there is a well-established theory. We refer to the survey papers of Sylvester and Uhlmann [249], and of Uhlmann [255, 256], as well as to the book of Isakov [158], since this problem is out of the scope of our monograph. When $d \geq 2$, many boundary measurements provide much more information about the conductivity of Ω than a finite number of measurements. Thus, the inverse conductivity problem with finite measurements is more difficult than the one with many boundary measurements and not much was known about it until recently. Fortunately, there has been over the last few years a considerable amount of interesting work and new techniques dedicated to both theoretical and numerical aspects of this problem. It is the purpose of this part to describe some of these fresh and promising techniques, in particular, those for the reconstruction of diametrically small inclusions.

Let us very briefly emphasize our general methodology for solving our inverse conductivity problem (with finite measurements). We first derive an asymptotic expansion of the boundary voltage difference $u - U$ to any order in ϵ. Then we apply this very explicit asymptotic behavior to the effective estimation of the location and some geometric features of the set of conductivity inclusions $\bigcup_{s=1}^{m} D_s$. To present these results we shall need a decomposition formula of u into a harmonic part and a refraction part, the Neumann function associated with the background conductor Ω, and the generalized polarization tensors (GPT's) associated with the scaled domains B_s and their conductivities k_s. The GPT's are in fact the basic building blocks for our full asymptotic expansion of $u - U$ on $\partial\Omega$ and contain significant information on the domains B_s and their conductivities k_s. Then it is important to precisely characterize these GPT's and derive their basic properties.

The problem we consider here occurs in many practical situations. The inclusions $\bigcup_{s=1}^{m} D_s$ might in a medical application represent potential tumors, in a material science application they might represent impurities, and finally in a war or post-war situation they could represent anti-personnel mines.

In medical applications, EIT is supported by the experimental evidence that different biological tissues have different electrical properties that change with cell concentration, cellular structure, molecular composition, and so on [165, 245]. Therefore, these properties manifest structural, functional, metabolic, and pathological conditions of the tissue providing valuable diagnostic information.

We conclude this introduction with a discussion of classical image reconstruction algorithms in EIT.

The most classical technique consists of a minimization approach. We assume an initial conductivity distribution for the model and iteratively update it until it minimizes the difference between measured and computed boundary voltages. This kind of method was first introduced in EIT by Yorkey, Webster, and Tompkins [266] following numerous variations and improvements. These include utilization of *a priori* information, various forms of regularization, and so on [264, 145, 257, 86]. Even though this approach is widely adopted for imaging by many researchers, it requires a large amount of computation time for producing images even with low spatial resolution and poor accuracy.

In the 1980's, Barber and Brown [45] introduced the back-projection algorithm for EIT that was the first fast and useful algorithm although it provides images with very low resolution. Since this algorithm is inspired from the computed tomography (CT) algorithm, it can be viewed as a generalized Radon Transform [240].

The third technique is the dynamical electrical impedance imaging. This interesting and sophisticated technique, developed by the Rensselaer impedance tomography group [78, 80, 79, 213, 244, 246, 108, 129, 124, 154, 153], is designed to produce images of a change of conductivity in the human body for purpose of applications in cardiac and respiratory imaging. The main idea is to decompose the conductivity into a static term, viewed as the background conductivity of human body, and a perturbing term, considered as the change of conductivity due to respiratory or heart function. The mathematical problem here is to visualize the perturbing term by an EIT system. Although this algorithm can provide accurate images when an initial guess of the background conductivity is reasonably good, it seems that new ideas are still needed to obtain good resolution images and completely satisfy practitioners, specially in screening for breast cancer.

Our main aim in this part is to propose a new mathematical direction of future EIT research mainly for biomedical applications. A new electronic system based on the mathematical modeling described in this book is being developed for breast cancer imaging at the Impedance Imaging Research Center by Jin Keun Seo and his group [26, 243, 192].

2

Transmission Problem

In this chapter we review some well-known results on the solvability and layer potentials for the conductivity problem, which we shall use frequently in subsequent chapters, and prove a decomposition formula of the steady-state voltage potential into a harmonic part and a refraction part. Our main aim here is to collect together the various concepts, basic definitions and key theorems on layer potentials on Lipschitz domains, with which the reader might not be familiar. This chapter gives a concise treatment of this subject where the technicalities are kept to a minimum.

2.1 Some Notations and Preliminaries

We begin with the concept of a Lipschitz domain. A bounded open connected domain D in $\mathbb{R}^d$ is called a Lipschitz domain with Lipschitz character (r_0, L, N) if for each point $x \in \partial D$ there is a coordinate system $(x', x_d), x' \in \mathbb{R}^{d-1}, x_d \in \mathbb{R}$, so that with respect to this coordinate system $x = (0, 0)$, and there are a double truncated cylinder Z (called a coordinate cylinder) centered at x with axis parallel to the x_d-axis and whose bottom and top are at a positive distance $r_0 < l < 2r_0$ from ∂D, and a Lipschitz function φ with $\|\nabla \varphi\|_{L^\infty(\mathbb{R}^{d-1})} \leq L$, so that $Z \cap D = Z \cap \{(x', x_d) : x_d > \varphi(x')\}$ and $Z \cap \partial D = Z \cap \{(x', x_d) : x_d = \varphi(x')\}$. The pair (Z, φ) is called a coordinate pair. By compactness it is possible to cover ∂D with a finite number of coordinate cylinders $Z_1, \ldots, Z_N$. Bounded Lipschitz domains satisfy both the interior and exterior cone conditions.

We say that $f \in W_1^2(\partial D)$ if $f \in L^2(\partial D)$ and for every cylinder Z as in the above definition with associated Lipschitz function φ, there are $L^2(\partial D \cap Z)$ functions g_p, $1 \leq p \leq d - 1$, such that

$$\int_{\mathbb{R}^{d-1}} h(x') g_p(x', \varphi(x')) \, dx' = -\int_{\mathbb{R}^{d-1}} \frac{\partial}{\partial x_p} h(x') f(x', \varphi(x')) \, dx'$$

for $1 \le p \le d - 1$, whenever $h \in C_0^\infty(\mathbb{R}^{d-1} \cap Z)$. Fixing a covering of ∂D by cylinders $Z_1, \ldots, Z_N$, $f \in W_1^2(\partial D)$ may be normed by the sum of L^2 norms of all the locally defined g_p together with the L^2 norm of f.

We define the Banach spaces $W^{1,p}(D), 1 < p < +\infty$, for an open set D by

$$W^{1,p}(D) = \left\{ f \in L^p(D) : \int_D |f|^p + \int_D |\nabla f|^p < +\infty \right\}.$$

A norm is introduced by defining

$$\|f\|_{W^{1,p}(D)} = \left(\int_D |f|^p + \int_D |\nabla f|^p \right)^{1/p}.$$

Another Banach space $W_0^{1,p}(D)$ arises by taking the closure of $C_0^\infty(D)$, the set of infinitely differentiable functions with compact support in D, in $W^{1,p}(D)$. The spaces $W^{1,p}(D)$ and $W_0^{1,p}(D)$ do not coincide for bounded D. The case $p = 2$ is special, since the spaces $W^{1,2}(D)$ and $W_0^{1,2}(D)$ are Hilbert spaces under the scalar product

$$(u, v) = \int_D u\,v + \int_D \nabla u \cdot \nabla v.$$

If D is a bounded Lipschitz domain, we will also need the space $W_{\text{loc}}^{1,2}(\mathbb{R}^d \setminus \overline{D})$ of functions $f \in L_{\text{loc}}^2(\mathbb{R}^d \setminus \overline{D})$ such that

$$hf \in W^{1,2}(\mathbb{R}^d \setminus \overline{D}), \forall\, h \in C_0^\infty(\mathbb{R}^d \setminus \overline{D}).$$

Further, we define $W^{2,2}(D)$ as the space of functions $f \in W^{1,2}(D)$ such that $\partial^2 f \in L^2(D)$ and the space $W^{3/2,2}(D)$ as the interpolation space $[W^{1,2}(D), W^{2,2}(D)]_{1/2}$. See, for example, the book of Bergh and Löfström [55].

We recall that if D is a bounded Lipschitz domain then the Poincaré inequality [128]

$$\int_D |u(x) - u_0|^2 \, dx \le C \int_D |\nabla u(x)|^2 \, dx,$$

holds for all $u \in W^{1,2}(D)$, where

$$u_0 = \frac{1}{|D|} \int_D u(x) \, dx.$$

It is known that the trace operator $u \mapsto u|_{\partial D}$ is a bounded linear surjective operator from $W^{1,2}(D)$ into $W_{\frac{1}{2}}^2(\partial D)$, where $f \in W_{\frac{1}{2}}^2(\partial D)$ if and only if $f \in L^2(\partial D)$ and

$$\int_{\partial D} \int_{\partial D} \frac{|f(x) - f(y)|^2}{|x - y|^d} \, d\sigma(x) \, d\sigma(y) < +\infty.$$

See [128] . Let $W^2_{-\frac{1}{2}}(\partial D) = (W^2_{\frac{1}{2}}(\partial D))^*$ and let $\langle , \rangle_{\frac{1}{2},-\frac{1}{2}}$ denote the duality pair between these dual spaces.

Let $T_1, \ldots, T_{d-1}$ be an orthonormal basis for the tangent plane to ∂D at x and let $\partial/\partial T = \sum_{p=1}^{d-1} \partial/\partial T_p \, T_p$ denote the tangential derivative on ∂D. The space $W_1^2(\partial D)$ is then the set of functions $f \in L^2(\partial D)$ such that $\partial f/\partial T \in L^2(\partial D)$. We also recall that according to [208] and [93] (Theorem 1.10) if D is a bounded Lipschitz domain (with connected boundary) then the following Poincaré inequality on ∂D holds

$$\|f - f_0\|_{L^2(\partial D)} \leq C \left\| \frac{\partial f}{\partial T} \right\|_{L^2(\partial D)} , \tag{2.1}$$

for any $f \in W_1^2(\partial D)$, where $f_0 = \frac{1}{|\partial D|} \int_{\partial D} f \, d\sigma$. Here the constant C depends only on the Lipschitz character of D.

The following lemma from [18] is of use to us.

Lemma 2.1 *Let $f(x) = \sum_{|\alpha| \leq n} a_\alpha x^\alpha$ be a harmonic polynomial and D be a bounded Lipschitz domain in $\mathbb{R}^d$. There is a constant C depending only on the Lipschitz character of D and n such that*

$$\|\nabla f\|_{L^2(\partial D)} \leq C \|\nabla f\|_{L^2(D)} . \tag{2.2}$$

Let us now turn to the concept of variational solutions. Let $(a_{pq})_{p,q=1}^d$ be a real symmetric $d \times d$ matrix with $a_{pq}(x) \in L^\infty(\mathbb{R}^d)$. We assume that $(a_{pq})_{p,q=1}^d$ is strongly elliptic, *i.e.*,

$$\frac{1}{C}|\xi|^2 \leq \sum_{p,q} a_{pq}(x)\xi_p\xi_q \leq C|\xi|^2$$

for all $\xi = (\xi_p)_{p=1}^d \in \mathbb{R}^d \setminus \{0\}$, where C is a positive constant. Let Ω be a bounded Lipschitz domain in $\mathbb{R}^d$. Given $g \in W^2_{-\frac{1}{2}}(\partial\Omega)$, with $\langle 1, g \rangle_{\frac{1}{2},-\frac{1}{2}} = 0$, we say that $u \in W^{1,2}(\Omega)$ is the (variational) solution to the Neumann problem

$$\begin{cases} \displaystyle\sum_{p,q=1}^d \frac{\partial}{\partial x_p} a_{pq} \frac{\partial}{\partial x_q} u = 0 & \text{in } \Omega , \\[2mm] \displaystyle\frac{\partial u}{\partial \widetilde{\nu}}\bigg|_{\partial\Omega} = g , \end{cases} \tag{2.3}$$

where $\widetilde{\nu}_p = \sum_q a_{pq}\nu_q$, if given any $\eta \in W^{1,2}(\Omega)$ we have

$$\int_\Omega \sum_{p,q=1}^d a_{pq} \frac{\partial u}{\partial x_p} \frac{\partial \eta}{\partial x_q} \, dx = \langle \eta, g \rangle_{\frac{1}{2},-\frac{1}{2}} .$$

The Lax-Milgram lemma shows that there exists a unique (modulo constants) $u \in W^{1,2}(\Omega)$ which solves (2.3).

2.2 Layer Potentials for the Laplacian

Let us first review some well-known properties of the layer potentials for the Laplacian and prove some useful identities. The theory of layer potentials has been developed in relation to boundary value problems in a Lipschitz domain.

To give a fundamental solution to the Laplacian with a general d, we denote the area of the $(d-1)$-dimensional unit sphere by ω_d. Even though the following result is elementary we give its proof for the reader's convenience.

Lemma 2.2 *A fundamental solution to the Laplacian is given by*

$$\Gamma(x) = \begin{cases} \dfrac{1}{2\pi} \ln|x| \,, & d = 2 \,, \\[2mm] \dfrac{1}{(2-d)\omega_d} |x|^{2-d} \,, & d \geq 3 \,. \end{cases} \tag{2.4}$$

Proof. The Laplacian is radially symmetric, so it is natural to seek Γ in the form $\Gamma(x) = w(r)$ where $r = |x|$. Since

$$\Delta w = \frac{d^2 w}{d^2 r} + \frac{(d-1)}{r} \frac{dw}{dr} = \frac{1}{r^{d-1}} \frac{d}{dr} \big(r^{d-1} \frac{dw}{dr} \big) \,,$$

$\Delta \Gamma = 0$ in $\mathbb{R}^d \setminus \{0\}$ shows that w must satisfy

$$\frac{1}{r^{d-1}} \frac{d}{dr} \big(r^{d-1} \frac{dw}{dr} \big) = 0 \quad \text{for } r > 0 \,,$$

so

$$w(r) \doteq \begin{cases} \dfrac{a_d}{(2-d)} \dfrac{1}{r^{d-2}} + b_d & \text{when } d \geq 3 \,, \\[2mm] a_2 \log r + b_2 & \text{when } d = 2 \,, \end{cases}$$

for some constants a_d and b_d. The choice of b_d is arbitrary, but a_d is fixed by the requirement that $\Delta \Gamma = \delta_0$ in $\mathbb{R}^d$, where δ_0 is the Dirac function at 0, or in other words

$$\int_{\mathbb{R}^d} \Gamma \Delta \phi = \phi(0) \quad \text{for } \phi \in \mathcal{C}_0^\infty(\mathbb{R}^d) \,. \tag{2.5}$$

Any test function $\phi \in \mathcal{C}_0^\infty(\mathbb{R}^d)$ has compact support, so we can apply Green's formula over the unbounded domain $\{x : |x| > \epsilon\}$ to arrive at

$$\int_{|x|>\epsilon} \Gamma(x)\Delta\phi(x)\, dx = \int_{|x|=\epsilon} \phi(x)\frac{\partial\Gamma}{\partial\nu}(x)\, d\sigma(x) - \int_{|x|=\epsilon} \Gamma(x)\frac{\partial\phi}{\partial\nu}(x)\, d\sigma(x) \,, \tag{2.6}$$

where $\nu = x/|x|$ on $\{|x| = \epsilon\}$. Since

$$\nabla\Gamma(x) = \frac{dw}{dr}\frac{x}{|x|} = \frac{a_d x}{|x|^d} \quad \text{for } d \geq 2 \,,$$

we have

$$\frac{\partial \Gamma}{\partial \nu}(x) = a_d \epsilon^{1-d} \quad \text{for } |x| = \epsilon \,.$$

Thus by the mean-value theorem for integrals,

$$\int_{|x|=\epsilon} \phi(x) \frac{\partial \Gamma}{\partial \nu}(x) \, d\sigma(x) = \frac{a_d}{\epsilon^{d-1}} \int_{|x|=\epsilon} \phi(x) \, d\sigma(x) = a_d \omega_d \phi(x_\epsilon)$$

for some x_ϵ satisfying $|x_\epsilon| = \epsilon$, whereas

$$\int_{|x|=\epsilon} \Gamma(x) \frac{\partial \phi}{\partial \nu}(x) \, d\sigma(x) = \begin{cases} O(\epsilon) & \text{if } d \geq 3 \,, \\ O(\epsilon |\log \epsilon|) & \text{if } d = 2 \,. \end{cases}$$

Thus, if $a_d = 1/\omega_d$, then (2.5) follows from (2.6) after sending $\epsilon \to 0$. $\square$

Now we prove Green's identity.

Lemma 2.3 *Suppose D is a bounded Lipschitz domain in $\mathbb{R}^d, d \geq 2$, and let $u \in W^{1,2}(D)$ be a harmonic function. Then for any $x \in D$ there holds*

$$u(x) = \int_{\partial D} \left(u(y) \frac{\partial \Gamma}{\partial \nu_y}(x-y) - \frac{\partial u}{\partial \nu_y}(y) \Gamma(x-y) \right) d\sigma(y) \,. \tag{2.7}$$

Proof. For $x \in D$ let $B_\epsilon(x)$ be the ball of center x and radius ϵ. We apply Green's formula to u and $\Gamma(x - \cdot)$ in the domain $D \setminus \overline{B_\epsilon}$ for small ϵ and get

$$\int_{D \setminus B_\epsilon(x)} \left(\Gamma \Delta u - u \Delta \Gamma \right) dy = \int_{\partial D} \left(\Gamma \frac{\partial u}{\partial \nu} - u \frac{\partial \Gamma}{\partial \nu} \right) d\sigma(y)$$

$$- \int_{\partial B_\epsilon(x)} \left(\Gamma \frac{\partial u}{\partial \nu} - u \frac{\partial \Gamma}{\partial \nu} \right) d\sigma(y) \,.$$

Note $\Delta \Gamma = 0$ in $D \setminus B_\epsilon(x)$. Then we have

$$\int_{\partial D} \left(\Gamma \frac{\partial u}{\partial \nu} - u \frac{\partial \Gamma}{\partial \nu} \right) d\sigma(y) = \int_{\partial B_\epsilon(x)} \left(\Gamma \frac{\partial u}{\partial \nu} - u \frac{\partial \Gamma}{\partial \nu} \right) d\sigma(y) \,.$$

For $d \geq 3$, we get by definition of Γ

$$\int_{\partial B_\epsilon(x)} \Gamma \frac{\partial u}{\partial \nu} \, d\sigma(y) = \frac{1}{(2-d)\omega_d} \epsilon^{2-d} \int_{\partial B_\epsilon(x)} \frac{\partial u}{\partial \nu} \, d\sigma(y) = 0$$

and

$$\int_{\partial B_\epsilon(x)} u \frac{\partial \Gamma}{\partial \nu} \, d\sigma(y) = \frac{1}{\omega_d \epsilon^{d-1}} \int_{\partial B_\epsilon(x)} u \, d\sigma(y) = u(x) \,,$$

by the mean value property. We get the same conclusion for $d = 2$ in the same way. $\square$

Given a bounded Lipschitz domain D in $\mathbb{R}^d, d \geq 2$, we will denote the single and double layer potentials of a function $\phi \in L^2(\partial D)$ as $\mathcal{S}_D \phi$ and $\mathcal{D}_D \phi$, respectively, where

$$\mathcal{S}_D\phi(x) := \int_{\partial D} \Gamma(x-y)\phi(y)\,d\sigma(y)\,, \quad x \in \mathbb{R}^d, \tag{2.8}$$

$$\mathcal{D}_D\phi(x) := \int_{\partial D} \frac{\partial}{\partial \nu_y}\Gamma(x-y)\phi(y)\,d\sigma(y)\,, \quad x \in \mathbb{R}^d \setminus \partial D\,. \tag{2.9}$$

For a function u defined on $\mathbb{R}^d \setminus \partial D$, we denote

$$u\Big|_\pm (x) := \lim_{t \to 0^+} u(x \pm t\nu_x), \quad x \in \partial D\,,$$

and

$$\frac{\partial}{\partial \nu_x}u\Big|_\pm (x) := \lim_{t \to 0^+} \langle \nabla u(x \pm t\nu_x), \nu_x \rangle\,, \quad x \in \partial D\,,$$

if the limits exist. Here ν_x is the outward unit normal to ∂D at x, and $\langle,\rangle$ denotes the scalar product in $\mathbb{R}^d$. For ease of notation we will sometimes use the dot for the scalar product in $\mathbb{R}^d$.

We now state without proofs the jump relations obeyed by the double layer potential and by the normal derivative of the single layer potential. The boundedness of these operators is not clear in the Lipschitz domains, because of the critical singularities of the kernel and the fact that we are dealing with non-convolution type operators. The following theorem can be proved using the deep results of Coifman-McIntosh-Meyer [87] on the boundedness of the Cauchy integral on Lipschitz curves, which together with the method of rotations of Calderón [68] allows one to produce patterns of arguments like those found in [97] for $\mathcal{C}^1$-domains. Complete proofs can be found in [258] (for smooth domains, see [115, 119, 217]). The reader is referred to Appendix A.1 for a statement of the theorem of Coifman-McIntosh-Meyer.

Theorem 2.4 *Let D be a bounded Lipschitz domain in $\mathbb{R}^d$. For $\phi \in L^2(\partial D)$*

$$\mathcal{S}_D\phi\big|_+ (x) = \mathcal{S}_D\phi\big|_- (x) \quad a.e.\ x \in \partial D\,, \tag{2.10}$$

$$\frac{\partial}{\partial T}\mathcal{S}_D\phi\Big|_+ (x) = \frac{\partial}{\partial T}\mathcal{S}_D\phi\Big|_- (x) \quad a.e.\ x \in \partial D\,, \tag{2.11}$$

$$\frac{\partial}{\partial \nu}\mathcal{S}_D\phi\Big|_\pm (x) = \left(\pm\frac{1}{2}I + \mathcal{K}_D^*\right)\phi(x) \quad a.e.\ x \in \partial D\,, \tag{2.12}$$

$$(\mathcal{D}_D\phi)\big|_\pm = \left(\mp\frac{1}{2}I + \mathcal{K}_D\right)\phi(x) \quad a.e.\ x \in \partial D\,, \tag{2.13}$$

where $\mathcal{K}_D$ is defined by

$$\mathcal{K}_D\phi(x) = \frac{1}{\omega_d}p.v.\int_{\partial D}\frac{\langle y-x, \nu_y\rangle}{|x-y|^d}\phi(y)\,d\sigma(y) \tag{2.14}$$

and $\mathcal{K}_D^*$ is the L^2-adjoint of $\mathcal{K}_D$, i.e.,

$$\mathcal{K}_D^* \phi(x) = \frac{1}{\omega_d} \text{p.v.} \int_{\partial D} \frac{\langle x - y, \nu_x \rangle}{|x - y|^d} \phi(y) \, d\sigma(y) \, . \tag{2.15}$$

Here p.v. denotes the Cauchy principal value. The operators $\mathcal{K}_D$ and $\mathcal{K}_D^*$ are singular integral operators and bounded on $L^2(\partial D)$.

Observe that if D is a two dimensional disk with radius r, then, as was observed in [168],

$$\frac{\langle x - y, \nu_x \rangle}{|x - y|^2} = \frac{1}{2r} \quad \forall \, x, y \in \partial D, x \neq y \, ,$$

and therefore, for any $\phi \in L^2(\partial D)$,

$$\mathcal{K}_D^* \phi(x) = \mathcal{K}_D \phi(x) = \frac{1}{4\pi r} \int_{\partial D} \phi(y) \, d\sigma(y) \, , \tag{2.16}$$

for all $x \in \partial D$. For $d \geq 3$, if D denotes a sphere with radius r, then, since

$$\frac{\langle x - y, \nu_x \rangle}{|x - y|^d} = \frac{1}{2r} \frac{1}{|x - y|^{d-2}} \quad \forall \, x, y \in \partial D, x \neq y \, ,$$

we have, as shown by Lemma 2.3 of [170], that for any $\phi \in L^2(\partial D)$,

$$\mathcal{K}_D^* \phi(x) = \mathcal{K}_D \phi(x) = \frac{(2 - d)}{2r} \mathcal{S}_D \phi(x) \tag{2.17}$$

for all $x \in \partial D$.

Let now D be a bounded Lipschitz domain, and let

$$L_0^2(\partial D) := \left\{ \phi \in L^2(\partial D) : \int_{\partial D} \phi \, d\sigma = 0 \right\} .$$

Let $\lambda \neq 0$ be a real number. Of particular interest for solving the transmission problem for the Laplacian would be the invertibility of the operator $\lambda I - \mathcal{K}_D^*$ on $L^2(\partial D)$ or $L_0^2(\partial D)$.

First, it was proved by Kellog in [175] that the eigenvalues of $\mathcal{K}_D^*$ on $L^2(\partial D)$ lie in $(-\frac{1}{2}, \frac{1}{2}]$ for smooth domains; but this argument goes through for Lipschitz domains [110]. The following injectivity result holds.

Lemma 2.5 *Let λ be a real number. The operator $\lambda I - \mathcal{K}_D^*$ is one to one on $L_0^2(\partial D)$ if $|\lambda| \geq \frac{1}{2}$, and for $\lambda \in (-\infty, -\frac{1}{2}] \cup (\frac{1}{2}, \infty)$, $\lambda I - \mathcal{K}_D^*$ is one to one on $L^2(\partial D)$.*

Proof. The argument is by contradiction. Let $\lambda \in (-\infty, -\frac{1}{2}] \cup (\frac{1}{2}, \infty)$, and suppose that $\phi \in L^2(\partial D)$ satisfies $(\lambda I - \mathcal{K}_D^*)\phi = 0$ and ϕ is not identically zero. Since by Green's formula $\mathcal{K}_D(1) = 1/2$, it follows by duality that ϕ has mean value zero on ∂D. Hence $\mathcal{S}_D \phi(x) = O(1/|x|^{d-1})$ and $\nabla \mathcal{S}_D \phi(x) = O(1/|x|^d)$

at infinity for $d \geq 2$. Since ϕ is not identically zero, the following numbers cannot be zero

$$A = \int_D |\nabla \mathcal{S}_D \phi|^2 \, dx \text{ and } B = \int_{\mathbb{R}^d \setminus \overline{D}} |\nabla \mathcal{S}_D \phi|^2 \, dx \;.$$

On the other hand, using the divergence theorem and (2.12), we have

$$A = \int_{\partial D} (-\frac{1}{2}I + \mathcal{K}_D^*)\phi \; \mathcal{S}_D \phi \, d\sigma \text{ and } B = -\int_{\partial D} (\frac{1}{2}I + \mathcal{K}_D^*)\phi \; \mathcal{S}_D \phi \, d\sigma \;.$$

Since $(\lambda I - \mathcal{K}_D^*)\phi = 0$, it follows that

$$\lambda = \frac{1}{2} \frac{B - A}{B + A} \;.$$

Thus, $|\lambda| < 1/2$, which is a contradiction and so, for $\lambda \in (-\infty, -\frac{1}{2}] \cup (\frac{1}{2}, \infty)$, $\lambda I - \mathcal{K}_D^*$ is one to one on $L^2(\partial D)$.

If $\lambda = 1/2$, then $A = 0$ and hence $\mathcal{S}_D \phi = $ constant in D. Thus $\mathcal{S}_D \phi$ is harmonic in $\mathbb{R}^d \setminus \partial D$, behaves like $O(|x|^{1-d})$ as $|x| \to +\infty$ (since $\phi \in L_0^2(\partial D)$), and is constant on ∂D. It then follows that $\mathcal{S}_D \phi = 0$ in $\mathbb{R}^d$, and hence $\phi = 0$. This proves that $(1/2) I - \mathcal{K}_D^*$ is one to one on $L_0^2(\partial D)$. $\square$

Let us now turn to the surjectivity of the operator $\lambda I - \mathcal{K}_D^*$ on $L^2(\partial D)$ or $L_0^2(\partial D)$. If D is a bounded $\mathcal{C}^{1+\alpha}$-domain for some $\alpha > 0$ then we have the bound

$$\left| \frac{\langle x - y, \nu_x \rangle}{|x - y|^d} \right| \leq C \frac{1}{|x - y|^{d-1-\alpha}} \quad \text{for } x, y \in \partial D, x \neq y \;,$$

which shows that the operators $\mathcal{K}_D$ and $\mathcal{K}_D^*$ are compact operators in $L^2(\partial D)$ [119]. From the above bound we can also deduce that there exists a constant C such that the estimate

$$||\mathcal{K}_D \phi||_{L^\infty(\partial D)} \leq C ||\phi||_{L^\infty(\partial D)} \tag{2.18}$$

holds for all $\phi \in L^\infty(\partial D)$. Moreover, for any $\lambda \in (-\infty, -\frac{1}{2}] \cup (\frac{1}{2}, \infty)$ there exists a constant C_λ such that

$$||\phi||_{L^\infty(\partial D)} \leq C_\lambda ||(\lambda I - \mathcal{K}_D)\phi||_{L^\infty(\partial D)}, \quad \forall \phi \in L^\infty(\partial D) \;. \tag{2.19}$$

If D is a bounded $\mathcal{C}^1$-domain, the operators $\mathcal{K}_D$ and $\mathcal{K}_D^*$ are still compact operators in $L^2(\partial D)$ [115] but more elaborate arguments are needed for a proof.

Hence, by Fredholm theory, it follows from Lemma 2.5 that $\lambda I - \mathcal{K}_D^*$ is invertible on $L_0^2(\partial D)$ if $|\lambda| \geq \frac{1}{2}$, and for $\lambda \in (-\infty, -\frac{1}{2}] \cup (\frac{1}{2}, \infty)$, $\lambda I - \mathcal{K}_D^*$ is invertible on $L^2(\partial D)$.

Unlike the $\mathcal{C}^1$-case, the operators $\mathcal{K}_D$ and $\mathcal{K}_D^*$ are not compact on a Lipschitz domain and thus, Fredholm theory is not applicable. This difficulty for the invertibility of $\lambda I - \mathcal{K}_D^*$ was overcome by Verchota [258] who made the key observation that the following Rellich identities, see [237, 231, 223, 162], are appropriate substitutes of compactness in the case of Lipschitz domains. We need to fix a notation first. For a vector field α and a function u let

$$\langle \alpha, \frac{\partial u}{\partial T} \rangle = \sum_{p=1}^{d-1} \langle \alpha, T_p \rangle \frac{\partial u}{\partial T_p} \, .$$

Here $T_1, \ldots, T_{d-1}$ is an orthonormal basis for the tangent plane to ∂D at x.

Lemma 2.6 (Rellich's identities) *Let D be a bounded Lipschitz domain in $\mathbb{R}^d, d \geq 2$. Let u be a function such that either*

(i) u is Lipschitz in $\overline{D}$ and $\Delta u = 0$ in D, or
(ii) u is Lipschitz in $\mathbb{R}^d \setminus D$, $\Delta u = 0$ in $\mathbb{R}^d \setminus \overline{D}$, and $|u(x)| = O(1/|x|^{d-2})$ when $d \geq 3$ and $|u(x)| = O(1/|x|)$ when $d = 2$ as $|x| \to +\infty$.

Let α be a C^1-vector field in $\mathbb{R}^d$ with compact support. Then

$$\int_{\partial D} \langle \alpha, \nu \rangle \left| \frac{\partial u}{\partial \nu} \right|^2 = \int_{\partial D} \langle \alpha, \nu \rangle \left| \frac{\partial u}{\partial T} \right|^2 - 2 \int_{\partial D} \langle \alpha, \frac{\partial u}{\partial T} \rangle \frac{\partial u}{\partial \nu}$$

$$+ \begin{cases} \displaystyle\int_D 2\langle \nabla \alpha \nabla u, \nabla u \rangle - (\nabla \cdot \alpha)|\nabla u|^2 & \text{if } u \text{ satisfies (i)}, \quad (2.20) \\ \displaystyle\int_{\mathbb{R}^d \setminus \overline{D}} 2\langle \nabla \alpha \nabla u, \nabla u \rangle - (\nabla \cdot \alpha)|\nabla u|^2 & \text{if } u \text{ satisfies (ii)}. \end{cases}$$

Proof. Suppose that u satisfies (i). Observe that

$$\nabla \cdot (\alpha |\nabla u|^2) = (\nabla \cdot \alpha)|\nabla u|^2 + \langle \alpha, \nabla |\nabla u|^2 \rangle ,$$
$$= (\nabla \cdot \alpha)|\nabla u|^2 + 2\langle \nabla^2 u \alpha, \nabla u \rangle ,$$

and

$$\nabla \cdot (\nabla u \langle \alpha, \nabla u \rangle) = \langle \alpha, \nabla u \rangle \Delta u + \langle \nabla \langle \alpha, \nabla u \rangle, \nabla u \rangle$$
$$= \langle \nabla \alpha \nabla u, \nabla u \rangle + \langle \nabla^2 u \alpha, \nabla u \rangle .$$

Here $\nabla^2 u$ is the Hessian of u. Combining these identities, we obtain

$$\nabla \cdot (\alpha |\nabla u|^2) = 2\nabla \cdot (\nabla u \langle \alpha, \nabla u \rangle) + (\nabla \cdot \alpha)|\nabla u|^2 - 2\langle \nabla \alpha \nabla u, \nabla u \rangle .$$

Stokes' formula shows that

$$\int_{\partial D} \langle \alpha, \nu \rangle |\nabla u|^2 = 2 \int_{\partial D} \frac{\partial u}{\partial \nu} \langle \alpha, \nabla u \rangle + \int_D (\nabla \cdot \alpha)|\nabla u|^2 - 2\langle \nabla \alpha \nabla u, \nabla u \rangle .$$

Since

$$\alpha = \langle \alpha, \nu \rangle \nu + \sum_{p=1}^{d-1} \langle \alpha, T_p \rangle T_p ,$$

we get

$$\langle \boldsymbol{\alpha}, \nabla u \rangle = \langle \boldsymbol{\alpha}, \nu \rangle \frac{\partial u}{\partial \nu} + \langle \boldsymbol{\alpha}, \frac{\partial u}{\partial T} \rangle \,.$$

We also get

$$|\nabla u|^2 = \left| \frac{\partial u}{\partial \nu} \right|^2 + \left| \frac{\partial u}{\partial T} \right|^2 \,.$$

Thus after rearranging, we find

$$\int_{\partial D} \langle \boldsymbol{\alpha}, \nu \rangle \left(\left| \frac{\partial u}{\partial \nu} \right|^2 + \left| \frac{\partial u}{\partial T} \right|^2 \right) = 2 \int_{\partial D} \langle \boldsymbol{\alpha}, \nu \rangle \left| \frac{\partial u}{\partial \nu} \right|^2 + 2 \int_{\partial D} \langle \boldsymbol{\alpha}, \frac{\partial u}{\partial T} \rangle \frac{\partial u}{\partial \nu}$$
$$+ \int_D (\nabla \cdot \boldsymbol{\alpha}) |\nabla u|^2 - 2 \langle \nabla \boldsymbol{\alpha} \nabla u, \nabla u \rangle \,.$$

Hence

$$\int_{\partial D} \langle \boldsymbol{\alpha}, \nu \rangle \left| \frac{\partial u}{\partial \nu} \right|^2 = \int_{\partial D} \langle \boldsymbol{\alpha}, \nu \rangle \left| \frac{\partial u}{\partial T} \right|^2 - 2 \int_{\partial D} \langle \boldsymbol{\alpha}, \frac{\partial u}{\partial T} \rangle \frac{\partial u}{\partial \nu}$$
$$+ \int_D 2 \langle \nabla \boldsymbol{\alpha} \nabla u, \nabla u \rangle - (\nabla \cdot \boldsymbol{\alpha}) |\nabla u|^2 \,,$$

and the identity (2.20) holds.

In order to establish the Rellich identity (2.20) when u satisfies (ii), we merely replace D by $\mathbb{R}^d \setminus \overline{D}$ in the above proof and use the decay estimate at infinity $|u(x)| = O(1/|x|^{d-2})$ when $d \geq 3$ and $|u(x)| = O(1/|x|)$ when $d = 2$ as $|x| \to +\infty$ to apply the Stokes' formula in all $\mathbb{R}^d \setminus \overline{D}$. □

As an easy consequence of the Rellich identities (2.20) the following important result holds.

Corollary 2.7 *Let u be as in Lemma 2.6. Then there exists a positive constant C depending only on the Lipschitz character of D such that*

$$\frac{1}{C} \left\| \frac{\partial u}{\partial T} \right\|_{L^2(\partial D)} \leq \left\| \frac{\partial u}{\partial \nu} \right\|_{L^2(\partial D)} \leq C \left\| \frac{\partial u}{\partial T} \right\|_{L^2(\partial D)} \,. \tag{2.21}$$

Proof. Let c_0 be a fixed positive number. Let $\boldsymbol{\alpha}$ be a vector field supported in the set $\mathrm{dist}(x, \partial D) < 2c_0$ such that $\boldsymbol{\alpha} \cdot \nu \geq \delta$ for some $\delta > 0$, $\forall \, x \in \partial D$ (here, δ depends only on the Lipschitz character of D). Applying (2.20) we obtain

$$\int_{\partial D} \langle \boldsymbol{\alpha}, \nu \rangle \left| \frac{\partial u}{\partial \nu} \right|^2 = \int_{\partial D} \langle \boldsymbol{\alpha}, \nu \rangle \left| \frac{\partial u}{\partial T} \right|^2 + O \left(\left\| \frac{\partial u}{\partial T} \right\|_{L^2(\partial D)} \left\| \frac{\partial u}{\partial \nu} \right\|_{L^2(\partial D)} + \| \nabla u \|_{L^2(D)} \right).$$

Since

$$\| \nabla u \|_{L^2(D)}^2 = \int_{\partial D} u \frac{\partial u}{\partial \nu} \, d\sigma \leq \| u - u_0 \|_{L^2(\partial D)} \left\| \frac{\partial u}{\partial \nu} \right\|_{L^2(\partial D)} \,,$$

(because $\int_{\partial D} \partial u/\partial \nu = 0$) where $u_0 = \frac{1}{|\partial D|} \int_{\partial D} u \, d\sigma$, the Poincaré inequality (2.1) yields

$$\|\nabla u\|^2_{L^2(D)} \leq C \left\|\frac{\partial u}{\partial T}\right\|_{L^2(\partial D)} \left\|\frac{\partial u}{\partial \nu}\right\|_{L^2(\partial D)},$$

where the constant C depends only on the Lipschitz character of D. Thus

$$\int_{\partial D} \langle \alpha, \nu \rangle \left|\frac{\partial u}{\partial \nu}\right|^2 = \int_{\partial D} \langle \alpha, \nu \rangle \left|\frac{\partial u}{\partial T}\right|^2 + O\left(\left\|\frac{\partial u}{\partial T}\right\|_{L^2(\partial D)} \left\|\frac{\partial u}{\partial \nu}\right\|_{L^2(\partial D)}\right).$$

Employing a small constant-large constant argument, we conclude that estimates (2.21) hold. □

The following results are due to Verchota [258] and Escauriaza, Fabes, and Verchota [110]. For proofs when ∂D is smooth, see [115, 119].

Theorem 2.8 *The operator $\lambda I - \mathcal{K}_D^*$ is invertible on $L_0^2(\partial D)$ if $|\lambda| \geq \frac{1}{2}$, and for $\lambda \in (-\infty, -\frac{1}{2}] \cup (\frac{1}{2}, \infty)$, $\lambda I - \mathcal{K}_D^*$ is invertible on $L^2(\partial D)$.*

Proof.   Let us first prove that the operators $\pm(1/2)I + \mathcal{K}_D^* : L_0^2(\partial D) \to L_0^2(\partial D)$ are invertible. Observe that $\pm(1/2)I + \mathcal{K}_D^*$ maps $L_0^2(\partial D)$ into $L_0^2(\partial D)$. In fact, since $\mathcal{K}_D(1) = 1/2$, we have

$$\int_{\partial D} \mathcal{K}_D^* f \, d\sigma = \frac{1}{2} \int_{\partial D} f \, d\sigma$$

for all $f \in L^2(\partial D)$.

Let $u(x) = \mathcal{S}_D f(x)$, where $f \in L_0^2(\partial D)$. Then u satisfies conditions (i) and (ii) in Lemma 2.6. Because of the second formula (2.11) in Theorem 2.4, $\partial u/\partial T$ is continuous across the boundary ∂D, and by the jump formula (2.12)

$$\frac{\partial u}{\partial \nu}\bigg|_{\pm} = (\pm \frac{1}{2} I + \mathcal{K}_D^*) f.$$

We now apply Corollary 2.7 in D and $\mathbb{R}^d \setminus \overline{D}$ to obtain that

$$\left\|\frac{\partial u}{\partial \nu}\bigg|_-\right\|_{L^2(\partial D)} \simeq \left\|\frac{\partial u}{\partial \nu}\bigg|_+\right\|_{L^2(\partial D)},$$

or equivalently

$$\frac{1}{C} \|(\frac{1}{2}I + \mathcal{K}_D^*)f\|_{L^2(\partial D)} \leq \|(\frac{1}{2}I - \mathcal{K}_D^*)f\|_{L^2(\partial D)}$$

$$\|(\frac{1}{2}I - \mathcal{K}_D^*)f\|_{L^2(\partial D)} \leq C \|(\frac{1}{2}I + \mathcal{K}_D^*)f\|_{L^2(\partial D)}.$$

$\qquad\qquad(2.22)$

Here the constant C depends only on the Lipschitz character of D. Since

$$f = (\frac{1}{2}I + \mathcal{K}_D^*)f + (\frac{1}{2}I - \mathcal{K}_D^*)f \,,$$

(2.22) shows that

$$\|(\frac{1}{2}I + \mathcal{K}_D^*)f\|_{L^2(\partial D)} \geq C\|f\|_{L^2(\partial D)} \,. \qquad (2.23)$$

In order to keep the technicalities to a minimum, we deal with the case when ∂D is given by a Lipschitz graph by localizing the situation. Assume

$$\partial D = \left\{(x', x_d) : x_d = \varphi(x')\right\},$$

where $\varphi : \mathbb{R}^{d-1} \to \mathbb{R}$ is a Lipschitz function. To show that $A = (1/2)\,I + \mathcal{K}_D^*$ is invertible we consider the Lipschitz graph corresponding to $t\varphi$,

$$\partial D_t = \left\{(x', x_d) : x_d = t\varphi(x')\right\} \quad \text{for } 0 < t < 1 \,,$$

and the corresponding operators $\mathcal{K}_{D_t}^*, A_t$. Then $A_0 = (1/2)\,I, A_1 = A$, and A_t are continuous in norm as a function of t. Moreover, by (2.23), $\|A_t f\|_{L^2(\partial D_t)} \geq C\|f\|_{L^2(\partial D_t)}$, with C independent of t in $(0,1)$ due to the fact that the constant in (2.23) depends only on the Lipschitz character of D. The invertibility of A now follows from the continuity method. See Appendix A.2. This establishes the invertibility of $(1/2)I + \mathcal{K}_D^*$ on $L_0^2(\partial D)$. Invertibility of $-(1/2)I + \mathcal{K}_D^*$ on $L_0^2(\partial D)$ can be proved in the same way starting from the inequality

$$\|(-\frac{1}{2}I + \mathcal{K}_D^*)f\|_{L^2(\partial D)} \geq C\|f\|_{L^2(\partial D)} \,.$$

We now show that $(1/2)\,I + \mathcal{K}_D^*$ is invertible on $L^2(\partial D)$. To do that, it suffices to show that it is onto on $L^2(\partial D)$. Since $\mathcal{K}_D(1) = 1/2$, we get

$$\int_{\partial D} (\frac{1}{2}I + \mathcal{K}_D^*)f \, d\sigma = \int_{\partial D} f \, d\sigma$$

for all $f \in L^2(\partial D)$. Let $h := ((1/2)\,I + \mathcal{K}_D^*)(1)$. For a given $g \in L^2(\partial D)$, let

$$g = g - ch + ch := g_0 + ch \,, \quad c = \frac{1}{|\partial D|} \int_{\partial D} g \, d\sigma \,.$$

Since

$$\int_{\partial D} h \, d\sigma = \int_{\partial D} (\frac{1}{2}I + \mathcal{K}_D^*)h \, d\sigma = |\partial D| \,,$$

one can easily see that $g_0 \in L_0^2(\partial D)$. Let $f_0 \in L_0^2(\partial D)$ be such that

$$((1/2)\,I + \mathcal{K}_D^*)f_0 = g_0 \,.$$

Then $f := f_0 + c$ satisfies $((1/2)I + \mathcal{K}_D^*)f = g$. Thus $(1/2)I + \mathcal{K}_D^*$ is onto on $L^2(\partial D)$.

Suppose now that $|\lambda| > 1/2$. Let $f \in L^2(\partial D)$ and set $u(x) = \mathcal{S}_D f(x)$. Let c_0 be a fixed positive number. Let $\boldsymbol{\alpha}$ be a vector field supported in the set $\operatorname{dist}(x, \partial D) < 2c_0$ such that $\boldsymbol{\alpha} \cdot \nu \geq \delta$ for some $\delta > 0$, $\forall\, x \in \partial D$. From the Rellich identity (2.20), we have

$$\int_{\partial D} \langle \boldsymbol{\alpha}, \nu \rangle \left| \frac{\partial u}{\partial \nu} \right|^2 = \int_{\partial D} \langle \boldsymbol{\alpha}, \nu \rangle \left| \frac{\partial u}{\partial T} \right|^2 - 2 \int_{\partial D} \langle \boldsymbol{\alpha}, \frac{\partial u}{\partial T} \rangle \frac{\partial u}{\partial \nu} \tag{2.24}$$
$$+ \int_D 2 \langle \nabla \boldsymbol{\alpha} \nabla u, \nabla u \rangle - (\nabla \cdot \boldsymbol{\alpha}) |\nabla u|^2 \,.$$

Observe that on ∂D

$$\left. \frac{\partial u}{\partial \nu} \right|_{-} = (-\frac{1}{2}I + \mathcal{K}_D^*)f = (\lambda - \frac{1}{2})f - (\lambda I - \mathcal{K}_D^*)f \,,$$

and

$$\langle \nabla u, \boldsymbol{\alpha} \rangle = \frac{\partial u}{\partial \nu} \langle \boldsymbol{\alpha}, \nu \rangle + \langle \boldsymbol{\alpha}, \frac{\partial u}{\partial T} \rangle$$
$$= -\frac{1}{2} \langle \boldsymbol{\alpha}, \nu \rangle f + \mathcal{K}_{\boldsymbol{\alpha}} f \,,$$

where

$$\mathcal{K}_{\boldsymbol{\alpha}}(f) = \frac{1}{\omega_d} \text{p.v.} \int_{\partial D} \frac{\langle x - y, \boldsymbol{\alpha}(x) \rangle}{|x - y|^d} f(y) \, d\sigma(y) \,.$$

We also have

$$\int_D |\nabla u|^2 \, dx = \int_{\partial D} u \frac{\partial u}{\partial \nu} \Big|_{-} \, d\sigma$$
$$= \int_{\partial D} \mathcal{S}_D(f) \left[(\lambda - \frac{1}{2})f - (\lambda I - \mathcal{K}_D^*)f \right] d\sigma \,.$$

By using

$$-2 \int_{\partial D} \langle \boldsymbol{\alpha}, \frac{\partial u}{\partial T} \rangle \frac{\partial u}{\partial \nu} = 2 \int_{\partial D} \langle \boldsymbol{\alpha}, \nu \rangle \left| \frac{\partial u}{\partial \nu} \right|^2 - 2 \int_{\partial D} \frac{\partial u}{\partial \nu} \left[-\frac{1}{2} \langle \boldsymbol{\alpha}, \nu \rangle f + \mathcal{K}_{\boldsymbol{\alpha}}(f) \right] \,,$$

we get from (2.24) that

$$\frac{1}{2} \int_{\partial D} \langle \boldsymbol{\alpha}, \nu \rangle \left| \frac{\partial u}{\partial \nu} \right|^2 = -\frac{1}{2} \int_{\partial D} \langle \boldsymbol{\alpha}, \nu \rangle \left| \frac{\partial u}{\partial T} \right|^2 + \int_{\partial D} \frac{\partial u}{\partial \nu} \left[-\frac{1}{2} \langle \boldsymbol{\alpha}, \nu \rangle f + \mathcal{K}_{\boldsymbol{\alpha}}(f) \right]$$
$$- \int_D \langle \nabla \boldsymbol{\alpha} \nabla u, \nabla u \rangle + \frac{1}{2} (\nabla \cdot \boldsymbol{\alpha}) |\nabla u|^2 \,.$$

Thus we obtain

$$\frac{1}{2}\left(\lambda - \frac{1}{2}\right)^2 \int_{\partial D} \langle \boldsymbol{\alpha}, \nu \rangle f^2 \, d\sigma$$

$$\leq \int_{\partial D} \left[-\frac{1}{2}\langle \boldsymbol{\alpha}, \nu \rangle f + \mathcal{K}_{\boldsymbol{\alpha}}(f) \right] \left[(\lambda - \frac{1}{2})f - (\lambda I - \mathcal{K}_D^*)(f) \right] d\sigma$$

$$+ C\|f\|_{L^2(\partial D)} \left(\|\mathcal{S}_D f\|_{L^2(\partial D)} + \|(\lambda I - \mathcal{K}_D^*)(f)\|_{L^2(\partial D)} \right)$$

$$+ C\|\mathcal{S}_D f\|_{L^2(\partial D)} \|(\lambda I - \mathcal{K}_D^*)(f)\|_{L^2(\partial D)} + C\|(\lambda I - \mathcal{K}_D^*)(f)\|^2_{L^2(\partial D)} \,,$$

where C denotes a constant depending on the Lipschitz character of D and λ. Multiplying out the integrand in the second integral above and taking to the left-hand side of the inequality the term involving f^2, we obtain

$$\frac{1}{2}\left(\lambda^2 - \frac{1}{4}\right) \int_{\partial D} \langle \boldsymbol{\alpha}, \nu \rangle f^2 \, d\sigma \leq \left(\lambda - \frac{1}{2}\right) \int_{\partial D} \mathcal{K}_{\boldsymbol{\alpha}}(f) f \, d\sigma$$

$$+ C\|f\|_{L^2(\partial D)} \left(\|\mathcal{S}_D f\|_{L^2(\partial D)} + \|(\lambda I - \mathcal{K}_D^*)(f)\|_{L^2(\partial D)} \right)$$

$$+ C\|\mathcal{S}_D f\|_{L^2(\partial D)} \|(\lambda I - \mathcal{K}_D^*)(f)\|_{L^2(\partial D)} + C\|(\lambda I - \mathcal{K}_D^*)(f)\|^2_{L^2(\partial D)} \,.$$

If $\mathcal{K}_{\boldsymbol{\alpha}}^*$ denotes the adjoint operator on $L^2(\partial D)$ of the operator $\mathcal{K}_{\boldsymbol{\alpha}}$, it is easy to see that $\mathcal{K}_{\boldsymbol{\alpha}} + \mathcal{K}_{\boldsymbol{\alpha}}^* = R_{\boldsymbol{\alpha}}$, where the operator $R_{\boldsymbol{\alpha}}$ is defined by

$$R_{\boldsymbol{\alpha}}(f) = \frac{1}{\omega_d} \text{p.v.} \int_{\partial D} \frac{\langle x - y, \boldsymbol{\alpha}(x) - \boldsymbol{\alpha}(y) \rangle}{|x - y|^d} f(y) \, d\sigma(y) \,.$$

By duality, we have

$$\int_{\partial D} \mathcal{K}_{\boldsymbol{\alpha}}(f) f \, d\sigma = \frac{1}{2} \int_{\partial D} R_{\boldsymbol{\alpha}}(f) f \, d\sigma \,.$$

Since $|\lambda| > 1/2$ and $\boldsymbol{\alpha} \cdot \nu \geq \delta > 0$, using the large constant-small constant argument, we can get from the above inequality that

$$\|f\|_{L^2(\partial D)} \leq C\left(\|(\lambda I - \mathcal{K}_D^*)(f)\|_{L^2(\partial D)} + \|\mathcal{S}_D f\|_{L^2(\partial D)} \right. \tag{2.25}$$

$$\left. + \|R_{\boldsymbol{\alpha}}(f)\|_{L^2(\partial D)} \right) \,.$$

Since $\mathcal{S}_D$ and $R_{\boldsymbol{\alpha}}$ are compact on $L^2(\partial D)$, we conclude from the above estimate that $\lambda I - \mathcal{K}_D^*$ has a closed range.

We now prove that $\lambda I - \mathcal{K}_D^*$ is surjective on $L^2(\partial D)$ and hence invertible on $L^2(\partial D)$ by Lemma 2.5.

Suppose on the contrary that for some λ real, $|\lambda| > 1/2$, $\lambda I - \mathcal{K}_D^*$ is not invertible on $L^2(\partial D)$. Then the intersection of the spectrum of $\mathcal{K}_D^*$ and the set $\{\lambda \in \mathbb{R} : |\lambda| > 1/2\}$ is not empty and so there exists a real number λ_0 that belongs to this intersection and is a boundary point of this set. To reach a contradiction it suffices to show that $\lambda_0 I - \mathcal{K}_D^*$ is invertible. By (2.25) we know that $\lambda_0 I - \mathcal{K}_D^*$ is injective and has a closed range. Hence there exists a constant C such that for all $f \in L^2(\partial D)$ the following estimate holds

$$\|f\|_{L^2(\partial D)} \leq C\|(\lambda_0 I - \mathcal{K}_D^*)(f)\|_{L^2(\partial D)} . \tag{2.26}$$

Since λ_0 is a boundary point of the intersection of spectrum of $\mathcal{K}_D^*$ and the real line there exists a sequence of real numbers λ_p with $|\lambda_p| > 1/2$, $\lambda_p \to \lambda_0$, as $p \to +\infty$, and $\lambda_p I - \mathcal{K}_D^*$ is invertible on $L^2(\partial D)$. Therefore, given $g \in L^2(\partial D)$ there exists a unique $f_p \in L^2(\partial D)$ such that $(\lambda_p I - \mathcal{K}_D^*)(f_p) = g$. If $\|f_p\|_{L^2(\partial D)}$ has a bounded subsequence then there exists another subsequence that converges weakly to some f_0 in $L^2(\partial D)$ and we have

$$\int_{\partial D} (\lambda_p I - \mathcal{K}_D^*)(f_0)h\, d\sigma = \lim_{p \to +\infty} \int_{\partial D} f_p(\lambda_0 I - \mathcal{K}_D)(h)\, d\sigma$$
$$= \lim_{p \to +\infty} \int_{\partial D} (\lambda_0 I - \mathcal{K}_D^*)(f_p)h\, d\sigma = \int_{\partial D} gh\, d\sigma .$$

Hence $(\lambda_0 I - \mathcal{K}_D^*)(f_0) = g$. In the opposite case we may assume that $\|f_p\|_{L^2(\partial D)} = 1$ and $(\lambda_0 I - \mathcal{K}_D^*)(f_p)$ converges to zero in $L^2(\partial D)$. However from (2.26)

$$1 = \|f_p\|_{L^2(\partial D)} \leq C\|(\lambda_0 I - \mathcal{K}_D^*)(f_p)\|_{L^2(\partial D)}$$
$$\leq C|\lambda_0 - \lambda_p| + C\|(\lambda_p I - \mathcal{K}_D^*)(f_p)\|_{L^2(\partial D)} .$$

Since the final two terms converge to zero as $p \to +\infty$, we arrive at a contradiction. We conclude that for each λ real, $|\lambda| > 1/2$, $\lambda I - \mathcal{K}_D^*$ is invertible. $\square$

Analogously to (2.19) we can deduce from Theorem 2.8 that for any $\lambda \in (-\infty, -\frac{1}{2}] \cup (\frac{1}{2}, \infty)$ there exists a constant C_λ such that

$$\|\phi\|_{L^2(\partial D)} \leq C_\lambda\|(\lambda I - \mathcal{K}_D)\phi\|_{L^2(\partial D)}, \quad \forall\, \phi \in L^2(\partial D) . \tag{2.27}$$

Moreover,

$$\|\phi\|_{L^2(\partial D)} \leq C\|(-\frac{1}{2}I + \mathcal{K}_D)\phi\|_{L^2(\partial D)}, \quad \forall\, \phi \in L_0^2(\partial D) , \tag{2.28}$$

for some positive constant C.

Suppose $D \subset B_1(0)$ is a star-shaped domain with respect to the origin in two-dimensional space, where $B_1(0)$ is the disk of radius 1 and center 0. We can quantify the constant C. For so doing, define

$$\delta(D) := \inf_{x \in \partial D} \langle x, \nu_x \rangle .$$

Note that since D is a star-shaped domain with respect to the origin, $\delta(D) > 0$. For $\phi \in L_0^2(\partial D)$, set $u := \mathcal{S}_D \phi$. It follows from the Rellich identity (2.20) with $\alpha(x) = x$ that

$$\int_{\partial D} \langle x, \nu \rangle \left|\frac{\partial u}{\partial T}\right|^2 d\sigma = \int_{\partial D} \langle x, \nu \rangle \left|\frac{\partial u}{\partial \nu}\right|_{\pm}^2 d\sigma + 2\int_{\partial D} \langle x, \frac{\partial u}{\partial T}\rangle \frac{\partial u}{\partial \nu}\Big|_{\pm} d\sigma ,$$

which leads to the following estimate:

$$\delta\left\|\frac{\partial u}{\partial T}\right\|_{L^2(\partial D)}^2 \leq \left\|\frac{\partial u}{\partial \nu}\right|_{\pm}\right\|_{L^2(\partial D)}^2 + 2\left\|\frac{\partial u}{\partial \nu}\right|_{\pm}\right\|_{L^2(\partial D)}\left\|\frac{\partial u}{\partial T}\right\|_{L^2(\partial D)},$$

$$\delta\left\|\frac{\partial u}{\partial \nu}\right|_{\pm}\right\|_{L^2(\partial D)}^2 \leq \left\|\frac{\partial u}{\partial T}\right\|_{L^2(\partial D)}^2 + 2\left\|\frac{\partial u}{\partial \nu}\right|_{\pm}\right\|_{L^2(\partial D)}\left\|\frac{\partial u}{\partial T}\right\|_{L^2(\partial D)}.$$

Therefore

$$\left\|\frac{\partial u}{\partial T}\right\|_{L^2(\partial D)}^2 \leq \frac{2\delta+4}{\delta^2}\left\|\frac{\partial u}{\partial \nu}\right|_{\pm}\right\|_{L^2(\partial D)}^2,$$

$$\left\|\frac{\partial u}{\partial \nu}\right|_{\pm}\right\|_{L^2(\partial D)}^2 \leq \frac{2\delta+4}{\delta^2}\left\|\frac{\partial u}{\partial T}\right\|_{L^2(\partial D)}^2.$$

Thus, by the jump formula (2.12), we get

$$\left\|(\pm\frac{1}{2}I + \mathcal{K}_D^*)\phi\right\|_{L^2(\partial D)} = \left\|\frac{\partial u}{\partial \nu}\right|_{\pm}\right\|_{L^2(\partial D)}$$

$$\leq \frac{2\delta+4}{\delta^2}\left\|\frac{\partial u}{\partial \nu}\right|_{\pm}\right\|_{L^2(\partial D)}$$

$$= \frac{2\delta+4}{\delta^2}\|(\mp\frac{1}{2}I + \mathcal{K}_D^*)\phi\|_{L^2(\partial D)},$$

to conclude that

$$\|\phi\|_{L^2(\partial D)} \leq \|(\pm\frac{1}{2}I + \mathcal{K}_D^*)\phi\|_{L^2(\partial D)} + \|(\mp\frac{1}{2}I + \mathcal{K}_D^*)\phi\|_{L^2(\partial D)}$$

$$\leq \frac{(\delta+2)^2}{\delta^2}\|(\pm\frac{1}{2}I + \mathcal{K}_D^*)\phi\|_{L^2(\partial D)}.$$

We have proved the following result from [29].

Lemma 2.9 Let $D \subset B_1(0)$ be a star-shaped domain with respect to the origin, where $B_1(0)$ is the disk of radius 1 and center 0. Define $\delta(D) := \inf_{x \in \partial D}\langle x, \nu_x \rangle$. Then, for any $\phi \in L_0^2(\partial D)$,

$$\|\phi\|_{L^2(\partial D)} \leq \frac{\left(\delta(D)+2\right)^2}{\delta(D)^2}\left\|(\pm\frac{1}{2}I + \mathcal{K}_D^*)\phi\right\|_{L^2(\partial D)}.$$

Estimate (2.28) will be useful in Chap. 4. A more refined one will be needed in Chap. 3.

Lemma 2.10 *There exists a constant C depending only on the Lipschitz character of D such that*

$$\|\phi\|_{L^2(\partial D)} \leq C \frac{|k-1|}{k+1} \left\| \left(\frac{k+1}{2(k-1)} I - \mathcal{K}_D^* \right) \phi \right\|_{L^2(\partial D)} \tag{2.29}$$

for all $\phi \in L_0^2(\partial D)$.

Proof.   By (2.28), there is a constant C depending only on the Lipschitz character of D such that

$$\|\phi\|_{L^2(\partial D)} \leq C \left\| \left(\frac{1}{2} I - \mathcal{K}_D^* \right) \phi \right\|_{L^2(\partial D)}$$

for all $\phi \in L_0^2(\partial D)$. Hence we get

$$\|\phi\|_{L^2(\partial D)} \leq C \left\| \left(\frac{k+1}{2(k-1)} I - \mathcal{K}_D^* \right) \phi \right\|_{L^2(\partial D)} + \frac{C}{|k-1|} \|\phi\|_{L^2(\partial D)} .$$

It then follows that for $k > C + 1$

$$\|\phi\|_{L^2(\partial D)} \leq \frac{C}{1 - \frac{C}{k-1}} \left\| \left(\frac{k+1}{2(k-1)} I - \mathcal{K}_D^* \right) \phi \right\|_{L^2(\partial D)} ,$$

and hence, if k is larger than $2C + 1$, then

$$\|\phi\|_{L^2(\partial D)} \leq 2C \left\| \left(\frac{k+1}{2(k-1)} I - \mathcal{K}_D^* \right) \phi \right\|_{L^2(\partial D)} .$$

When k is smaller than $1/(C+1)$, we can proceed in the same way starting from the estimate

$$\|\phi\|_{L^2(\partial D)} \leq C \left\| \left(-\frac{1}{2} I - \mathcal{K}_D^* \right) \phi \right\|_{L^2(\partial D)} .$$

Now suppose that $|k-1|$ is small, or, equivalently, λ is large. Then

$$\|\phi\|_{L^2(\partial D)} \leq \frac{1}{\lambda} \|\lambda \phi\|_{L^2(\partial D)} \leq \frac{1}{\lambda} \|(\lambda I - \mathcal{K}_D^*)\phi\|_{L^2(\partial D)} + \frac{1}{\lambda} \|\mathcal{K}_D^* \phi\|_{L^2(\partial D)} .$$

Since $\|\mathcal{K}_D^* \phi\|_{L^2(\partial D)} \leq C \|\phi\|_{L^2(\partial D)}$ for some C then, if $\lambda > 2C$, we get

$$\|\phi\|_{L^2(\partial D)} \leq \frac{2}{\lambda} \|(\lambda I - \mathcal{K}_D^*)\phi\|_{L^2(\partial D)} .$$

Since the norm on the right-hand side of (2.29) depends continuously on k, by a compactness argument, the proof is complete. $\square$

We will also need the following theorem due to Verchota [258].

Theorem 2.11 *Let D be a bounded Lipschitz domain in $\mathbb{R}^d$. Then the single layer potential $\mathcal{S}_D$ maps $L^2(\partial D)$ into $W_1^2(\partial D)$ boundedly and $\mathcal{K}_D$: $W_1^2(\partial D) \to W_1^2(\partial D)$ is a bounded operator.*

Proof.   That $\mathcal{S}_D$ maps $L^2(\partial D)$ into $W_1^2(\partial D)$ boundedly is clear. In fact, by Corollary 2.7, (2.12), and Theorem 2.8, we get

$$\left\| \frac{\partial(\mathcal{S}_D f)}{\partial T} \right\|_{L^2(\partial D)} \approx \left\| \frac{\partial(\mathcal{S}_D f)}{\partial \nu} \Big|_- \right\|_{L^2(\partial D)}$$

$$\approx \left\| (-\frac{1}{2}I + \mathcal{K}_D^*)f \right\|_{L^2(\partial D)} \leq C \|f\|_{L^2(\partial D)} .$$

Thus we have

$$\|\mathcal{S}_D f\|_{W_1^2(\partial D)} \leq C \|f\|_{L^2(\partial D)} .$$

Given $h \in W_1^2(\partial D)$, let v be the solution to the problem $\Delta v = 0$ in D and $v = h$ on ∂D. Then $v \in W^{1,2}(D)$. By the Green representation, we get

$$\mathcal{S}_D \left(\frac{\partial v}{\partial \nu} \Big|_- \right) (x) = \mathcal{D}_D(v|_-)(x), \quad x \in \mathbb{R}^d \setminus \overline{D} .$$

It then follows from (2.13) that

$$(-\frac{1}{2}I + \mathcal{K}_D)h = \mathcal{S}_D \left(\frac{\partial v}{\partial \nu} \Big|_- \right) \quad \text{on } \partial D .$$

Therefore we get

$$\|\mathcal{K}_D h\|_{W_1^2(\partial D)} \leq \frac{1}{2}\|h\|_{W_1^2(\partial D)} + \left\| \mathcal{S}_D \left(\frac{\partial v}{\partial \nu} \Big|_+ \right) \right\|_{W_1^2(\partial D)}$$

$$\leq \frac{1}{2}\|h\|_{W_1^2(\partial D)} + C \left\| \frac{\partial v}{\partial \nu} \Big|_+ \right\|_{L^2(\partial D)} \leq C \|h\|_{W_1^2(\partial D)} ,$$

where the last inequality follows from Corollary 2.7. Thus we obtain that $\mathcal{K}_D : W_1^2(\partial D) \to W_1^2(\partial D)$ is bounded. $\square$

We conclude this section by investigating the invertibility of the single layer potential. We shall see that complications arise when $d = 2$.

Lemma 2.12 *Let D be a bounded Lipschitz domain in $\mathbb{R}^d$. Let $\phi \in L^2(\partial D)$ satisfy $\mathcal{S}_D\phi = 0$ on ∂D.*

(i) If $d \geq 3$, then $\phi = 0$.
(ii) If $d = 2$ and $\int_{\partial D} \phi = 0$, then $\phi = 0$.

Proof.   The single layer potential $u = \mathcal{S}_D\phi$ satisfies $\Delta u = 0$ in $\mathbb{R}^d \setminus \overline{D}, u = 0$ on ∂D and as $|x| \to +\infty$, we have $u(x) = O(|x|^{2-d})$ when $d \geq 3$, but

$$u(x) = \frac{1}{2\pi}\Big(\int_{\partial D} \phi\Big) \log |x| + O(|x|^{-1}) \quad \text{when } d = 2 \,.$$

Thus, provided we assume that $\int_{\partial D} \phi = 0$ when $d = 2$, we have $u(x) = 0(|x|^{-1})$ as $|x| \to +\infty$. Therefore, for large R,

$$\int_{B_R(0) \setminus \overline{D}} |\nabla u|^2 = \int_{\partial B_R(0)} \frac{\partial u}{\partial \nu} u = \begin{cases} O(R^{2-d}) & \text{if } d \geq 3 \,, \\ O(R^{-2}) & \text{if } d = 2 \,. \end{cases}$$

Sending $R \to +\infty$, we deduce that $\nabla u = 0$ in $\mathbb{R}^d \setminus \overline{D}$, and thus u is constant in $\mathbb{R}^d \setminus \overline{D}$. Since $u = 0$ on ∂D, it follows that $u = 0$ in $\mathbb{R}^d \setminus \overline{D}$. But $\mathcal{S}_D \phi = 0$ in D, and hence $\phi = \partial \mathcal{S}_D \phi / \partial \nu|_+ - \partial \mathcal{S}_D \phi / \partial \nu|_- = 0$ on ∂D. □

Theorem 2.13 *Let D be a bounded Lipschitz domain in $\mathbb{R}^d$.*

(i) If $d \geq 3$, then $\mathcal{S}_D : L^2(\partial D) \to W_1^2(\partial D)$ has a bounded inverse.
(ii) If $d = 2$, then the operator $A : L^2(\partial D) \times \mathbb{R} \to W_1^2(\partial D) \times \mathbb{R}$ defined by

$$A(\phi, a) = \Big(\mathcal{S}_D \phi + a, \int_{\partial D} \phi\Big)$$

has a bounded inverse.
(iii) Suppose $d = 2$ and let $(\phi_e, a) \in L^2(\partial D) \times \mathbb{R}$ denote the solution of the system

$$\begin{cases} \mathcal{S}_D \phi_e + a = 0 \,, \\ \int_{\partial D} \phi_e = 1 \,, \end{cases}$$

then $\mathcal{S}_D : L^2(\partial D) \to W_1^2(\partial D)$ has a bounded inverse if and only if $a \neq 0$.

Proof. Since $W_1^2(\partial D) \hookrightarrow L^2(\partial D)$ is compact, it follows from Theorem 2.11 that the operator $\mathcal{S}_D : L^2(\partial D) \to W_1^2(\partial D)$ is Fredholm with zero index. But, by Lemma 2.12, we have $Ker \mathcal{S}_D = \{0\}$ when $d \geq 3$, and therefore $\mathcal{S}_D$ has a bounded inverse.

We now establish that A has a bounded inverse. Since $\mathcal{S}_D : L^2(\partial D) \to W_1^2(\partial D)$ is Fredholm with zero index we need only to prove injectivity. In fact, if $\mathcal{S}_D \phi + a = 0$ and $\int_{\partial D} \phi = 0$, then $\int_{\partial D} \mathcal{S}_D \phi \phi = 0$. But $\int_{\partial D} \mathcal{S}_D \phi \phi = \int_{\mathbb{R}^d} |\nabla \phi|^2$, and consequently, $\mathcal{S}_D \phi = 0$ since $\mathcal{S}_D \phi \to 0$ as $|x| \to +\infty$. According to Lemma 2.12 this implies $\phi = 0$ and in turn $a = 0$.

Turning to part (iii), we note that if $a = 0$, then $\mathcal{S}_D$ cannot be invertible because $\mathcal{S}_D \phi_e = 0$. Thus, suppose that $a \neq 0$ and there exists $\phi \in L^2(\partial D)$ such that $\mathcal{S}_D \phi = 0$. Define $\phi_0 = \phi - (\int_{\partial D} \phi) \phi_e$, and observe that

$$\mathcal{S}_D \phi_0 = -\Big(\int_{\partial D} \phi\Big) \mathcal{S}_D \phi_e = a \int_{\partial D} \phi \text{ and } \int_{\partial D} \phi_0 = 0 \,.$$

Hence $\int_{\partial D} \mathcal{S}_D \phi_0 \phi_0 = 0$ and therefore $\phi_0 = 0$. In turn, $\int_{\partial D} \phi = 0$ because $a \neq 0$, giving $\psi = 0$ by Lemma 2.12. Thus, the homogeneous equation $\mathcal{S}_D \phi = 0$ has only the trivial solution, and $\mathcal{S}_D$ is invertible. □

2.3 Neumann and Dirichlet Functions

Let Ω be a bounded Lipschitz domain in $\mathbb{R}^d, d \geq 2$. Let $N(x,z)$ be the Neumann function for Δ in Ω corresponding to a Dirac mass at z. That is, N is the solution to

$$\begin{cases} \Delta_x N(x,z) = -\delta_z & \text{in } \Omega\,, \\ \frac{\partial N}{\partial \nu_x}\big|_{\partial\Omega} = -\frac{1}{|\partial\Omega|}\,, \int_{\partial\Omega} N(x,z)\,d\sigma(x) = 0 & \text{for } z \in \Omega\,. \end{cases} \tag{2.30}$$

Note that the Neumann function $N(x,z)$ is defined as a function of $x \in \overline{\Omega}$ for each fixed $z \in \Omega$. The operator defined by $N(x,z)$ is the solution operator for the Neumann problem

$$\begin{cases} \Delta U = 0 & \text{in } \Omega\,, \\ \frac{\partial U}{\partial \nu}\big|_{\partial\Omega} = g\,, \end{cases} \tag{2.31}$$

namely, the function U defined by

$$U(x) := \int_{\partial\Omega} N(x,z)g(z)d\sigma(z)$$

is the solution to (2.31) satisfying $\int_{\partial\Omega} U\,d\sigma = 0$.

Now we discuss some properties of N as a function of x and z.

Lemma 2.14 *The Neumann function N is symmetric in its arguments, that is, $N(x,z) = N(z,x)$ for $x \neq z \in \Omega$. It furthermore has the form*

$$N(x,z) = \begin{cases} -\frac{1}{2\pi}\log|x-z| + R_2(x,z) & \text{if } d = 2\,, \\ \frac{1}{(d-2)\omega_d}\frac{1}{|x-z|^{d-2}} + R_d(x,z) & \text{if } d \geq 3\,, \end{cases} \tag{2.32}$$

where $R_d(\cdot,z)$ belongs to $W^{\frac{3}{2},2}(\Omega)$ for any $z \in \Omega, d \geq 2$ and solves

$$\begin{cases} \Delta_x R_d(x,z) = 0 & \text{in } \Omega\,, \\ \frac{\partial R_d}{\partial \nu_x}\big|_{\partial\Omega} = -\frac{1}{|\partial\Omega|} + \frac{1}{\omega_d}\frac{\langle x-z, \nu_x\rangle}{|x-z|^d} & \text{for } x \in \partial\Omega\,. \end{cases}$$

Proof. Pick $z_1, z_2 \in \Omega$ with $z_1 \neq z_2$. Let $B_r(z_p) = \{|x-z_p| < r\}$, $p = 1,2$. Choose $r > 0$ so small that $B_r(z_1) \cap B_r(z_2) = \emptyset$. Set $N_1(x) = N(x,z_1)$ and $N_2(x) = N(x,z_2)$. We apply Green's formula in $\Omega' = \Omega \setminus B_r(z_1) \cup B_r(z_2)$ to get

$$\int_{\Omega'} \left(N_1 \Delta N_2 - N_2 \Delta N_1\right) dx = \int_{\partial\Omega} \left(N_1 \frac{\partial N_2}{\partial \nu} - N_2 \frac{\partial N_1}{\partial \nu}\right) d\sigma$$

$$-\int_{\partial B_r(z_1)} \left(N_1 \frac{\partial N_2}{\partial \nu} - N_2 \frac{\partial N_1}{\partial \nu}\right) d\sigma - \int_{\partial B_r(z_2)} \left(N_1 \frac{\partial N_2}{\partial \nu} - N_2 \frac{\partial N_1}{\partial \nu}\right) d\sigma\,.$$

Since N_p is harmonic for $x \neq z_p, p = 1, 2$, $\partial N_1 / \partial \nu = \partial N_2 / \partial \nu = -1/|\partial \Omega|$, and $\int_{\partial \Omega} (N_1 - N_2) \, d\sigma = 0$, we have

$$\int_{\partial B_r(z_1)} \left(N_1 \frac{\partial N_2}{\partial \nu} - N_2 \frac{\partial N_1}{\partial \nu} \right) d\sigma + \int_{\partial B_r(z_2)} \left(N_1 \frac{\partial N_2}{\partial \nu} - N_2 \frac{\partial N_1}{\partial \nu} \right) d\sigma = 0 \ .$$

Thanks to (2.32) which will be proved shortly, the left hand side of the above has the same limit as the left hand side of the following as $r \to 0$:

$$\int_{\partial B_r(z_1)} \left(\Gamma \frac{\partial N_2}{\partial \nu} - N_2 \frac{\partial \Gamma}{\partial \nu} \right) d\sigma + \int_{\partial B_r(z_2)} \left(N_1 \frac{\partial \Gamma}{\partial \nu} - \Gamma \frac{\partial N_1}{\partial \nu} \right) d\sigma = 0 \ .$$

Since

$$\int_{\partial B_r(z_1)} \Gamma \frac{\partial N_2}{\partial \nu} \, d\sigma \to 0 \ , \int_{\partial B_r(z_2)} \Gamma \frac{\partial N_1}{\partial \nu} \, d\sigma \to 0 \quad \text{as } r \to 0$$

and

$$\int_{\partial B_r(z_1)} N_2 \frac{\partial \Gamma}{\partial \nu} \, d\sigma \to N_2(z_1) \ , \int_{\partial B_r(z_2)} N_1 \frac{\partial \Gamma}{\partial \nu} \, d\sigma \to N_1(z_2) \quad \text{as } r \to 0 \ ,$$

we obtain $N_2(z_1) - N_1(z_2) = 0$, or equivalently $N(z_2, z_1) = N(z_1, z_2)$ for any $z_1 \neq z_2 \in \Omega$.

Now let $R_d, d \geq 2$, be defined by

$$R_d(x, z) = \begin{cases} N(x, z) + \dfrac{1}{2\pi} \log |x - z| & \text{if } d = 2 \ , \\[2mm] N(x, z) + \dfrac{1}{(2 - d)\omega_d} \dfrac{1}{|x - z|^{d-2}} & \text{if } d \geq 3 \ . \end{cases}$$

Since $R_d(\cdot, z)$ is harmonic in Ω and $\partial R_d(\cdot, z)/\partial \nu \in L^2(\partial \Omega)$ then it follows from [162] that $R_d(\cdot, z) \in W^{\frac{3}{2}, 2}(\Omega)$ for any $z \in \Omega$. □

For D, a subset of Ω, let

$$N_D f(x) := \int_{\partial D} N(x, y) f(y) \, d\sigma(y) \ .$$

The following lemma from [17] relates the fundamental solution with the Neumann function.

Lemma 2.15 *For $z \in \Omega$ and $x \in \partial \Omega$, let $\Gamma_z(x) := \Gamma(x - z)$ and $N_z(x) := N(x, z)$. Then*

$$\left(-\frac{1}{2} I + \mathcal{K}_\Omega \right)(N_z)(x) = \Gamma_z(x) \quad \text{modulo constants,} \quad x \in \partial \Omega \ , \qquad (2.33)$$

or, to be more precise, for any simply connected Lipschitz domain D compactly contained in Ω and for any $g \in L_0^2(\partial D)$, we have for any $x \in \partial \Omega$

$$\int_{\partial D} \left(-\frac{1}{2} I + \mathcal{K}_\Omega \right)(N_z)(x) g(z) \, d\sigma(z) = \int_{\partial D} \Gamma_z(x) g(z) \, d\sigma(z) \ . \qquad (2.34)$$

Proof. Let $f \in L_0^2(\partial\Omega)$ and define

$$u(z) := \int_{\partial\Omega} \left(-\frac{1}{2}I + \mathcal{K}_\Omega\right)(N_z)(x)f(x)\,d\sigma(x), \quad z \in \Omega .$$

Then

$$u(z) = \int_{\partial\Omega} N(x, z)\left(-\frac{1}{2}I + \mathcal{K}_\Omega^*\right)f(x)\,d\sigma(x) .$$

Therefore, $\Delta u = 0$ in Ω and

$$\left.\frac{\partial u}{\partial \nu}\right|_{\partial\Omega} = (-\frac{1}{2}I + \mathcal{K}_\Omega^*)f .$$

Hence by the uniqueness modulo constants of a solution to the Neumann problem we have

$$u(z) - \mathcal{S}_\Omega f(z) = \text{constant}, \quad z \in \Omega .$$

Thus if $g \in L_0^2(\partial D)$, then we obtain

$$\int_{\partial\Omega} \int_{\partial D} \left(-\frac{1}{2}I + \mathcal{K}_\Omega\right)(N_z)(x)g(z)f(x)\,d\sigma(z)\,d\sigma(x)$$
$$= \int_{\partial\Omega} \int_{\partial D} \Gamma_z(x)g(z)f(x)\,d\sigma(z)\,d\sigma(x) .$$

Since f is arbitrary, we have equation (2.33) or, equivalently, (2.34). This completes the proof. $\square$

In Chap. 4 we will be dealing with the inclusions of the form $D = \epsilon B + z$ where B is a bounded Lipschitz domain in $\mathbb{R}^d$. For the purpose of using in Chap. 4, we now expand $N(x, \epsilon y + z)$ asymptotically for $x \in \partial\Omega$, $z \in \Omega$, and $y \in \partial B$, and as $\epsilon \to 0$. By (2.33) we have the following relation:

$$\left(-\frac{1}{2}I + \mathcal{K}_\Omega\right)\left[N(\cdot, \epsilon y + z)\right](x) = \Gamma(x - z - \epsilon y) \quad \text{modulo constants}, \quad x \in \partial\Omega .$$

Using the Taylor expansion

$$\Gamma(x - \epsilon y) = \sum_{|j|=0}^{+\infty} \frac{(-1)^j}{j!} \epsilon^{|j|} \partial^j \Gamma(x) y^j ,$$

we obtain

$$\left(-\frac{1}{2}I + \mathcal{K}_\Omega\right)\left[N(\cdot, \epsilon y + z)\right](x)$$

$$= \sum_{|j|=0}^{+\infty} \frac{(-1)^j}{j!} \epsilon^{|j|} \partial^j (\Gamma(x-z)) y^j$$

$$= \sum_{|j|=0}^{+\infty} \frac{(-1)^j}{j!} \epsilon^{|j|} \partial_x^j \left(\left(-\frac{1}{2}I + \mathcal{K}_\Omega\right) N(\cdot, z)(x)\right) y^j$$

$$= \sum_{|j|=0}^{+\infty} \frac{1}{j!} \epsilon^{|j|} \left(\left(-\frac{1}{2}I + \mathcal{K}_\Omega\right) \partial_z^j N(\cdot, z)(x)\right) y^j$$

$$= \left(-\frac{1}{2}I + \mathcal{K}_\Omega\right)\left[\sum_{|j|=0}^{+\infty} \frac{1}{j!} \epsilon^{|j|} \partial_z^j N(\cdot, z) y^j\right](x) .$$

Since $\int_{\partial\Omega} N(x, w) \, d\sigma(x) = 0$ for all $w \in \Omega$, we have the following asymptotic expansion of the Neumann function.

Lemma 2.16 *For $x \in \partial\Omega$, $z \in \Omega$, and $y \in \partial B$, and as $\epsilon \to 0$,*

$$N(x, \epsilon y + z) = \sum_{|j|=0}^{+\infty} \frac{1}{j!} \epsilon^{|j|} \partial_z^j N(x, z) y^j . \qquad (2.35)$$

Now we turn to the properties of the Dirichlet function. Let $G(x, z)$ be the Green's function for the Dirichlet problem in Ω, that is, the unique solution to

$$\begin{cases} \Delta_x G(x, z) = -\delta_z & \text{in } \Omega , \\ G(x, z) = 0 & \text{on } \partial\Omega , \end{cases}$$

and let $G_z(x) = G(x, z)$. Then for any $x \in \partial\Omega$, and $z \in \Omega$ we can prove that

$$\left(\frac{1}{2}I + \mathcal{K}_\Omega^*\right)^{-1} \left(\frac{\partial \Gamma_z(y)}{\partial \nu_y}\right)(x) = -\frac{\partial G_z}{\partial \nu_x}(x) . \qquad (2.36)$$

Moreover, we would like to mention the following important properties of G [136]:

(i) the Green's function G is symmetric in $\Omega \times \Omega$;
(ii) the maximum principle implies that for $x, z \in \Omega$ with $x \neq z$

$$0 > G(x, z) > -\Gamma(x - z) \quad \text{for } d \geq 3 ,$$

$$0 > G(x, z) > -\Gamma(x - z) + \frac{1}{2\pi} \log \operatorname{diam}(\Omega) \quad \text{for } d = 2 ;$$

(iii) the Green's function for the ball $B_R(0)$ is given by

$$G(x, z) = \frac{1}{(2-d)\omega_d} \left(|x-z|^{2-d} - \left| \frac{R}{|x|}x - \frac{|x|}{R}z \right|^{2-d} \right) \quad \text{for } d \geq 3 \,,$$

$$G(x, z) = \frac{1}{2\pi} \left(\log |x-z| - \log \left| \frac{R}{|x|}x - \frac{|x|}{R}z \right| \right) \quad \text{for } d = 2 \,;$$

(iv) the normal derivative of the Green's function on the sphere $\partial B_R(0)$ is given by

$$\frac{\partial G}{\partial \nu}(x, z) = \frac{R^2 - |z|^2}{\omega_d R |x-z|^d} \quad \text{for any } z \in B_R(0) \text{ and } x \in \partial B_R(0) \,.$$

2.4 Representation Formula

Let Ω be a bounded domain in $\mathbb{R}^d$ with a connected Lipschitz boundary and conductivity equal to 1. Consider a bounded domain $D \subset\subset \Omega$ with a connected Lipschitz boundary and conductivity $0 < k \neq 1 < +\infty$. Let $g \in L_0^2(\partial\Omega)$, and let u and U be the (variational) solutions of the Neumann problems

$$\begin{cases} \nabla \cdot \left(1 + (k-1)\chi(D) \right) \nabla u = 0 & \text{in } \Omega \,, \\ \dfrac{\partial u}{\partial \nu} \bigg|_{\partial\Omega} = g \,, \\ \displaystyle\int_{\partial\Omega} u(x)\, d\sigma(x) = 0 \,, \end{cases} \tag{2.37}$$

and

$$\begin{cases} \Delta U = 0 & \text{in } \Omega \,, \\ \dfrac{\partial U}{\partial \nu} \bigg|_{\partial\Omega} = g \,, \\ \displaystyle\int_{\partial\Omega} U(x)\, d\sigma(x) = 0 \,, \end{cases} \tag{2.38}$$

where $\chi(D)$ is the characteristic function of D. Clearly, the Lax-Milgram lemma shows that, given $g \in L_0^2(\partial\Omega)$, there exist unique u and U in $W^{1,2}(\Omega)$ which solve (2.37) and (2.38), respectively.

At this point we have all the necessary ingredients to state and prove a decomposition formula of the steady-state voltage potential u into a harmonic part and a refraction part which will be the main tool for both deriving the asymptotic expansion in Chap. 4 and providing efficient reconstruction algorithms in Chap. 5. This decomposition formula is unique and seems to inherit geometric properties of the inclusion D, as it is shown in Chap. 3.

The following theorem was proved in [168, 169, 171].

Theorem 2.17 *Suppose that D is a domain compactly contained in Ω with a connected Lipschitz boundary and conductivity $0 < k \neq 1 < +\infty$. Then the solution u of the Neumann problem (2.37) is represented as*

$$u(x) = H(x) + \mathcal{S}_D\phi(x), \quad x \in \Omega, \tag{2.39}$$

where the harmonic function H is given by

$$H(x) = -\mathcal{S}_\Omega(g)(x) + \mathcal{D}_\Omega(f)(x), \quad x \in \Omega, \quad f := u|_{\partial\Omega}, \tag{2.40}$$

and $\phi \in L_0^2(\partial D)$ satisfies the integral equation

$$\left(\frac{k+1}{2(k-1)}I - \mathcal{K}_D^* \right)\phi = \left.\frac{\partial H}{\partial\nu}\right|_{\partial D} \quad \text{on } \partial D. \tag{2.41}$$

The decomposition (2.39) into a harmonic part and a refraction part is unique. Moreover, $\forall\, n \in \mathbb{N}$, there exists a constant $C_n = C(n, \Omega, \text{dist}(D, \partial\Omega))$ independent of $|D|$ and the conductivity k such that

$$\|H\|_{C^n(\overline{D})} \leq C_n\|g\|_{L^2(\partial\Omega)}. \tag{2.42}$$

Furthermore, the following holds

$$H(x) + \mathcal{S}_D\phi(x) = 0, \quad \forall\, x \in \mathbb{R}^d \setminus \overline{\Omega}. \tag{2.43}$$

Proof.   Consider the following two phase transmission problem:

$$\begin{cases} \nabla \cdot \left(1 + (k-1)\chi(D) \right)\nabla v = 0 & \text{in } \mathbb{R}^d \setminus \partial\Omega, \\[2mm] v\big|_- - v\big|_+ = f & \text{on } \partial\Omega, \\[2mm] \dfrac{\partial v}{\partial\nu}\bigg|_- - \dfrac{\partial v}{\partial\nu}\bigg|_+ = g & \text{on } \partial\Omega, \\[2mm] v(x) = O(|x|^{1-d}) & \text{as } |x| \to \infty. \end{cases} \tag{2.44}$$

Let $v_1 := -\mathcal{S}_\Omega g + \mathcal{D}_\Omega f + \mathcal{S}_D\phi$ in $\mathbb{R}^d$. Since $\phi \in L_0^2(\partial D)$ and $g \in L_0^2(\partial\Omega)$, $v_1(x) = O(|x|^{1-d})$ and hence v_1 is a solution of (2.44) by the jump formulae (2.12) and (2.13). If we put $v_2 = u$ in Ω and $v_2 = 0$ in $\mathbb{R}^d \setminus \overline{\Omega}$, then v_2 is also a solution of (2.44). Therefore, in order to prove (2.39) and (2.43), it suffices to show that the problem (2.44) has a unique solution in $W^{1,2}_{\text{loc}}(\mathbb{R}^d \setminus \partial\Omega)$.

Suppose that $v \in W^{1,2}_{\text{loc}}(\mathbb{R}^d \setminus \partial\Omega)$ is a solution of (2.44) with $f = g = 0$. Then v is a weak solution of $\nabla \cdot (1 + (k-1)\chi(D))\nabla v = 0$ in the entire domain $\mathbb{R}^d$. Therefore, for a large R,

$$\begin{aligned} \int_{B_R(0)} |\nabla v|^2 &\leq \frac{1+k}{k} \int_{B_R(0)} \left(1 + (k-1)\chi(D) \right)|\nabla v|^2 \\ &\leq \frac{1+k}{k} \int_{\partial B_R(0)} v\frac{\partial v}{\partial\nu} \\ &\leq -\frac{1+k}{k} \int_{\mathbb{R}^d \setminus \overline{B_R(0)}} |\nabla v|^2 \leq 0, \end{aligned}$$

where $B_R(0) = \{|x| < R\}$. This inequality holds for all R and hence v is constant. Since $v(x) \to 0$ at the infinity, we conclude that $v \equiv 0$.

To prove the uniqueness of the representation, suppose that H' is harmonic in Ω and

$$H + S_D\phi = H' + S_D\phi' \text{ in } \Omega .$$

Then $S_D(\phi - \phi')$ is harmonic in Ω and hence

$$\left.\frac{\partial}{\partial\nu}S_D(\phi - \phi')\right|_- = \left.\frac{\partial}{\partial\nu}S_D(\phi - \phi')\right|_+ \quad \text{on } \partial D.$$

It then follows from (2.12) that $\phi - \phi' = 0$ on ∂D and $H = H'$.

We finally prove estimate (2.42). Suppose that $\text{dist}(D, \partial\Omega) > c_0$ for some constant $c_0 > 0$. From the definition of H in (2.40) it is easy to see that

$$\|H\|_{C^n(\overline{D})} \leq C_n\left(\|g\|_{L^2(\partial\Omega)} + \|u|_{\partial\Omega}\|_{L^2(\partial\Omega)}\right), \tag{2.45}$$

where C_n depends only on n, $\partial\Omega$, and c_0. It suffices then to show as in Corollary 2.7 that

$$\|u|_{\partial\Omega}\|_{L^2(\partial\Omega)} \leq C\|g\|_{L^2(\partial\Omega)} .$$

To do so, we use the Rellich identity. Let $\boldsymbol{\alpha}$ be a vector field supported in the set $\text{dist}(x, \partial\Omega) < c_0$ such that $\boldsymbol{\alpha} \cdot \nu_x \geq \delta$ for some $\delta > 0$, $\forall\, x \in \partial\Omega$. Using the Rellich identity (2.20) with this $\boldsymbol{\alpha}$, we can show that

$$\left\|\frac{\partial u}{\partial T}\right\|_{L^2(\partial\Omega)} \leq C\left(\|g\|_{L^2(\partial\Omega)} + \|\nabla u\|_{L^2(\Omega\setminus\overline{D})}\right),$$

where C depends only on $\partial\Omega$ and c_0. Observe that

$$\|\nabla u\|_{L^2(\Omega\setminus\overline{D})}^2 \leq C\int_\Omega \left(1 + (k-1)\chi(D)\right)\nabla u \cdot \nabla u\, dx$$

$$\leq C\int_{\partial\Omega} gu\, d\sigma$$

$$\leq C\|g\|_{L^2(\partial\Omega)}\|u|_{\partial\Omega}\|_{L^2(\partial\Omega)} .$$

Since $\int_{\partial\Omega} u\, d\sigma = 0$, it follows from the Poincaré inequality (2.1) that

$$\|u|_{\partial\Omega}\|_{L^2(\partial\Omega)} \leq C\left\|\frac{\partial u}{\partial T}\right\|_{L^2(\partial\Omega)} .$$

Thus we obtain

$$\|u|_{\partial\Omega}\|_{L^2(\partial\Omega)}^2 \leq C\left(\|g\|_{L^2(\partial\Omega)}^2 + \|g\|_{L^2(\partial\Omega)}\|u|_{\partial\Omega}\|_{L^2(\partial\Omega)}\right),$$

and hence

$$\|u|_{\partial\Omega}\|_{L^2(\partial\Omega)} \leq C \|g\|_{L^2(\partial\Omega)} .$$

From (2.45) we finally obtain (2.42). □

It is important to note that based on this representation formula, Kang and Seo proved global uniqueness results for the inverse conductivity problem with one measurement when the conductivity inclusion D is a disk or a ball in the three-dimensional space [168, 170]. See Appendix A.4.2.

Another useful expression of the harmonic part H of u is given in the following lemma.

Lemma 2.18 *We have*

$$H(x) = \begin{cases} u(x) - (k-1) \displaystyle\int_D \nabla_y \Gamma(x-y) \cdot \nabla u(y)\, dy & x \in \Omega , \\[3mm] -(k-1) \displaystyle\int_D \nabla_y \Gamma(x-y) \cdot \nabla u(y)\, dy & x \in \mathbb{R}^d \setminus \overline{\Omega} . \end{cases} \tag{2.46}$$

Proof. We claim that

$$\phi = (k-1)\frac{\partial u}{\partial \nu}\bigg|_- . \tag{2.47}$$

In fact, it follows from the jump formula (2.12) and the equation (2.42) that

$$\frac{\partial u}{\partial \nu}\bigg|_- = \frac{\partial H}{\partial \nu} + \frac{\partial}{\partial \nu} \mathcal{S}_D \phi \bigg|_- = \frac{\partial H}{\partial \nu} + (-\frac{1}{2}I + \mathcal{K}_D^*)\phi = \frac{1}{k-1}\phi .$$

Then (2.46) follows from (2.43) and (2.47) by Green's formula. □

Let $g \in L_0^2(\partial\Omega)$ and

$$U(y) := \int_{\partial\Omega} N(x,y)g(x)\, d\sigma(x) .$$

Then U is the solution to the Neumann problem (2.38) and the following representation holds.

Theorem 2.19 *The solution u of (2.37) can be represented as*

$$u(x) = U(x) - N_D\phi(x), \quad x \in \partial\Omega , \tag{2.48}$$

where ϕ is defined in (2.41).

Proof. By substituting (2.39) into (2.40), we obtain

$$H(x) = -\mathcal{S}_\Omega(g)(x) + \mathcal{D}_\Omega\left(H|_{\partial\Omega} + (\mathcal{S}_D\phi)|_{\partial\Omega} \right)(x), \quad x \in \Omega.$$

It then follows from (2.13) that

$$\left(\frac{1}{2}I - \mathcal{K}_\Omega\right)(H|_{\partial\Omega}) = -(\mathcal{S}_\Omega g)|_{\partial\Omega} + \left(\frac{1}{2}I + \mathcal{K}_\Omega\right)((\mathcal{S}_D\phi)|_{\partial\Omega}) \quad \text{on } \partial\Omega. \quad (2.49)$$

Since by Green's formula $U = -\mathcal{S}_\Omega(g) + \mathcal{D}_\Omega(U|_{\partial\Omega})$ in Ω, we have

$$\left(\frac{1}{2}I - \mathcal{K}_\Omega\right)(U|_{\partial\Omega}) = -(\mathcal{S}_\Omega g)|_{\partial\Omega}. \quad (2.50)$$

Since $\phi \in L_0^2(\partial D)$, it follows from (2.33) that

$$-\left(\frac{1}{2}I - \mathcal{K}_\Omega\right)((N_D\phi)|_{\partial\Omega}) = (\mathcal{S}_D\phi)|_{\partial\Omega}. \quad (2.51)$$

Recall that we have established in the course of the proof of Lemma 2.15 that

$$\left(\frac{1}{2}I - \mathcal{K}_\Omega\right)f = 0 \,, f \in L^2(\partial\Omega) \Rightarrow f = \text{ constant.} \quad (2.52)$$

Then, from (2.49), (2.50), and (2.51), we conclude that

$$\left(\frac{1}{2}I - \mathcal{K}_\Omega\right)\left(H|_{\partial\Omega} - U|_{\partial\Omega} + \left(\frac{1}{2}I + \mathcal{K}_\Omega\right)((N_D\phi)|_{\partial\Omega})\right) = 0.$$

Therefore, we have

$$H|_{\partial\Omega} - U|_{\partial\Omega} + \left(\frac{1}{2}I + \mathcal{K}_\Omega\right)((N_D\phi)|_{\partial\Omega}) = C \text{ (constant).} \quad (2.53)$$

Note that

$$(\frac{1}{2}I + \mathcal{K}_\Omega)((N_D\phi)|_{\partial\Omega}) = (N_D\phi)|_{\partial\Omega} + (\mathcal{S}_D\phi)|_{\partial\Omega}.$$

Thus we get from (2.39) and (2.53) that

$$u|_{\partial\Omega} = U|_{\partial\Omega} - (N_D\phi)|_{\partial\Omega} + C. \quad (2.54)$$

Since all the functions entering in (2.54) belong to $L_0^2(\partial\Omega)$, we conclude that $C = 0$, and the theorem is proved. $\square$

We have a similar representation for solutions of the Dirichlet problem. Let $f \in W_{\frac{1}{2}}^2(\partial\Omega)$, and let v and V be the (variational) solutions of the Dirichlet problems:

$$\begin{cases} \nabla \cdot \left(1 + (k-1)\chi(D)\right)\nabla v = 0 & \text{in } \Omega, \\ v = f & \text{on } \partial\Omega, \end{cases} \quad (2.55)$$

and

$$\begin{cases} \Delta V = 0 & \text{in } \Omega, \\ V = f & \text{on } \partial\Omega. \end{cases} \quad (2.56)$$

The following representation theorem holds.

Theorem 2.20 *Let v and V be the solution of the Dirichlet problems (2.55) and (2.56). Then $\partial v/\partial \nu$ on ∂D can be represented as*

$$\frac{\partial v}{\partial \nu}(x) = \frac{\partial V}{\partial \nu}(x) - \frac{\partial}{\partial \nu}G_D\phi(x), \quad x \in \partial\Omega, \tag{2.57}$$

where ϕ is defined in (2.41) with H given by (2.40) and $g = \partial v/\partial \nu$ on $\partial\Omega$, and

$$G_D\phi(x) := \int_{\partial D} G(x,y)\phi(y)\,d\sigma(y) \ .$$

Theorem 2.20 can be proved in the same way as Theorem 2.19. In fact, it is simpler because of the solvability of the Dirichlet problem or, equivalently, the invertibility of $(1/2)\,I + \mathcal{K}_\Omega^*$. So we omit the proof.

2.5 Energy Identities

The following energy identities hold [172, 10].

Lemma 2.21 *The solutions u and U of (2.37) and (2.38) satisfy*

$$\int_\Omega |\nabla(u-U)|^2\,dx + (k-1)\int_D |\nabla u|^2\,dx = \int_{\partial\Omega}(U-u)g\,d\sigma, \tag{2.58}$$

$$\int_\Omega \Big(1 + (k-1)\chi(D)\Big)|\nabla(u-U)|^2\,dx - (k-1)\int_D |\nabla U|^2\,dx \tag{2.59}$$

$$= -\int_{\partial\Omega}(U-u)g\,d\sigma \ .$$

Proof. From the weak formulations of the Neumann problems (2.37) and (2.38), it follows that

$$\int_\Omega \nabla(u-U)\cdot\nabla\eta\,dx + (k-1)\int_D \nabla u\cdot\nabla\eta\,dx = 0, \tag{2.60}$$

for every test function $\eta \in W^{1,2}(\Omega)$. Substituting $\eta = u$ in (2.60) and integrating by parts, we have

$$\int_\Omega |\nabla(u-U)|^2\,dx + (k-1)\int_D |\nabla u|^2\,dx = \int_{\partial\Omega}(U-u)g\,d\sigma,$$

while substituting $\eta = u - U$ yields

$$\int_\Omega (1+(k-1)\chi(D))|\nabla(u-U)|^2\,dx - (k-1)\int_D |\nabla U|^2\,dx = -\int_{\partial\Omega}(U-u)g\,d\sigma \ .$$

Then Lemma 2.21 immediately follows from the above two identities. $\quad\square$

3

Generalized Polarization Tensors

In this chapter we introduce the notion of generalized polarization tensors (GPT's) associated with a bounded Lipschitz domain B and a conductivity k and study their basic properties. The GPT's are the basic building blocks for the full asymptotic expansions of the boundary voltage perturbations due to the presence of a small conductivity inclusion D of the form $D = \epsilon B + z$ with conductivity k inside a conductor Ω with conductivity 1. See Chap. 4.

The use of these GPT's leads to stable and accurate algorithms for the numerical computations of the steady-state voltage in the presence of small conductivity inclusions. It is known that small size features cause difficulties in the numerical solution of the conductivity problem by the finite element or finite difference methods. This is because such features require refined meshes in their neighborhoods, with their attendant problems [174].

On the other hand, it is important from an imaging point of view to precisely characterize these GPT's and derive some of their properties, such as symmetry, positivity, and optimal bounds on their elements, for developing efficient algorithms for reconstructing conductivity inclusions of small volume. The GPT's seem to contain significant information on the domain B and its conductivity k which is yet to be investigated.

The concepts of higher-order polarization tensors generalize those of classical Pólya–Szegö polarization tensors which have been extensively studied in the literature by many authors for various purposes [72, 24, 73, 105, 200, 198, 123, 180, 186, 231, 232, 241, 95]. The notion of Pólya–Szegö polarization tensor appeared in problems of potential theory related to certain problems arising in hydrodynamics and in electrostatics. If the conductivity k is zero, namely, if B is insulated, the polarization tensor of Pólya–Szegö is called the virtual mass. The concept of polarization tensors also occurs in several other interesting contexts, in particular in asymptotic models of dilute composites [212, 24, 70, 104] and in low-frequency scattering of acoustic and electromagnetic waves [180, 94].

Our plan of this chapter is as follows. We first give two slightly different but equivalent definitions of the GPT's. We then prove that the knowledge of all the GPT's uniquely determines the domain and the constitutive parameter. Furthermore, we show important symmetric properties and positivity of the GPT's and derive isoperimetric inequalities satisfied by the tensor elements of the GPT's. We also establish relations that can be used to provide bounds on the weighted volume. We understand an isoperimetric inequality to be any inequality which relates two or more geometric and/or physical quantities associated with the same domain. The inequality must be optimal in the sense that the equality sign holds for some domain or in the limit as the domain degenerates [229]. The classical isoperimetric inequality–the one after which all such inequalities are named–states that of all plane curves of given perimeter the circle encloses the largest area. This inequality was known already to the Greeks. The reader is referred to [232, 40, 229, 230] for a variety of important isoperimetric inequalities.

We conclude the chapter by considering the polarization tensors associated with multiple inclusions. We prove their symmetry and positivity. We also estimate their eigenvalues in terms of the total volume of the inclusions. Explicit formulae for the GPT's in the multi-disk case are given in [21].

The results presented here will be applied in Chap. 5 to obtain accurate reconstructions of small conductivity inclusions from a small number of boundary measurements. Note that similar results have been established for the (generalized) anisotropic polarization tensors in [167]. These tensors are defined in the same way as the GPT's. However, they occur due to not only the presence of discontinuity, but also the difference of the anisotropy.

3.1 Definition

Let B be a Lipschitz bounded domain in $\mathbb{R}^d$ and let the conductivity of B be k, $0 < k \neq 1 < +\infty$. Denote $\lambda := (k+1)/(2(k-1))$.

Definition 3.1 *For a multi-index $i = (i_1, \ldots, i_d) \in \mathbb{N}^d$, let $\partial^i f = \partial_1^{i_1} \cdots \partial_d^{i_d} f$ and $x^i := x_1^{i_1} \cdots x_d^{i_d}$. For $i, j \in \mathbb{N}^d$, we define the generalized polarization tensor M_{ij} by*

$$M_{ij} := \int_{\partial B} y^j \phi_i(y) \, d\sigma(y) \,, \tag{3.1}$$

where ϕ_i is given by

$$\phi_i(y) := (\lambda I - \mathcal{K}_B^*)^{-1} \left(\nu_x \cdot \nabla x^i \right)(y), \quad y \in \partial B \,. \tag{3.2}$$

If $|i| = |j| = 1$, we denote M_{ij} by $(m_{pq})_{p,q=1}^d$ and call $M = (m_{pq})_{p,q=1}^d$ the polarization tensor of Pólya–Szegö.

Lemma 3.2 *For any multi-index $i = (i_1, \ldots, i_d) \in \mathbb{N}^d$ there is a unique solution ψ_i of the following transmission problem:*

$$\begin{cases} \Delta\psi_i(x) = 0, \quad x \in B \cup (\mathbb{R}^d \setminus \overline{B}) , \\ \psi_i\Big|_+(x) - \psi_i\Big|_-(x) = 0 \quad x \in \partial B , \\ \dfrac{\partial\psi_i}{\partial\nu}\Big|_+(x) - k\dfrac{\partial\psi_i}{\partial\nu}\Big|_-(x) = \nu \cdot \nabla x^i \quad x \in \partial B , \\ \psi_i(x) \to 0 \ as \ |x| \to \infty \quad if \ d = 3 , \\ \psi_i(x) - \dfrac{1}{2\pi}\log|x| \displaystyle\int_{\partial B} \nu \cdot \nabla y^i \, d\sigma(y) \to 0 \ as \ |x| \to \infty \quad if \ d = 2 . \end{cases} \quad (3.3)$$

Moreover, ψ_i satisfies the following decay estimate at infinity

$$\psi_i(x) - \Gamma(x)\int_{\partial B} \nu \cdot \nabla y^i \, d\sigma(y) = O\left(\frac{1}{|x|^{d-1}}\right) \quad as \ |x| \to \infty . \quad (3.4)$$

Proof.   Existence and uniqueness of ψ_i can be established using single layer potentials with suitably chosen densities. Fairly simple manipulations show that $\partial\psi_i/\partial\nu|_-$ satisfies the integral equation

$$(\lambda I - \mathcal{K}_B^*)\left(\frac{\partial\psi_i}{\partial\nu}\Big|_-\right)(x) = \frac{1}{k-1}\left(-\frac{I}{2} + \mathcal{K}_B^*\right)(\nu \cdot \nabla y^i)(x) , \quad x \in \partial B . \quad (3.5)$$

Since $K_B(1) = 1/2$ then

$$\int_{\partial B}(-\frac{I}{2} + \mathcal{K}_B^*)(\nu \cdot \nabla y^i)(x) \, d\sigma(x) = \int_{\partial B}(\nu \cdot \nabla x^i)(-\frac{I}{2} + \mathcal{K}_B)(1) \, d\sigma(x) = 0 ,$$

and consequently, according to Theorem 2.8, there exists a unique solution $\partial\psi_i/\partial\nu|_- \in L_0^2(\partial B)$ to the integral equation (3.5). Furthermore, we can express $\psi_i(x)$ for all $x \in \mathbb{R}^d$ as follows

$$\psi_i(x) = \frac{1}{k-1}\mathcal{S}_B(\lambda I - \mathcal{K}_B^*)^{-1}(\nu \cdot \nabla y^i)(x) , \quad x \in \mathbb{R}^d . \quad (3.6)$$

To obtain the behavior at infinity of ψ_i we write

$$\psi_i(x) = \frac{1}{k-1}\int_{\partial B}\left(\Gamma(x-y) - \Gamma(x)\right)(\lambda I - \mathcal{K}_B^*)^{-1}(\nu \cdot \nabla z^i)(y) \, d\sigma(y)$$
$$+ \Gamma(x)\frac{1}{k-1}\int_{\partial B}(\lambda I - \mathcal{K}_B^*)^{-1}(\nu \cdot \nabla z^i)(y) \, d\sigma(y) .$$

Since

$$\int_{\partial B}(\lambda I - \mathcal{K}_B^*)^{-1}(\nu \cdot \nabla y^i)(z) \, d\sigma(z) = (k-1)\int_{\partial B}\nu \cdot \nabla y^i \, d\sigma(y) ,$$

we obtain

$$\psi_i(x) - \Gamma(x) \int_{\partial B} \nu \cdot \nabla y^i \, d\sigma(y)$$

$$= \frac{1}{k-1} \int_{\partial B} \Big(\Gamma(x-y) - \Gamma(x) \Big) (\lambda I - \mathcal{K}_B^*)^{-1} (\nu \cdot \nabla z^i)(y) \, d\sigma(y)$$

and therefore by the Cauchy–Schwarz inequality

$$\left| \psi_i(x) - \Gamma(x) \int_{\partial B} \nu \cdot \nabla y^i \, d\sigma(y) \right| \leq C_i \, \| \Gamma(x-y) - \Gamma(x) \|_{L^2(\partial B)} \,,$$

which yields the desired decay estimate (3.4) due to the fact that

$$\| \Gamma(x-y) - \Gamma(x) \|_{L^2(\partial B)} = O \Big(\frac{1}{|x|^{d-1}} \Big) \quad \text{as } |x| \to +\infty \,.$$

Observe that uniqueness of a solution ψ_i to (3.3) can be proved in a straightforward way from the decay estimate (3.4). Let θ be the difference of two solutions, so that

$$\begin{cases} \nabla \cdot \Big(1 + (k-1)\chi(B) \Big) \nabla \theta = 0 \quad \text{in } \mathbb{R}^d, \\ \theta(x) = O \Big(1/|x|^{d-1} \Big) \quad \text{as } |x| \to +\infty \,. \end{cases}$$

Integrating by parts yields the energy identity

$$\int_{|y|<R} (1 + (k-1)\chi(B)) |\nabla \theta|^2 = \int_{|y|=R} \frac{\partial \theta}{\partial \nu} \, \theta \,.$$

Now let $R \to +\infty$; we have

$$\frac{\partial \theta}{\partial \nu} \theta = O(R^{-2d+1}) \quad \text{for } |y| = R \,,$$

so that

$$\int_{\mathbb{R}^d} (1 + (k-1)\chi(B)) |\nabla \theta|^2 = 0 \,.$$

This implies that θ is constant in $\mathbb{R}^d$, and, in fact $\theta = 0$ in $\mathbb{R}^d$ because $\theta(y)$ goes to 0 as $|y| \to +\infty$. $\square$

Lemma 3.3 *For all $i, j \in \mathbb{N}^d, M_{ij}$ can be rewritten in the following form:*

$$M_{ij} = (k-1) \int_{\partial B} x^j \frac{\partial x^i}{\partial \nu} \, d\sigma(x) + (k-1)^2 \int_{\partial B} x^j \frac{\partial \psi_i}{\partial \nu} \Big|_- (x) \, d\sigma(x) \,, \qquad (3.7)$$

where ψ_i is the unique solution of the transmission problem (3.3).

Proof. From the expression (3.6) of ψ_i and the identity

$$-\frac{1}{2}I + \mathcal{K}_B^* = -(\lambda I - \mathcal{K}_B^*) + (\lambda - \frac{1}{2})I ,$$

we compute by using the jump relation (2.12)

$$\int_{\partial B} x^j \frac{\partial \psi_i}{\partial \nu}\bigg|_{-}(x) = \frac{1}{k-1}\int_{\partial B} x^j \left[(\lambda I - \mathcal{K}_B^*)^{-1}(-\frac{I}{2} + \mathcal{K}_B^*)(\nu \cdot \nabla y^i)(x)\right] d\sigma(x)$$

$$= \frac{1}{k-1}\int_{\partial B} x^j \left[(\lambda - \frac{1}{2})(\lambda I - \mathcal{K}_B^*)^{-1}(\nu \cdot \nabla y^i)(x) - \nu \cdot \nabla x^i\right] d\sigma(x) ,$$

which immediately leads to (3.7). $\square$

Note that the definition (3.1) of GPT's is valid even when $k = 0$ or ∞. If $k = 0$, namely, if B is insulated, then

$$M_{ij} := \int_{\partial B} y^j \left(-\frac{1}{2}I - \mathcal{K}_B^*\right)^{-1} (\nu_y \cdot \nabla y^i)(y)\, d\sigma(y) ,$$

while if $k = \infty$, namely, if B is perfectly conducting, then

$$M_{ij} := \int_{\partial B} y^j \left(\frac{1}{2}I - \mathcal{K}_B^*\right)^{-1} (\nu_y \cdot \nabla y^i)(y)\, d\sigma(y) .$$

When $|i| = |j| = 1$, these definitions exactly match those introduced by Pólya–Szegö [232] and Schiffer and Szegö [241].[1]

The polarization tensor M of Pólya–Szegö can be explicitly computed for disks and ellipses in the plane and balls and ellipsoids in three-dimensional space [180]. If, for example, B is an ellipse whose focal line is on either the x_1- or the x_2-axis, its semi-major axis is of length a, and its semi-minor axis is of length b then its polarization tensor of Pólya–Szegö M takes the form

$$M = (k-1)|B| \begin{pmatrix} \dfrac{a+b}{a+kb} & 0 \\ 0 & \dfrac{a+b}{b+ka} \end{pmatrix} ,$$

where $|B|$ denotes the volume of B.

The GPT's seem to carry important geometric and potential theoretic properties of the domain B. In the following sections we investigate these properties.

3.2 Uniqueness Result

In this section we prove that the knowledge of all the GPT's uniquely determines the geometry and the constitutive parameter of the inclusion. To do so,

[1] When $k = 0$, it is called the virtual mass.

we relate the GPT's to the Dirichlet-to-Neumann (DtN) map. We prove that we can recover the DtN map from all the GPT's and hence, by a uniqueness result due to Isakov [156] (see also Druskin [106] and Appendix A.4.1), B and k are uniquely determined from all the GPT's.

Let Ω be a bounded Lipschitz domain in $\mathbb{R}^d$ compactly containing B. Recall that the DtN map $\Lambda : W_{\frac{1}{2}}^2(\partial\Omega) \rightarrow W_{-\frac{1}{2}}^2(\partial\Omega)$ corresponding to k and B is defined by, for $f \in W_{\frac{1}{2}}^2(\partial\Omega)$,

$$\Lambda(f) := \frac{\partial u}{\partial \nu}\Big|_{\partial\Omega},$$

where u is the unique weak solution of

$$\begin{cases} \nabla \cdot \left(1 + (k-1)\chi(B)\right)\nabla u = 0 & \text{in } \Omega, \\ u|_{\partial\Omega} = f. \end{cases}$$

Let $M_{ij}(k, B)$ denote the GPT's associated with the domain B and conductivity k. The following theorem asserts that we can recover the DtN map and hence B and k from all the GPT's.

Theorem 3.4 *Let k_1, k_2 be numbers different from 1, and let B_1, B_2 be bounded Lipschitz domains in $\mathbb{R}^d$. Let Ω be a domain compactly containing $\overline{B_1 \cup B_2}$, and let Λ_p be the DtN map corresponding to k_p and B_p, $p = 1, 2$, on $\partial\Omega$. If $M_{ij}(k_1, B_1) = M_{ij}(k_2, B_2)$ for all multi-indices i and j, then $\Lambda_1 = \Lambda_2$, and hence $k_1 = k_2$ and $B_1 = B_2$.*

Proof. Let $\lambda_p = (k_p + 1)/(2(k_p - 1))$, $p = 1, 2$. Let H be an entire harmonic function in $\mathbb{R}^d$. Since

$$\Gamma(x - y) = \sum_{|j|=0}^{\infty} \frac{1}{j!}\partial^j \Gamma(x)y^j, \quad |x| \rightarrow \infty, \tag{3.8}$$

we obtain, for all sufficiently large x,

$$\mathcal{S}_{B_p}(\lambda_p I - \mathcal{K}_{B_p}^*)^{-1}(\nu \cdot \nabla H|_{\partial B_p})(x)$$

$$= \int_{\partial B_p} \Gamma(x - y)(\lambda_p I - \mathcal{K}_{B_p}^*)^{-1}(\nu \cdot \nabla H|_{\partial B_p})(y)\, d\sigma(y)$$

$$= \sum_{|i|=1}^{\infty} \sum_{|j|=1}^{\infty} \frac{\partial^i H(0)}{i!j!}\partial^j \Gamma(x) \int_{\partial B_p} y^j(\lambda_p I - \mathcal{K}_{B_p}^*)^{-1}(\nu \cdot \nabla y^i|_{\partial B_p})(y)\, d\sigma(y)$$

$$= \sum_{|i|=1}^{\infty} \sum_{|j|=1}^{\infty} \frac{\partial^i H(0)}{i!j!}\partial^j \Gamma(x)M_{ij}(k_p, B_p).$$

If $M_{ij}(k_1, B_1) = M_{ij}(k_2, B_2)$ for all i and j, then

$$\mathcal{S}_{B_1}(\lambda_1 I - \mathcal{K}_{B_1}^*)^{-1}(\nu \cdot \nabla H|_{\partial B_1})(x) = \mathcal{S}_{B_2}(\lambda_2 I - \mathcal{K}_{B_2}^*)^{-1}(\nu \cdot \nabla H|_{\partial B_2})(x) \quad (3.9)$$

for all large x. By the unique continuation property of harmonic functions, we conclude that (3.9) holds for $x \in \mathbb{R}^d \setminus \overline{B_1 \cup B_2}$ and entire harmonic functions H.

Let $f \in W_{\frac{1}{2}}^2(\partial \Omega)$ and u_1 be the $W^{1,2}(\Omega)$ solution of the boundary value problem $\nabla \cdot ((1 + (k_1 - 1)\chi(B_1))\nabla u) = 0$ in Ω with $u|_{\partial \Omega} = f$. Let $H(x) := -\mathcal{S}_\Omega(\Lambda_1(f))(x) + \mathcal{D}_\Omega(f)(x)$, $x \in \Omega$. Then by the representation formula in Theorem 2.17, we have

$$u_1(x) = H(x) + \mathcal{S}_{B_1}(\lambda_1 I - \mathcal{K}_{B_1}^*)^{-1}(\nu \cdot \nabla H|_{\partial B_1})(x), \quad x \in \Omega .$$

Define u_2 by

$$u_2(x) = H(x) + \mathcal{S}_{B_2}(\lambda_2 I - \mathcal{K}_{B_2}^*)^{-1}(\nu \cdot \nabla H|_{\partial B_2})(x), \quad x \in \Omega .$$

Then u_2 is a $W^{1,2}(\Omega)$ solution of the equation $\nabla \cdot ((1 + (k_2 - 1)\chi(B_2))\nabla u) = 0$ in Ω. Since H is harmonic in Ω, there is a sequence H_n of entire harmonic functions converging to H uniformly on every compact subset of Ω. Since $\overline{B_1 \cup B_2}$ is a compact subset of Ω, it follows from (3.9) that $u_1 = u_2$ in $\Omega' \setminus \overline{B_1 \cup B_2}$, where Ω' is any relatively compact subset of Ω. By the unique continuation property of harmonic functions, we get $u_1 = u_2$ in $\Omega \setminus \overline{B_1 \cup B_2}$. Therefore, we obtain

$$\Lambda_1(f) = \left.\frac{\partial u_1}{\partial \nu}\right|_{\partial \Omega} = \left.\frac{\partial u_2}{\partial \nu}\right|_{\partial \Omega} = \Lambda_2(f) .$$

Since f is arbitrary, we conclude that $\Lambda_1 = \Lambda_2$.

In order to prove that $k_1 = k_2$ and $B_1 = B_2$, it suffices to refer to a result of Isakov [156] that asserts that B and k are uniquely determined from the Dirichlet-to-Neumann map Λ. See Appendix A.4.1. This completes the proof. $\square$

More physical properties of GPT's are investigated in Sects. 3.3 and 3.5.

3.3 Symmetry and Positivity of GPT's

We now consider the symmetry and positivity of GPT's. When $|i| = |j| = 1$, these properties were proved first in [73]. For symmetry we have the following theorem.

Theorem 3.5 *Suppose that a_i and b_j are constants such that $\sum_i a_i y^i$ and $\sum_j b_j y^j$ are harmonic polynomials. Then*

$$\sum_{i,j} a_i b_j M_{ij} = \sum_{i,j} a_i b_j M_{ji} . \tag{3.10}$$

Proof. Note that

$$\sum_{i,j} a_i b_j M_{ij} = \int_{\partial B} \sum_j b_j y^j \sum_i a_i \phi_i(y) \, d\sigma(y) \, .$$

Put $f(y) = \sum_i a_i y^i$, $g(y) = \sum_j b_j y^j$, $\phi = \sum_i a_i \phi_i = (\lambda I - \mathcal{K}_B^*)^{-1}(\frac{\partial f}{\partial \nu})$, and $\psi = (\lambda I - \mathcal{K}_B^*)^{-1}(\frac{\partial g}{\partial \nu})$. Then $\mathcal{S}_B \phi$ and $\mathcal{S}_B \psi$ satisfy the transmission conditions

$$\frac{\partial}{\partial \nu} \mathcal{S}_B \phi|_+ - k \frac{\partial}{\partial \nu} \mathcal{S}_B \phi|_- = (k-1) \frac{\partial f}{\partial \nu}$$

and

$$\frac{\partial}{\partial \nu} \mathcal{S}_B \psi|_+ - k \frac{\partial}{\partial \nu} \mathcal{S}_B \psi|_- = (k-1) \frac{\partial g}{\partial \nu}$$

on ∂B. Recall that

$$\sum_{i,j} a_i b_j M_{ij} = \int_{\partial B} g \phi \, d\sigma \quad \text{and} \quad \sum_{i,j} a_i b_j M_{ji} = \int_{\partial B} f \psi \, d\sigma \, .$$

By (2.12) and the transmission condition, we have

$$\begin{aligned}
\int_{\partial B} g \phi \, d\sigma &= \int_{\partial B} g \left[\frac{\partial \mathcal{S}_B \phi}{\partial \nu} \Big|_+ - \frac{\partial \mathcal{S}_B \phi}{\partial \nu} \Big|_- \right] d\sigma \\
&= (k-1) \int_{\partial B} g \frac{\partial}{\partial \nu} (\mathcal{S}_B \phi + f) \Big|_- \, d\sigma \, .
\end{aligned} \tag{3.11}$$

We then immediately obtain

$$\begin{aligned}
\int_{\partial B} g \phi \, d\sigma &= (k-1) \int_{\partial B} (\mathcal{S}_B \psi + g) \frac{\partial}{\partial \nu} (\mathcal{S}_B \phi + f) \Big|_- \, d\sigma \\
&\quad - \int_{\partial B} \mathcal{S}_B \psi \frac{\partial}{\partial \nu} \mathcal{S}_B \phi \Big|_+ \, d\sigma + \int_{\partial B} \mathcal{S}_B \psi \frac{\partial}{\partial \nu} \mathcal{S}_B \phi \Big|_- \, d\sigma \\
&= (k-1) \int_B \nabla(\mathcal{S}_B \psi + g) \cdot \nabla(\mathcal{S}_B \phi + f) \, dx \\
&\quad + \int_{\mathbb{R}^d \setminus B} \nabla \mathcal{S}_B \psi \cdot \nabla \mathcal{S}_B \phi \, dx + \int_B \nabla \mathcal{S}_B \psi \cdot \nabla \mathcal{S}_B \phi \, dx \, .
\end{aligned}$$

Symmetry property (3.10) follows from the above identity.   □

Suppose that $f = g$ in the proof of Theorem 3.5. It then follows from (3.11) that

$$\int_{\partial B} f \phi \, d\sigma = (k-1) \int_{\partial B} \frac{\partial f}{\partial \nu} (\mathcal{S}_B \phi + f) \, d\sigma \, . \tag{3.12}$$

On the other hand, it follows from the transmission condition that

$$\int_{\partial B} f\phi\, d\sigma = (k-1)\int_{\partial B}(\mathcal{S}_B\phi + f)\frac{\partial}{\partial\nu}(\mathcal{S}_B\phi + f)\bigg|_{-}\, d\sigma$$

$$- (k-1)\int_{\partial B}\mathcal{S}_B\phi\frac{\partial}{\partial\nu}\mathcal{S}_B\phi\bigg|_{-}\, d\sigma - (k-1)\int_{\partial B}\mathcal{S}_B\phi\frac{\partial f}{\partial\nu}\, d\sigma$$

$$= (k-1)\int_{\partial B}(\mathcal{S}_B\phi + f)\frac{\partial}{\partial\nu}(\mathcal{S}_B\phi + f)\bigg|_{-}\, d\sigma$$

$$- \left(1 - \frac{1}{k}\right)\int_{\partial B}\mathcal{S}_B\phi\frac{\partial}{\partial\nu}\mathcal{S}_B\phi\bigg|_{+}\, d\sigma - \left(1 - \frac{1}{k}\right)\int_{\partial B}\mathcal{S}_B\phi\frac{\partial f}{\partial\nu}\, d\sigma\,.$$

$$(3.13)$$

Define quadratic forms $Q_D(u)$ by

$$Q_D(u) := \int_D |\nabla u|^2\, dx\,, \tag{3.14}$$

where D is a Lipschitz domain in $\mathbb{R}^d$. Then, by equating (3.12) and (3.13), we obtain

$$\int_{\partial B}\mathcal{S}_B\phi\frac{\partial f}{\partial\nu}\, d\sigma = \frac{k}{k+1}Q_B(\mathcal{S}_B\phi + f) + \frac{1}{k+1}Q_{\mathbb{R}^d\setminus B}(\mathcal{S}_B\phi) - \frac{k}{k+1}Q_B(f)\,.$$

Substituting this identity into (3.12), we get

$$\int_{\partial B} f\phi\, d\sigma = \frac{k(k-1)}{k+1}Q_B(\mathcal{S}_B\phi + f) + \frac{k-1}{k+1}Q_{\mathbb{R}^d\setminus B}(\mathcal{S}_B\phi)$$

$$+ \frac{k-1}{k+1}Q_B(f)\,.$$

So we obtain the following theorem of positivity.

Theorem 3.6 *Suppose that a_i, $i \in I$, where I is a finite index set, are constants such that $f(y) = \sum_{i\in I} a_i y^i$ is a harmonic polynomial. Let $\phi = (\lambda I - \mathcal{K}_B^*)^{-1}(\frac{\partial f}{\partial\nu})$. Then*

$$\sum_{i,j\in I} a_i a_j M_{ij} = \frac{k-1}{k+1}\left[kQ_B(\mathcal{S}_B\phi + f) + Q_{\mathbb{R}^d\setminus B}(\mathcal{S}_B\phi) + Q_B(f)\right]\,. \tag{3.15}$$

Theorem 3.6 says that if $k > 1$, then GPT's are positive definite, and they are negative definite if $0 < k < 1$.

3.4 Bounds for the Polarization Tensor of Pólya–Szegö

Our aim of this section is to derive important isoperimetric inequalities satisfied by the elements of the polarization tensor of Pólya–Szegö.

Let $\{e_p\}_{p=1}^d$ be an orthonormal basis of $\mathbb{R}^d$. Denote

$$\phi_p = (\lambda I - \mathcal{K}_B^*)^{-1}(\frac{\partial x_p}{\partial \nu}), p = 1, \ldots, d .$$

Consider the polarization tensor $M = (m_{pq})_{p,q=1}^d$ of Pólya–Szegö associated with B and k. From Theorem 3.6 it follows that

$$m_{pq} = \frac{k-1}{k+1}\left[k \int_B (\nabla \mathcal{S}_B \phi_p + e_p)\cdot(\nabla \mathcal{S}_B \phi_q + e_q) + \int_{\mathbb{R}^d \backslash \overline{B}} \nabla \mathcal{S}_B \phi_p \cdot \nabla \mathcal{S}_B \phi_q + \delta_{pq}|B| \right],$$

or equivalently

$$\begin{aligned} \frac{k+1}{k-1} m_{pq} - \delta_{pq}|B| &= k \int_B (\nabla \mathcal{S}_B \phi_p + e_p) \cdot (\nabla \mathcal{S}_B \phi_q + e_q) \\ &+ \int_{\mathbb{R}^d \backslash \overline{B}} \nabla \mathcal{S}_B \phi_p \cdot \nabla \mathcal{S}_B \phi_q . \end{aligned} \tag{3.16}$$

From the Schwarz' inequality

$$\left(\int_B (\nabla \mathcal{S}_B \phi_p + e_p) \cdot (\nabla \mathcal{S}_B \phi_q + e_q) \right)^2 \leq \int_B |\nabla \mathcal{S}_B \phi_p + e_p|^2 \int_B |\nabla \mathcal{S}_B \phi_q + e_q|^2 ,$$

and similarly for the second integral in (3.16). By squaring both sides of (3.16) and then applying the inequality to the new right-hand side, we obtain the following inequality

$$\left(\frac{k+1}{k-1} m_{pq} - \delta_{pq}|B| \right)^2 \leq \left(\frac{k+1}{k-1} m_{pp} - |B| \right)\left(\frac{k+1}{k-1} m_{qq} - |B| \right) . \tag{3.17}$$

We can also develop upper and lower bounds on the diagonal elements $(m_{pp})_{p=1,\ldots,d}$. We have

$$m_{pp} = \frac{k-1}{k+1}\left[k \int_B |\nabla \mathcal{S}_B \phi_p + e_p|^2 + \int_{\mathbb{R}^d \backslash \overline{B}} |\nabla \mathcal{S}_B \phi_p|^2 + |B| \right] ,$$

where $\phi_p = (\lambda I - \mathcal{K}_B^*)^{-1}(\nu_p)$. $\forall\, \tau \in \mathbb{R}$, we compute

$$\int_B |\tau \nabla(\mathcal{S}_B \phi_p + y_p) + e_p|^2$$

$$= \tau^2 \int_B |\nabla \mathcal{S}_B \phi_p + e_p|^2 + 2\tau \int_B \nabla(\mathcal{S}_B \phi_p + y_p) \cdot e_p + |B|$$

$$= \tau^2 \int_B |\nabla \mathcal{S}_B \phi_p + e_p|^2 + 2\tau \int_{\partial B} \left(\frac{\partial}{\partial \nu}\mathcal{S}_B \phi_p \Big|_- + \nu_p \right) y_p + |B| .$$

Since

$$\frac{\partial}{\partial \nu}\mathcal{S}_B \phi_p \Big|_- = (-\frac{1}{2}I + \mathcal{K}_B^*)\phi_p = (\lambda - \frac{1}{2})\phi_p - \nu_p$$

then

$$\int_B |\nabla \mathcal{S}_B \phi_p + e_p|^2 = \frac{1}{\tau^2} \int_B |\tau \nabla(\mathcal{S}_B \phi_p + y_p) + e_p|^2 - \frac{2}{\tau}(\lambda - \frac{1}{2})m_{pp} - \frac{1}{\tau^2}|B| \,,$$

and hence

$$\frac{m_{pp}}{k-1}\left(1 + \frac{2k}{\tau(k+1)}\right) = |B|(1 - \frac{k}{\tau})\frac{1}{k+1}$$
$$+ \frac{1}{k+1}\left[\frac{k}{\tau^2}\int_B |\tau \nabla(\mathcal{S}_B \phi_p + y_p) + e_p|^2 + \int_{\mathbb{R}^d \setminus \overline{B}} |\nabla \mathcal{S}_B \phi_p|^2\right].$$

Taking $\tau = -1$ in the above identity we arrive at

$$\frac{m_{pp}}{k-1} = |B| + \frac{1}{1-k}\left[k\int_B |-\nabla(\mathcal{S}_B \phi_p + y_p) + e_p|^2 + \int_{\mathbb{R}^d \setminus \overline{B}} |\nabla \mathcal{S}_B \phi_p|^2\right],$$

and therefore

$$\frac{m_{pp}}{k-1} \geq |B| \quad \text{if } k < 1\,,$$

and

$$\frac{m_{pp}}{k-1} \leq |B| \quad \text{if } k > 1\,.$$

Taking $\tau = -k$ yields

$$\frac{m_{pp}}{k-1} \leq \frac{1}{k}|B| \quad \text{if } k < 1\,,$$

and

$$\frac{m_{pp}}{k-1} \geq \frac{1}{k}|B| \quad \text{if } k > 1\,.$$

The following optimal upper and lower bounds for the diagonal elements of the Polarization Tensor of Pólya–Szegö hold.

Lemma 3.7 *If $M = (m_{pq})_{p,q=1}^d$ is the polarization tensor of Pólya–Szegö associated with the bounded Lipschitz domain B and the conductivity $0 < k \neq 1 < +\infty$ then*

$$\min(1, \frac{1}{k})|B| \leq \frac{m_{pp}}{k-1} \leq \max(1, \frac{1}{k})|B|, \quad p = 1, \ldots, d\,. \tag{3.18}$$

These bounds are optimal in the sense that they are achieved by the diagonal elements of thin ellipses (for $d = 2$) and thin spheroids (for $d = 3$).

Note that the bounds $d|B|\min(1, 1/k)$ and $d|B|\max(1, 1/k)$ on the trace $\text{Tr}(M)$ of the matrix M which follow directly from (3.18) are not optimal. Indeed, the optimal bounds are given by the following lemma [71, 72].

Lemma 3.8 *If M is the polarization tensor of Pólya–Szegö associated with the bounded Lipschitz domain B and the conductivity $0 < k \neq 1 < +\infty$ then*

$$\frac{d^2}{d-1+k}|B| \leq \frac{\text{Tr}(M)}{k-1} \leq (d-1+\frac{1}{k})|B|\,.$$

The domain of possible values of the polarization tensor of Pólya–Szegö for a domain of a given volume and a given conductivity can be found in [186, 72]. As shown by Capdeboscq and Vogelius in [72], any point in this domain is achieved by coated ellipses. As has been pointed out by Kozlov in [185, 186], the derivation of optimal bounds for the polarization tensor of Pólya–Szegö and the estimates of its possible values are direct analogues of the corresponding estimates for the effective conductivity matrix known in the theory of composite materials [139, 199, 215, 205].

The proof of Lemma 3.8 relies on a variational principle associated with the representation (3.16) of the elements of the matrix M. This variational principle can be formulated as follows [22]:

$$\frac{1}{k-1} \sum_{p,q=1}^{d} m_{pq}\xi_p\xi_q = (k-1) \min_{w \in W_d} \int_{\mathbb{R}^d} \left(1 + (k-1)\chi(B)\right) \left|\nabla w + \frac{1}{k}\chi(B)\xi\right|^2$$

$$+ \frac{|B|}{k}|\xi|^2, \quad \forall \, \xi = (\xi_p)_{p=1}^d \in \mathbb{R}^d, d = 2, 3 \,.$$

(3.19)

Here

$$W_3 := \left\{ w \in W_{\text{loc}}^{1,2}(\mathbb{R}^3) : \frac{w}{r} \in L^2(\mathbb{R}^3), \nabla w \in L^2(\mathbb{R}^3) \right\}$$

and

$$W_2 := \left\{ w \in W_{\text{loc}}^{1,2}(\mathbb{R}^2) : \frac{w}{\sqrt{1 + r^2} \log(2 + r^2)} \in L^2(\mathbb{R}^2), \nabla w \in L^2(\mathbb{R}^2) \right\} \,.$$

Another interesting result that can be obtained by using the variational principle (3.19) is the following: $(1/(k-1)) M$ is a monotonically increasing positive definite matrix if we replace the given domain B by another B' which contains B.

Finally we would like to mention the following important unproven conjecture of Pólya–Szegö that is related to Lemma 3.8: if

$$\text{Tr}(M) = (k-1)|B|\frac{d^2}{(d-1+k)}$$

then B is a disk in the plane and a ball in three-dimensional space.

3.5 Estimates of the Weighted Volume and the Center of Mass

If $f(x) = \sum a_i x^i$ is a harmonic polynomial, then $Q_B(f) = \int_B |\nabla(\sum a_i x^i)|^2 \, dx$, where Q_B is defined by (3.14). In particular, if $f(x) = x_p$, $p = 1, \ldots, d$, then $Q_B(f) = |B|$. One can observe from (3.15) that if $\sum_{i \in I} a_i x^i$ is a harmonic polynomial, then

$$\left| \sum_{i,j \in I} a_i a_j M_{ij} \right| \geq \frac{|k-1|}{k+1} \int_B \left| \nabla \left(\sum a_i x^i \right) \right|^2 dx \; .$$

We now derive an upper bound for $\sum_{i,j \in I} a_i a_j M_{ij}$ in terms of the weighted volume.

Theorem 3.9 *There exists a constant C depending only on the Lipschitz character of B such that if $f(x) = \sum_{i \in I} a_i x^i$ is a harmonic polynomial, then*

$$\int_B |\nabla f|^2 \, dx \leq \frac{k+1}{|k-1|} \left| \sum_{i,j \in I} a_i a_j M_{ij} \right| \leq C \int_B |\nabla f|^2 \, dx \; . \tag{3.20}$$

Proof. By the definition of GPT's, we have

$$\sum_{i,j \in I} a_i a_j M_{ij} = \int_{\partial B} f(y) (\lambda I - \mathcal{K}_B^*)^{-1} \left(\frac{\partial f}{\partial \nu} \Big|_{\partial B} \right) (y) d\sigma(y) \; .$$

Since $\int_{\partial B} (\lambda I - \mathcal{K}_B^*)^{-1} (\frac{\partial f}{\partial \nu}|_{\partial B}) \, d\sigma = 0$, we get

$$\sum_{i,j \in I} a_i a_j M_{ij} = \int_{\partial B} (f(y) - f_0)(\lambda I - \mathcal{K}_B^*)^{-1} \left(\frac{\partial f}{\partial \nu} \Big|_{\partial B} \right) (y) d\sigma(y) \; ,$$

where $f_0 := \frac{1}{|\partial B|} \int_{\partial B} f \, d\sigma$. It thus follows from Lemma 2.10 that

$$\left| \sum_{i,j \in I} a_i a_j M_{ij} \right| \leq C \frac{|k-1|}{k+1} \|f - f_0\|_{L^2(\partial B)} \left\| \frac{\partial f}{\partial \nu} \right\|_{L^2(\partial B)} \; .$$

By the Poincaré inequality,

$$\|f - f_0\|_{L^2(\partial B)} \leq C \|\nabla f\|_{L^2(\partial B)} \; .$$

Thus the proof is complete by Lemma 2.1. $\square$

We now investigate the relation of GPT's with the centroid of B. Suppose that B is a two dimensional disk with radius r, then, (2.16) yields

$$\mathcal{K}_B^* \phi(x) = \mathcal{K}_B \phi(x) = \frac{1}{4\pi r} \int_{\partial B} \phi(y) \, d\sigma(y) \; ,$$

which gives that $\mathcal{K}_B^*(\phi) = 0$ for all $\phi \in L_0^2(\partial B)$. Thus, if $f(y) = \sum_i a_i y^i$ is harmonic, then

$$\sum_i a_i (\lambda I - \mathcal{K}_B^*)^{-1} (\nu_y \cdot \nabla y^i)(x) = \frac{1}{\lambda} \nu_x \cdot \nabla f \; .$$

Therefore, we have

$$\sum_i a_i M_{ij} = \frac{1}{\lambda} \int_{\partial B} y^j \nu_y \cdot \nabla f \, d\sigma(y) = \frac{1}{\lambda} \int_B \nabla y^j \cdot \nabla f \, dy \,.$$

Thus, if $i = j = e_p$, $p = 1, \ldots, d$, then $M_{ij} = \lambda^{-1}|B|$, and if $i = e_p$ and $j = 2e_p$, then $M_{ij} = 2\lambda^{-1}|B|x_p^*$, where x^* is the center of the ball. Here $\{e_p\}_{p=1}^d$ is an orthonormal basis of $\mathbb{R}^d$.

Suppose now that $d = 3$ and $B = B_r(x^*)$ is a ball of center x^* and radius r. Then, by (2.17), $\mathcal{K}_B^* \phi(x) = -\frac{1}{2r}\mathcal{S}_B\phi(x)$ for all $x \in \partial B$.

Let f be a harmonic polynomial homogeneous of degree n with respect to the center x^*. Let

$$\varphi(x) = \mathcal{S}_B\left(\frac{\partial f}{\partial \nu}\Big|_{\partial B}\right)(x), \quad x \in \mathbb{R}^d \setminus B \,.$$

By (3.8) we have

$$\varphi(x) = \sum_{p=1}^{\infty} \int_{\partial B} \sum_{|j|=p} \frac{1}{j!} \partial^j \Gamma(x - x^*)(y - x^*)^j \frac{\partial f}{\partial \nu}(y) \, d\sigma(y)$$

$$= \int_{\partial B} \sum_{|j|=n} \frac{1}{j!} \partial^j \Gamma(x - x^*)(y - x^*)^j \frac{\partial f}{\partial \nu}(y) \, d\sigma(y) \,.$$

In particular, $\varphi(x)$, $x \in \mathbb{R}^3 \setminus B$, is homogeneous of degree $-n - 1$ with respect to x^*.

By (2.12), we get

$$\frac{\partial \varphi}{\partial \nu}\Big|_+ (x) = \left(\frac{1}{2}I + \mathcal{K}_B^*\right)\left(\frac{\partial f}{\partial \nu}\Big|_{\partial B}\right)(x)$$

$$= \frac{1}{2}\frac{\partial f}{\partial \nu}(x) - \frac{1}{2r}\mathcal{S}_B\left(\frac{\partial f}{\partial \nu}\Big|_{\partial B}\right)(x), \quad x \in \partial B \,.$$

Therefore,

$$\frac{x - x^*}{r} \cdot \nabla\varphi + \frac{1}{2r}\varphi = \frac{1}{2}\frac{\partial f}{\partial \nu}(x) \quad \text{on } \partial B \,.$$

It then follows from the homogeneity of φ and f that $(x - x^*) \cdot \nabla\varphi = -(n+1)\varphi$, and hence

$$\varphi = -\frac{r}{2n+1}\frac{\partial f}{\partial \nu} \quad \text{on } \partial B \,.$$

So far we proved that if f is a harmonic polynomial homogeneous of degree n with respect to x^*, then

$$\mathcal{K}_B^*\left(\frac{\partial f}{\partial \nu}\Big|_{\partial B}\right)(x) = -\frac{1}{2r}\mathcal{S}_B\left(\frac{\partial f}{\partial \nu}\Big|_{\partial B}\right)(x) = \frac{1}{2(2n+1)}\frac{\partial f}{\partial \nu}(x), \quad x \in \partial B \,.$$

It then follows that

$$(\lambda I - \mathcal{K}_B^*)^{-1}\left(\frac{\partial f}{\partial \nu}\bigg|_{\partial B}\right) = \frac{(k-1)(2n+1)}{kn+n+1}\frac{\partial f}{\partial \nu} \quad \text{on } \partial B . \tag{3.21}$$

In particular, if $f(x) = x_p$, $p = 1, 2, 3$, then

$$\frac{\partial f}{\partial \nu} = \frac{\partial}{\partial \nu}(x_p - x_p^*) .$$

Thus by (3.21) we get

$$(\lambda I - \mathcal{K}_B^*)^{-1}\left(\frac{\partial f}{\partial \nu}\bigg|_{\partial B}\right) = \frac{3(k-1)}{k+2}\frac{\partial f}{\partial \nu} \quad \text{on } \partial B .$$

Therefore, if $|i| = 1$, then

$$M_{ij} = \frac{3(k-1)}{k+2}\int_B \nabla y^j \cdot \nabla y^i \, dy. \tag{3.22}$$

Observe that if $j = 2e_p$ and $i = e_p$, $p = 1, \ldots, d$, then

$$\int_B \nabla y^j \cdot \nabla y^i \, dy = 2\int_B y_j \, dy = 2x_p^*|B| .$$

So far we have proved the following theorem.

Theorem 3.10 *Suppose that $B = B_r(x^*)$ is a ball in $\mathbb{R}^d$, $d = 2, 3$. Let $i_l := e_l$ and $j_l := 2e_l$, $l = 1, \ldots, d$. Then*

$$M_{i_l i_l} = \frac{d(k-1)}{k+d-1}|B|, \quad l = 1, \ldots, d ,$$

and

$$(M_{i_1 j_1}, \ldots, M_{i_d j_d}) = \frac{2d(k-1)}{k+d-1}|B|x^* .$$

For a general bounded Lipschitz domain B, we have the following theorem.

Theorem 3.11 *Let B be a bounded Lipschitz domain and x^* the center of mass of B. Let $i_l := e_l$ and $j_l := 2e_l$, $l = 1, \ldots, d$. Then there exists C which depends only on the Lipschitz character of B such that*

$$\left|\frac{M_{i_l j_l}}{M_{i_l i_l}} - 2x_l^*\right| \leq C\frac{|k-1|}{k+1}\operatorname{diam}(B) . \tag{3.23}$$

Proof. Since

$$(\lambda I - \mathcal{K}_B^*)^{-1}(\nu_l) = \lambda^{-1}\nu_l + \lambda^{-1}(\lambda I - \mathcal{K}_B^*)^{-1}\mathcal{K}_B^*(\nu_l) ,$$

it follows from (2.29) that

$$\|(\lambda I - \mathcal{K}_B^*)^{-1}(\nu_l) - \lambda^{-1}\nu_l\|_{L^2(\partial B)} \leq C|\lambda|^{-1}\|(\lambda I - \mathcal{K}_B^*)^{-1}\mathcal{K}_B^*(\nu_l)\|_{L^2(\partial B)}$$

$$\leq C|\lambda|^{-2}\|\mathcal{K}_B^*(\nu_l)\|_{L^2(\partial B)} \leq C|\lambda|^{-2}|\partial B|^{1/2} .$$

Note that

$$M_{i_l j_l} - 2x_l^* M_{i_l i_l} = \int_{\partial B} (y_l - x_l^*)^2 (\lambda I - \mathcal{K}_B^*)^{-1}(\nu_l)(y)\, d\sigma(y) .$$

We also note that

$$\int_{\partial B} (y_l - x_l^*)^2 \nu_l(y) d\sigma(y) = 0 .$$

It then follows from the Cauchy–Schwarz inequality that

$$|M_{i_l j_l} - 2x_l^* M_{i_l i_l}| = \left| \int_{\partial B} (y_l - x_l^*)^2 \left[(\lambda I - \mathcal{K}_B^*)^{-1}(\nu_l)(y) - \nu_l(y) \right] d\sigma(y) \right|$$

$$\leq C\text{diam}(B)^2|\partial B||\lambda|^{-2} .$$

Then (3.23) follows from (3.20). This completes the proof. □

Theorem 3.11 says that if either k is close to 1 or the diameter of B is small, then $(M_{i_l j_l}/M_{i_l i_l})_{l=1,\ldots,d}$, where $j_l = 2e_l$, is a good approximation of the centroid of B.

We note that even when the conductivity of the inclusion and the background is anisotropic, the polarization tensor shares the same properties, symmetry and positivity. For this, see [167].

3.6 Polarization Tensors of Multiple Inclusions

Our goal in this section is to investigate properties of polarization tensors associated with multiple inclusions such as symmetry and positivity, which, in a most natural way, generalize those already derived for a single inclusion in the above sections. We also estimate their eigenvalues in terms of the total volume of the inclusions and explicitly compute them in the multi-disk case. These results are from [21].

Let B_s for $s = 1, \ldots, m$ be a bounded Lipschitz domain in $\mathbb{R}^d$. Throughout this section we suppose that:

(H1) there exist positive constants C_1 and C_2 such that

$$C_1 \leq \text{diam}\, B_s \leq C_2, \quad \text{and} \quad C_1 \leq \text{dist}(B_s, B_{s'}) \leq C_2, \quad s \neq s' ;$$

(H2) the conductivity of the inclusion B_s for $s = 1, \ldots, m$ is equal to some positive constant $k_s \neq 1$.

3.6.1 Definition

Theorem 3.12 *Let H be a harmonic function in $\mathbb{R}^d$ for $d = 2$ or 3. Let u be the solution of the transmission problem:*

$$\begin{cases} \nabla \cdot \left(\chi(\Omega \setminus \bigcup_{s=1}^{m} \overline{B_s}) + \sum_{s=1}^{m} k_s \chi(B_s) \right) \nabla u = 0 \quad in \ \mathbb{R}^d \ , \\ u(x) - H(x) = O(|x|^{1-d}) \quad as \ |x| \to \infty \ . \end{cases} \tag{3.24}$$

There are unique functions $\varphi^{(l)} \in L_0^2(\partial B_l)$, $l = 1, \ldots, m$, such that

$$u(x) = H(x) + \sum_{l=1}^{m} \mathcal{S}_{B_l} \varphi^{(l)}(x) \ . \tag{3.25}$$

The potentials $\varphi^{(l)}$, $l = 1, \ldots, m$, satisfy

$$(\lambda_l I - \mathcal{K}_{B_l}^*)\varphi^{(l)} - \sum_{s \neq l} \frac{\partial(\mathcal{S}_{B_s}\varphi^{(s)})}{\partial \nu^{(l)}}\Big|_{\partial B_l} = \frac{\partial H}{\partial \nu^{(l)}}\Big|_{\partial B_l} \quad on \ \partial B_l \ , \tag{3.26}$$

where $\nu^{(l)}$ denotes the outward unit normal to ∂B_l and

$$\lambda_l = \frac{k_l + 1}{2(k_l - 1)} \ .$$

Proof. It is easy to see from (2.12) that u defined by (3.25) and (3.26) is the solution of (3.24). Thus it is enough to show that the integral equation (3.26) has a unique solution.

Let $X := L_0^2(\partial B_1) \times \cdots \times L_0^2(\partial B_m)$. We prove that the operator $T : X \to X$ defined by

$$T(\varphi^{(1)}, \cdots, \varphi^{(m)}) = T_0(\varphi^{(1)}, \cdots, \varphi^{(m)}) + T_1(\varphi^{(1)}, \cdots, \varphi^{(m)})$$

$$:= \left((\lambda_1 I - \mathcal{K}_{B_1}^*)\varphi^{(1)}, \cdots, (\lambda_m I - \mathcal{K}_{B_m}^*)\varphi^{(m)} \right)$$

$$- \left(\sum_{s \neq 1} \frac{\partial(\mathcal{S}_{B_s}\varphi^{(s)})}{\partial \nu^{(1)}}\Big|_{\partial B_1}, \cdots, \sum_{s \neq m} \frac{\partial(\mathcal{S}_{B_s}\varphi^{(s)})}{\partial \nu^{(m)}}\Big|_{\partial B_m} \right)$$

is invertible. By Theorem 2.8, T_0 is invertible on X. On the other hand, it is easy to see that T_1 is a compact operator on X. Thus, by the Fredholm alternative, it suffices to show that T is injective on X. If $T(\varphi^{(1)}, \cdots, \varphi^{(m)}) = 0$, then $u(x) := \sum_{l=1}^{m} \mathcal{S}_{B_l}\varphi^{(l)}(x)$, $x \in \mathbb{R}^d$ is the solution of (3.24) with $H = 0$. By the uniqueness of the solution to (3.24), we get $u \equiv 0$. In particular, $\mathcal{S}_{B_l}\varphi^{(l)}$ is smooth across ∂B_l, $l = 1, \ldots, m$. Therefore,

$$\varphi^{(l)} = \frac{\partial(\mathcal{S}_{B_l}\varphi^{(l)})}{\partial \nu^{(l)}}\Big|_{+} - \frac{\partial(\mathcal{S}_{B_l}\varphi^{(l)})}{\partial \nu^{(l)}}\Big|_{-} = 0 \ .$$

This completes the proof. $\square$

Definition 3.13 *Let* $i = (i_1, \ldots, i_d), j = (j_1, \ldots, j_d) \in \mathbb{N}^d$ *be multi-indices. For* $l = 1, \ldots, m$, *let* $\varphi_i^{(l)}$ *be the solution of*

$$(\lambda_l I - \mathcal{K}_{B_l}^*)\varphi_i^{(l)} - \sum_{s \neq l} \frac{\partial(\mathcal{S}_{B_s}\varphi_i^{(s)})}{\partial\nu^{(l)}}\Big|_{\partial B_l} = \frac{\partial x^i}{\partial\nu^{(l)}}\Big|_{\partial B_l} \quad on \ \partial B_l \,. \tag{3.27}$$

Then the polarization tensor M_{ij} *is defined to be*

$$M_{ij} = \sum_{l=1}^{m} \int_{\partial B_l} x^j \varphi_i^{(l)}(x) \, d\sigma(x) \,. \tag{3.28}$$

If $|i| = |j| = 1$, *we denote* M_{ij} *by* $(m_{pq})_{p,q=1}^d$. *We call* $M = (m_{pq})_{p,q=1}^d$ *the first-order polarization tensor.*

3.6.2 Properties

Theorem 3.14 *Suppose that* a_i *and* b_j *are constants such that* $\sum_i a_i y^i$ *and* $\sum_j b_j y^j$ *are harmonic polynomials. Then*

$$\sum_{i,j} a_i b_j M_{ij} = \sum_{i,j} a_i b_j M_{ji} \,. \tag{3.29}$$

Proof. Reasoning as in the proof of Theorem 3.5 we put $f(y) := \sum_i a_i y^i$, $g(y) := \sum_j b_j y^j$, $\varphi^{(l)} := \sum_i a_i \varphi_i^{(l)}$, and $\psi^{(l)} := \sum_j b_j \varphi_j^{(l)}$ to easily see that

$$\sum_{i,j} a_i b_j M_{ij} = \sum_{l=1}^{m} \int_{\partial B_l} g\varphi^{(l)} \, d\sigma \quad and \quad \sum_{i,j} a_i b_j M_{ji} = \sum_{l=1}^{m} \int_{\partial B_l} f\psi^{(l)} \, d\sigma \,.$$

We also put

$$\Phi(x) := \sum_{l=1}^{m} \mathcal{S}_{B_l}\varphi^{(l)} \quad and \quad \Psi(x) := \sum_{l=1}^{m} \mathcal{S}_{B_l}\psi^{(l)} \,.$$

From the definition of $\varphi_i^{(l)}$, one can readily get

$$k_l \frac{\partial(f+\Phi)}{\partial\nu^{(l)}}\Big|_- = \frac{\partial(f+\Phi)}{\partial\nu^{(l)}}\Big|_+ \quad on \ \partial B_l \,, \tag{3.30}$$

and the same relation for $g + \Psi$ holds. From (3.27) we obtain

$$\frac{\partial(\mathcal{S}_{B_l}\varphi^{(l)})}{\partial\nu^{(l)}}\Big|_+ - k_l\frac{\partial(\mathcal{S}_{B_l}\varphi^{(l)})}{\partial\nu^{(l)}}\Big|_- = \sum_i a_i \left[\frac{\partial(\mathcal{S}_{B_l}\varphi_i^{(l)})}{\partial\nu^{(l)}}\Big|_+ - k_l\frac{\partial(\mathcal{S}_{B_l}\varphi_i^{(l)})}{\partial\nu^{(l)}}\Big|_-\right]$$

$$= (k_l - 1)\sum_i a_i \frac{\partial}{\partial\nu^{(l)}}\left[x^i + \sum_{s \neq l} \mathcal{S}_{B_s}\varphi_i^{(s)}\right]$$

$$= (k_l - 1)\frac{\partial}{\partial\nu^{(l)}}\left[f + \sum_{s \neq l} \mathcal{S}_{B_s}\varphi_i^{(s)}\right] \,.$$

Thus, it follows from (3.30) that

$$\varphi^{(l)} = \frac{\partial(\mathcal{S}_{B_l}\varphi^{(l)})}{\partial\nu^{(l)}}\bigg|_+ - \frac{\partial(\mathcal{S}_{B_l}\varphi^{(l)})}{\partial\nu^{(l)}}\bigg|_- = (k_l - 1)\frac{\partial(f + \varPhi)}{\partial\nu^{(l)}}\bigg|_- \quad \text{on } \partial B_l . \quad (3.31)$$

Therefore, we get

$$\sum_{i,j} a_i b_j M_{ij} = \sum_{l=1}^{m}(k_l - 1)\int_{\partial B_l} g\frac{\partial(f + \varPhi)}{\partial\nu}\bigg|_- d\sigma$$

$$= \sum_{l=1}^{m}(k_l - 1)\int_{\partial B_l}(g + \varPsi)\frac{\partial(f + \varPhi)}{\partial\nu}\bigg|_- d\sigma$$

$$- \sum_{l=1}^{m}(k_l - 1)\int_{\partial B_l}\varPsi\frac{\partial(f + \varPhi)}{\partial\nu}\bigg|_- d\sigma$$

$$= \sum_{l=1}^{m}(k_l - 1)\int_{\partial B_l}(g + \varPsi)\frac{\partial(f + \varPhi)}{\partial\nu}\bigg|_- d\sigma$$

$$- \sum_{l=1}^{m}\int_{\partial B_l}\varPsi\left[\frac{\partial(\mathcal{S}_{B_l}\varphi^{(l)})}{\partial\nu}\bigg|_+ - \frac{\partial(\mathcal{S}_{B_l}\varphi^{(l)})}{\partial\nu}\bigg|_-\right]d\sigma .$$

Observe now that

$$\sum_{l=1}^{m}\int_{\partial B_l}\varPsi\frac{\partial(\mathcal{S}_{B_l}\varphi^{(l)})}{\partial\nu}\bigg|_+ d\sigma = \sum_{s,l}\int_{\partial B_l}\mathcal{S}_{B_s}\psi^{(s)}\frac{\partial(\mathcal{S}_{B_l}\varphi^{(l)})}{\partial\nu}\bigg|_+ d\sigma$$

$$= -\sum_{l=1}^{m}\int_{\mathbb{R}^d\setminus\overline{B_l}}\nabla\mathcal{S}_{B_l}\psi^{(l)} \cdot \nabla\mathcal{S}_{B_l}\varphi^{(l)}\, dx$$

$$- \frac{1}{2}\sum_{l\neq s}\int_{\mathbb{R}^d\setminus\overline{B_l\cup B_s}}\nabla\mathcal{S}_{B_s}\psi^{(s)} \cdot \nabla\mathcal{S}_{B_l}\varphi^{(l)}\, dx ,$$

and on the other hand

$$\sum_{l=1}^{m}\int_{\partial B_l}\varPsi\frac{\partial(\mathcal{S}_{B_l}\varphi^{(l)})}{\partial\nu}\bigg|_- d\sigma = \sum_{s,l}\int_{B_l}\nabla\mathcal{S}_{B_s}\psi^{(s)} \cdot \nabla\mathcal{S}_{B_l}\varphi^{(l)}\, dx$$

$$= \sum_{l=1}^{m}\int_{B_l}\nabla\mathcal{S}_{B_l}\psi^{(l)} \cdot \nabla\mathcal{S}_{B_l}\varphi^{(l)}\, dx + \frac{1}{2}\sum_{s\neq l}\int_{B_l\cup B_s}\nabla\mathcal{S}_{B_s}\psi^{(s)} \cdot \nabla\mathcal{S}_{B_l}\varphi^{(l)}\, dx ,$$

to finally arrive at

$$\sum_{i,j} a_i b_j M_{ij} = \sum_{l=1}^{m}(k_l - 1)\langle(g + \varPsi), (f + \varPhi)\rangle_{B_l}$$

$$+ \sum_{l=1}^{m}\langle\mathcal{S}_{B_l}\psi^{(l)}, \mathcal{S}_{B_l}\varphi^{(l)}\rangle_{\mathbb{R}^d} + \frac{1}{2}\sum_{s\neq l}\langle\mathcal{S}_{B_s}\psi^{(s)}, \mathcal{S}_{B_l}\varphi^{(l)}\rangle_{\mathbb{R}^d} .$$

$$(3.32)$$

Here, the notation $\langle u, v \rangle_D := \int_D \nabla u \cdot \nabla v \, dx$ has been used. The symmetry (3.29) follows immediately from (3.32) and the proof is complete.   □

Theorem 3.15 *Suppose that either $k_l - 1 > 0$ or $k_l - 1 < 0$ for all $l = 1, \ldots, m$. Let*

$$\kappa := \max_{1 \le l \le m} \left| 1 - \frac{1}{k_l} \right| .$$

For any a_i such that $\sum_i a_i y^i$ is harmonic,

$$\left| \sum_{i,j} a_i a_j M_{ij} \right| \ge \frac{|\kappa - 1|}{m + 1} \sum_{l=1}^{m} |k_l - 1| \int_{B_l} \left| \nabla (\sum_i a_i y^i) \right|^2 dy . \tag{3.33}$$

In particular, if $k_l - 1 > 0$ (resp. < 0) for all $l = 1, \ldots, m$, then $M = (m_{pq})_{p,q=1}^{d}$ is positive (resp. negative) definite and if $\sum_{p=1}^{d} a_p^2 = 1$, then

$$\left| \sum_{p,q=1}^{d} a_p a_q m_{pq} \right| \ge \frac{|\kappa - 1|}{m + 1} \sum_{l=1}^{m} |k_l - 1| \, |B_l| .$$

Proof.   Suppose that either $k_l - 1 > 0$ or $k_l - 1 < 0$ for all $l = 1, \ldots, m$. Recall that the quadratic form $Q_D(u)$ is defined by $Q_D(u) := \langle u, u \rangle_D$. It then follows from (3.32) that

$$\sum_{i,j} a_i a_j M_{ij} = \sum_{l=1}^{m} (k_l - 1) Q_{B_l}(f + \Phi) + \sum_{l=1}^{m} Q_{\mathbb{R}^d}(\mathcal{S}_{B_l} \varphi^{(l)})$$

$$+ \frac{1}{2} \sum_{s \ne l} \langle \mathcal{S}_{B_s} \varphi^{(s)}, \mathcal{S}_{B_l} \varphi^{(l)} \rangle_{\mathbb{R}^d} \tag{3.34}$$

$$= \sum_{l=1}^{m} (k_l - 1) Q_{B_l}(f + \Phi) + Q_{\mathbb{R}^d}(\Phi) .$$

On the other hand, because of (3.30), we get

$$(k_l - 1) \frac{\partial f}{\partial \nu^{(l)}} = \left. \frac{\partial \Phi}{\partial \nu^{(l)}} \right|_{+} - k_l \left. \frac{\partial \Phi}{\partial \nu^{(l)}} \right|_{-} \quad \text{on } \partial B_l, \quad l = 1, \ldots, d .$$

Thus, it follows from (3.31) that

$$\sum_{i,j} a_i a_j M_{ij} \tag{3.35}$$

$$= \sum_{l=1}^{m} (k_l - 1) \int_{\partial B_l} f \frac{\partial (f + \Phi)}{\partial \nu} \Big|_{-} \, d\sigma$$

$$= \sum_{l=1}^{m} (k_l - 1) Q_{B_l}(f) + \sum_{l=1}^{m} (k_l - 1) \int_{\partial B_l} \frac{\partial f}{\partial \nu} \Phi \, d\sigma$$

$$= \sum_{l=1}^{m} (k_l - 1) Q_{B_l}(f) + \sum_{l=1}^{m} \int_{\partial B_l} \frac{\partial \Phi}{\partial \nu} \Big|_{+} \Phi \, d\sigma - \sum_{l=1}^{m} k_l \int_{\partial B_l} \frac{\partial \Phi}{\partial \nu} \Big|_{-} \Phi \, d\sigma$$

$$= \sum_{l=1}^{m} (k_l - 1) Q_{B_l}(f) - \sum_{l=1}^{m} Q_{\mathbb{R}^d}(\Phi) - \sum_{l=1}^{m} (k_l - 1) Q_{B_l}(\Phi) . \tag{3.36}$$

By equating (3.34) and (3.36) we have

$$\sum_{l=1}^{m} (k_l - 1) Q_{B_l}(f + \Phi) + Q_{\mathbb{R}^d}(\Phi)$$

$$= \sum_{l=1}^{m} (k_l - 1) Q_{B_l}(f) - \sum_{l=1}^{m} Q_{\mathbb{R}^d}(\Phi) - \sum_{l=1}^{m} (k_l - 1) Q_{B_l}(\Phi) . \tag{3.37}$$

and consequently, one can claim that

$$\sum_{l=1}^{m} (k_l - 1) Q_{B_l}(f) \geq \sum_{l=1}^{m} k_l Q_{B_l}(\Phi) , \tag{3.38}$$

since the left-hand side of (3.37) is positive. It also follows from (3.37) that

$$Q_{\mathbb{R}^d}(\Phi) = \frac{1}{m+1} \sum_{l=1}^{m} (k_l - 1) \Big[Q_{B_l}(f) - Q_{B_l}(f + \Phi) - Q_{B_l}(\Phi) \Big] . \tag{3.39}$$

Substituting (3.39) into (3.34), we obtain

$$\sum_{i,j} a_i a_j M_{ij} = \frac{m}{m+1} \sum_{l=1}^{m} (k_l - 1) Q_{B_l}(f + \Phi)$$

$$+ \frac{1}{m+1} \sum_{l=1}^{m} (k_l - 1) \Big[Q_{B_l}(f) - Q_{B_l}(\Phi) \Big] ,$$

and hence

$$\sum_{i,j} a_i a_j M_{ij} \geq \frac{1}{m+1} \sum_{l=1}^{m} (k_l - 1) \Big[Q_{B_l}(f) - Q_{B_l}(\Phi) \Big] . \tag{3.40}$$

But by (3.38) we get

$$\sum_{l=1}^{m}(k_l - 1)Q_{B_l}(\varPhi) = \sum_{l=1}^{m}\frac{(k_l - 1)}{k_l}k_l Q_{B_l}(\varPhi)$$

$$\leq \kappa\sum_{l=1}^{m} k_l Q_{B_l}(\varPhi) \leq \kappa\sum_{l=1}^{m}(k_l - 1)Q_{B_l}(f) ,$$

and hence (3.33) follows immediately from (3.40). This completes the proof.
□

Based on the definition (3.28), polarizations tensors associated with multiple disks and balls are explicitly computed in [21, 197]. It should also be noted that Cheng and Greengard gave in Theorem 2.2 of their interesting paper [81] a solution to the two- and three-disk conductivity problem based on a method of images.

3.6.3 Representation by Equivalent Ellipses

Suppose $d = 2$, and let $M = (m_{pq})_{p,q=1}^2$ be the first-order polarization tensor of the inclusions $\cup_{s=1}^m B_s$. We define the overall conductivity $\bar{k}$ of $B = \cup_{s=1}^m B_s$ by

$$\frac{\bar{k} - 1}{\bar{k} + 1}\sum_{s=1}^{m}|B_s| := \sum_{s=1}^{m}\frac{k_s - 1}{k_s + 1}|B_s| , \qquad (3.41)$$

and its *center* $\bar{z}$ by

$$\frac{\bar{k} - 1}{\bar{k} + 1}\bar{z}\sum_{s=1}^{m}|B_s| = \sum_{s=1}^{m}\frac{k_s - 1}{k_s + 1}\int_{B_s} x\, dx . \qquad (3.42)$$

Note that if k_s is the same for all s then $\bar{k} = k_s$ and $\bar{z}$ is the center of mass of B.

In this section we represent and visualize the multiple inclusions $\cup_{s=1}^m B_s$ by means of an ellipse, $\mathcal{E}$, of center $\bar{z}$ with the same polarization tensor. We call $\mathcal{E}$ the equivalent ellipse of $\cup_{s=1}^m B_s$.

At this point let us review a method to find an ellipse from a given first-order polarization tensor. This method is due to Brühl, Hanke, and Vogelius [64]. Let $\mathcal{E}'$ be an ellipse whose focal line is on either the x_1- or the x_2-axis. We suppose that its semi-major axis is of length a and its semi-minor axis is of length b. Let $\mathcal{E} = \mathcal{R}\mathcal{E}'$ where $R = \begin{pmatrix} \cos\theta & -\sin\theta \\ \sin\theta & \cos\theta \end{pmatrix}$ and $\theta \in [0, \pi]$. Let M be the polarization tensor of $\mathcal{E}$. We want to recover a, b, and θ from M knowing the conductivity $k = \bar{k}$.

Recall that the polarization tensor M' for $\mathcal{E}'$ takes the form

$$M' = (k - 1)|\mathcal{E}'|\begin{pmatrix} \frac{a+b}{a+kb} & 0 \\ 0 & \frac{a+b}{b+ka} \end{pmatrix} ,$$

and that of $\mathcal{E}$ is given by $M = RM'R^T$. Suppose that the eigenvalues of M are λ_1 and λ_2, and corresponding eigenvectors of unit length are $(e_{11}, e_{12})^T$ and $(e_{21}, e_{22})^T$. Then it can be shown that

$$a = \sqrt{\frac{p}{\pi q}}, \quad b = \sqrt{\frac{pq}{\pi}}, \quad \theta = \arctan \frac{e_{21}}{e_{11}},$$

where

$$\frac{1}{p} = \frac{k-1}{k+1}\left(\frac{1}{\lambda_1} + \frac{1}{\lambda_2}\right) \quad \text{and} \quad q = \frac{\lambda_2 - k\lambda_1}{\lambda_1 - k\lambda_2}.$$

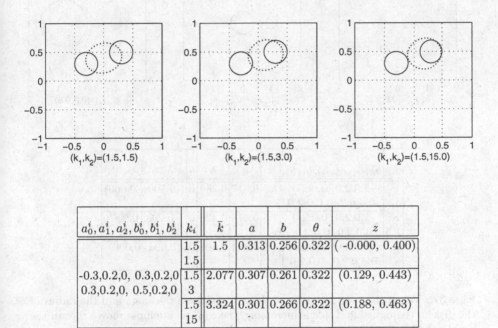

$a_0^i, a_1^i, a_2^i, b_0^i, b_1^i, b_2^i$	k_i	$\bar{k}$	a	b	θ	z
	1.5	1.5	0.313	0.256	0.322	(-0.000, 0.400)
	1.5					
-0.3,0.2,0, 0.3,0.2,0	1.5	2.077	0.307	0.261	0.322	(0.129, 0.443)
0.3,0.2,0, 0.5,0.2,0	3					
	1.5	3.324	0.301	0.266	0.322	(0.188, 0.463)
	15					

Fig. 3.1. When the two disks have the same radius and the conductivity of the one on the right-hand side is increasing, the equivalent ellipse moves toward the right inclusion. In the table $\bar{k}$ and $\bar{z}$ are the overall conductivity and center defined by (3.41) and (3.42) and a, b, θ are the semi-axes lengths and angle of orientation measured in radians of the equivalent ellipse.

We now show some numerical examples of equivalent ellipses. We represent the set of inclusions $B = \cup_{s=1}^m B_s$ by an equivalent ellipse of center $\bar{z}$ and conductivity $\bar{k}$. We assume that the inclusion B_s takes the following form:

$$\partial B_s = \left\{ \left(a_0^s + a_1^s \cos(t) + a_2^s \cos(2t), b_0^s + b_1^s \sin(t) + b_2^s \sin(2t)\right), 0 \le t < 2\pi \right\}.$$

In order to evaluate the first-order polarization tensor of multiple inclusions, we solve the integral equation (2.41) with $H(x) = x_p$ to find $\varphi_p^{(s)}$ for $p = 1, 2$ and $s = 1, \ldots, m$, and then calculate $m_{pq} = \sum_{s=1}^{m} \int_{\partial B_s} x_q \varphi_p^{(s)}(x) \, d\sigma(x)$.

Figures 3.1 and 3.2 show how the equivalent ellipse changes as the conductivities and the sizes of the inclusions B_s vary. The solid line represents the actual inclusions and the dashed lines are the effective ellipses.

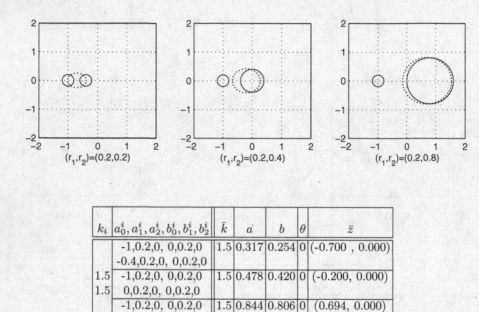

k_i	$a_0^i, a_1^i, a_2^i, b_0^i, b_1^i, b_2^i$	$\bar{k}$	a	b	θ	$\bar{z}$
	-1,0.2,0, 0,0.2,0 -0.4,0.2,0, 0,0.2,0	1.5	0.317	0.254	0	(-0.700 , 0.000)
1.5 1.5	-1,0.2,0, 0,0.2,0 0,0.2,0, 0,0.2,0	1.5	0.478	0.420	0	(-0.200, 0.000)
	-1,0.2,0, 0,0.2,0 0.8,0.2,0, 0,0.2,0	1.5	0.844	0.806	0	(0.694, 0.000)

Fig. 3.2. When the conductivities of the two disks are the same and the radius of the disk on the right-hand side is increasing, the equivalent ellipse moves toward the right inclusion.

4

Derivation of the Full Asymptotic Formula

In this chapter we derive full asymptotic expansions of the steady-state voltage potentials in the presence of a finite number of diametrically small inclusions with conductivities different from the background conductivity. The derivation is rigorous and based on layer potential techniques and the decomposition formula (2.39) of the steady-state voltage potential into a harmonic part and a refraction part. The asymptotic expansions in this chapter are valid for inclusions with Lipschitz boundaries and those with extreme conductivities (zero or infinite conductivity).

The main result of this chapter is the following full asymptotic expansion of the solution u of

$$\begin{cases} \nabla \cdot \left(\chi\left(\Omega \setminus \bigcup_{s=1}^{m} \overline{D_s} \right) + \sum_{s=1}^{m} k_s \chi(D_s) \right) \nabla u = 0 & \text{in } \Omega \,, \\ \dfrac{\partial u}{\partial \nu} \bigg|_{\partial \Omega} = g \,, \end{cases} \tag{4.1}$$

for the case $m = 1$. The leading-order term in this asymptotic formula which expresses the fact that the conductivity inclusion can be modeled by a dipole has been derived by Cedio-Fengya, Moskow, and Vogelius [73]; see also the prior work of Friedman and Vogelius [123] for the case of perfectly conducting or insulating inclusions. A very general representation formula for the boundary voltage perturbations caused by internal conductivity inclusions of small volume fraction has been obtained by Capdeboscq and Vogelius in their interesting paper [70].

Theorem 4.1 *Suppose that the inclusion consists of a single component, $D = \epsilon B + z$, and let u be the solution of (4.1). The following pointwise asymptotic expansion on $\partial \Omega$ holds for $d = 2, 3$:*

$$u(x) = U(x) - \epsilon^{d-2} \sum_{|i|=1}^{n} \sum_{|j|=1}^{n-|i|+1} \frac{\epsilon^{|i|+|j|}}{i!j!}$$

$$\times \left[\left(\left(I + \sum_{p=1}^{n+2-|i|-|j|-d} \epsilon^{d+p-1} \mathcal{Q}_p \right) (\partial^l U(z)) \right)_i M_{ij} \partial_z^j N(x,z) \right] \tag{4.2}$$

$$+ O(\epsilon^{d+n}),$$

where the remainder $O(\epsilon^{d+n})$ is dominated by $C\epsilon^{d+n}\|g\|_{L^2(\partial\Omega)}$ for some C independent of $x \in \partial\Omega$. Here $N(x,z)$ is the Neumann function, that is, the solution to (2.30), M_{ij}, $i,j \in \mathbb{N}^d$, are the generalized polarization tensors introduced in (3.1), and the matrix $\mathcal{Q}_p$ is defined in (4.17).

In particular, if $n = d$ then we simplify formula (4.2) to obtain:

$$u(x) = U(x) - \sum_{|i|=1}^{d} \sum_{|j|=1}^{d-|i|+1} \frac{\epsilon^{|i|+|j|+d-2}}{i!j!} \partial^i U(z) M_{ij} \partial_z^j N(x,z) + O(\epsilon^{2d}). \tag{4.3}$$

We have a similar expansion for the solutions of the Dirichlet problem (Theorem 4.7).

In the expression (4.3), the remainder $O(\epsilon^{2d})$ is dominated by $C'\epsilon^{2d}$, where the constant C' can be precisely quantified in terms of the Lipschitz character of B and dist$(D, \partial\Omega)$ (see [29, 13]).

The constant C' blows up if dist$(D, \partial\Omega) \to 0$ or B has a "bad" Lipschitz character, *i.e.*, the constant C in (2.28) goes to $+\infty$ (or, in view of Lemma 2.9, $\delta(B) = \min_{x \in \partial B}\langle x, \nu_x \rangle \to 0$ if B is a star-shaped domain with respect to the origin in two-dimensional space).

When B has a "bad" Lipschitz character we must use, in place of (4.3), the asymptotic formula corresponding to a small thin inclusion, which has been formally derived by Beretta, Mukherjee, and Vogelius in [53] and rigorously justified by Beretta, Francini, and Vogelius in their recent paper [52]; see also [51].

In the case where the small inclusion is nearly touching the boundary (dist$(D, \partial\Omega) \to 0$) a more complicated asymptotic formula established in [13] should be used instead of (4.3). The dipole-type expansion (4.3) is valid when the potential u within the inclusion D is nearly constant. On decreasing dist$(D, \partial\Omega)$ this assumption begins to fail because higher-order multi-poles become significant due to the interaction between D and $\partial\Omega$. Our approximation in [13] provides some essential insight for understanding this interaction.

The derivation of the asymptotic expansions for any fixed number m of well-separated inclusions (these are a fixed distance apart) follows by iteration of the arguments that we will present for the case $m = 1$. In other words, we may develop asymptotic formulae involving the difference between the fields u and U on $\partial\Omega$ with s inclusions and those with $s-1$ inclusions, $s = m, \ldots, 1$, and then at the end essentially form the sum of these m formulae (the reference fields change, but that may easily be remedied). The derivation of each of the

m formulae is virtually identical. If D takes the form $D = \cup_{s=1}^m (\epsilon B_s + z_s)$. The conductivity of the inclusion $\epsilon B_s + z_s$ is k_s, $s = 1, \ldots, m$. By iterating the formula (4.3) we can derive the following expansion in the case when there are several well separated inclusions:

$$u(x) = U(x) - \sum_{s=1}^m \sum_{|i|=1}^d \sum_{|j|=1}^{d-|i|+1} \frac{\epsilon^{|i|+|j|+d-2}}{i!j!} \partial^i U(z_s) M_{ij}(k_s, B_s) \partial_z^j N(x, z_s)$$

$$+ O(\epsilon^{2d}) .$$

$$(4.4)$$

As stated in the above theorem, we restrict our derivation to the case of a single inclusion ($m = 1$). We only give the details when considering the difference between the fields corresponding to one and zero inclusions. In order to further simplify notation we assume that the single inclusion D has the form $D = \epsilon B + z$, where $z \in \Omega$ and B is a bounded Lipschitz domain in $\mathbb{R}^d$ containing the origin. Suppose that the conductivity of D is a positive constant $k \neq 1$. Let $\lambda := k + 1/(2(k-1))$ as before. Then by (2.39) and (2.41), the solution u of (2.37) takes the form

$$u(x) = U(x) - N_D(\lambda I - \mathcal{K}_D^*)^{-1}\left(\frac{\partial H}{\partial \nu}\Big|_{\partial D}\right)(x) , \quad x \in \partial \Omega ,$$

where U is the background potential given in (2.38).

4.1 Energy Estimates

Let us begin with the following estimate of the trace of $u - U$ on the boundary $\partial \Omega$.

Proposition 4.2 *If $\partial \Omega$ and ∂D are Lipschitz then there exists a positive constant C independent of ϵ, k, and g such that, for ϵ small enough,*

$$\|u - U\|_{L^2(\partial \Omega)} \leq C(k-1)\|g\|_{L^2(\partial \Omega)}\epsilon^d \quad \text{if } k > 1 ,$$

$$\leq C(\frac{1}{k} - 1)\|g\|_{L^2(\partial \Omega)}\epsilon^d \quad \text{if } 0 < k < 1 .$$

Proof. We first observe that

$$\mathcal{D}_\Omega(u - U)(x) = H(x) , \quad x \in \mathbb{R}^d \setminus \overline{\Omega} , \qquad (4.5)$$

which follows immediately from the fact that $\mathcal{D}_\Omega(U)(x) - \mathcal{S}_\Omega(g)(x) = 0$ for $x \in \mathbb{R}^d \setminus \overline{\Omega}$.

Recall that according to (2.28) there exists a positive constant C that depends only on the Lipschitz character of Ω such that

$$\|f\|_{L^2(\partial\Omega)} \leq C \left\|(-\frac{1}{2}I + \mathcal{K}_\Omega)f\right\|_{L^2(\partial\Omega)} \qquad \forall f \in L_0^2(\partial\Omega) . \qquad (4.6)$$

Employing this inequality, we write

$$\|u - U\|_{L^2(\partial\Omega)}^2 \leq C\|(-\frac{1}{2}I + \mathcal{K}_\Omega)(u - U)\|_{L^2(\partial\Omega)}^2 .$$

It then follows from the jump formula (2.13) that

$$\|u - U\|_{L^2(\partial\Omega)}^2 \leq C \lim_{t \to 0^+} \int_{\partial\Omega} \left| \mathcal{D}_\Omega(u - U)(x + t\nu_x) \right|^2 d\sigma(x) ,$$

which gives with the help of (2.46) and (4.5) that

$$\|u - U\|_{L^2(\partial\Omega)}^2 \leq C(k - 1)^2 \int_{\partial\Omega} \left| \int_D \nabla_y \Gamma(x - y) \cdot \nabla u(y) \, dy \right|^2 d\sigma(x) .$$

Thus we get by the Cauchy–Schwarz inequality

$$\|u - U\|_{L^2(\partial\Omega)}^2 \leq C(k - 1)^2 (\int_D |\nabla u(y)|^2 \, dy) \int_{\partial\Omega} (\int_D |\nabla_y \Gamma(x - y)|^2 \, dy) \, d\sigma(x) .$$

$$(4.7)$$

If $k > 1$, then using the energy identity (2.58) we arrive at

$$\|u - U\|_{L^2(\partial\Omega)} \leq C(k - 1)\|g\|_{L^2(\partial\Omega)} \int_{\partial\Omega} (\int_D \cdot |\nabla_y \Gamma(x - y)|^2 \, dy) \, d\sigma(x) .$$

But there exists a positive constant C depending only on $|B|$ and $\mathrm{dist}(D, \partial\Omega)$ such that

$$\int_{\partial\Omega} (\int_D |\nabla_y \Gamma(x - y)|^2 \, dy) \, d\sigma(x) \leq C\epsilon^d ,$$

for ϵ small enough. Inserting this into the above inequality immediately yields the desired estimate for $k > 1$.

If $0 < k < 1$, then by using (2.59) we have

$$\int_D |\nabla u(y)|^2 \, dy \leq 2 \int_D |\nabla(u - U)(y)|^2 \, dy + 2 \int_D |\nabla U(y)|^2 \, dy$$

$$\leq \frac{2}{k} \int_\Omega \left(1 + (k - 1)\chi(D) \right) |\nabla(u - U)(y)|^2 \, dy$$

$$+ 2 \int_D |\nabla U(y)|^2 \, dy$$

$$\leq \frac{2}{k(1 - k)} \|u - U\|_{L^2(\partial\Omega)} \|g\|_{L^2(\partial\Omega)} .$$

Here we have used the energy identity (2.59). Combining (4.7) with the above estimate, we deduce that for $0 < k < 1$ the desired estimate holds and the proof of the proposition is then complete. $\square$

As a direct consequence of Proposition 4.2 and its proof we get the following corollary.

Corollary 4.3 Let $0 < k \neq 1 < +\infty$. There exists a constant $C(k)$ independent of ϵ such that

$$\|\nabla u\|_{L^2(D)} \leq C(k)\,\epsilon^{\frac{d}{2}}\,.$$

Next we employ the Rellich identity stated in Lemma 2.6 to estimate the L^2-norm of the tangential derivative of $u - U$ on the boundary $\partial\Omega$ as ϵ goes to zero.

Lemma 4.4 Let T_x be the tangent vector to $\partial\Omega$ at x. If $\partial\Omega$ is Lipschitz then there exists a positive constant C depending only on the Lipschitz character of $\partial\Omega$ such that

$$\left\|\frac{\partial}{\partial T}(u - U)\right\|_{L^2(\partial\Omega)} \leq C\,\|g\|_{L^2(\partial\Omega)}\epsilon^{\frac{d}{2}}\,. \tag{4.8}$$

Proof.   Let

$$\Omega_\epsilon = \left\{ x \in \Omega, \operatorname{dist}(x, \partial\Omega) > \left(C - \max_{x \in \partial B} |x|\right)\epsilon \right\}$$

and $\boldsymbol{\alpha}$ be a smooth vector field such that the support of $\boldsymbol{\alpha}$ lies in $\mathbb{R}^d \setminus \overline{\Omega_\epsilon}$ and $\langle \boldsymbol{\alpha}, \nu \rangle > c_1 > 0$ on $\partial\Omega$ (here, c_1 depends only on the Lipschitz character of $\partial\Omega$). Using the Rellich identity (2.20) with this $\boldsymbol{\alpha}$ we obtain

$$\int_{\partial\Omega} \langle \boldsymbol{\alpha}, \nu \rangle \left|\frac{\partial}{\partial T}(u-U)\right|^2 = \int_{\Omega} -2\langle \nabla\boldsymbol{\alpha}\nabla(u-U), \nabla(u-U)\rangle + (\nabla\cdot\boldsymbol{\alpha})|\nabla(u-U)|^2\,,$$

since $\partial(u - U)/\partial\nu = 0$ on $\partial\Omega$. Hence

$$\int_{\partial\Omega} \langle \boldsymbol{\alpha}, \nu \rangle \left|\frac{\partial}{\partial T}(u - U)\right|^2 \leq C \int_{\Omega} |\nabla(u-U)|^2. \tag{4.9}$$

Combining the energy identity (2.58) together with Proposition 4.2 leads us to the estimates

$$\begin{aligned}
\int_{\Omega} |\nabla(u - U)|^2 &\leq \int_{\partial\Omega} (U - u)g \\
&\leq \|U - u\|_{L^2(\partial\Omega)}\,\|g\|_{L^2(\partial\Omega)} \\
&\leq C\epsilon^d \|g\|_{L^2(\partial\Omega)}\,.
\end{aligned}$$

Therefore (4.9) implies that the estimate (4.8) holds.   □

Proposition 4.5 If $\partial\Omega$ is of class C^2 then there exists a positive constant C that is independent of ϵ, k, and g such that

$$\|u - U\|_{L^\infty(\partial\Omega)} \leq C\,|k - 1|\,\|g\|_{L^2(\partial\Omega)}\epsilon^d\,,$$

for ϵ small enough.

Proof. Since $\partial\Omega$ is of class $\mathcal{C}^2$ then

$$\|u - U\|_{L^\infty(\partial\Omega)} \leq C \left\|(-\frac{1}{2}I + \mathcal{K}_\Omega)(u - U)\right\|_{L^\infty(\partial\Omega)},$$

where C depends only on the $\mathcal{C}^2$ character of Ω and therefore

$$\|u - U\|_{L^\infty(\partial\Omega)} \leq C \lim_{t\to 0^+} \sup_{x\in\partial\Omega} \left|\mathcal{D}_\Omega(u - U)(x + t\nu_x)\right|.$$

Using (2.46), we readily get

$$\|u - U\|_{L^\infty(\partial\Omega)} \leq C|k - 1| \sup_{x\in\partial\Omega} \left|\int_D \nabla_y \Gamma(x - y) \cdot \nabla u(y)\, dy\right|$$

$$\leq C|k - 1| \sup_{x\in\partial\Omega} \left(\int_D |\nabla_y \Gamma(x - y)|^2\, dy\right)^{\frac{1}{2}} \|\nabla u\|_{L^2(D)}.$$

Since

$$\|\nabla u\|_{L^2(D)} \leq C\epsilon^{d/2}$$

by Corollary 4.3 and $\sup_{x\in\partial\Omega}(\int_D |\nabla_y \Gamma(x - y)|^2\, dy)^{1/2}$ is bounded by $C\epsilon^{d/2}$, we obtain the desired result. $\square$

Next, the following estimates hold.

Proposition 4.6 *(i) If Ω is Lipschitz, then*

$$\|H - U\|_{L^2(\partial\Omega)} \leq C\|u - U\|_{L^2(\partial\Omega)} \leq C|k - 1| \|g\|_{L^2(\partial\Omega)}\epsilon^d.$$

(ii) If Ω is of class $\mathcal{C}^2$, then

$$\|H - U\|_{L^\infty(\partial\Omega)} \leq C\|u - U\|_{L^\infty(\partial\Omega)} \leq C|k - 1| \|g\|_{L^2(\partial\Omega)}\epsilon^d.$$

(iii) If Ω is of class $\mathcal{C}^2$, then

$$\|H - U\|_{L^\infty(\overline{\Omega})} \leq C|k - 1| \|g\|_{L^2(\partial\Omega)}\epsilon^d.$$

(iv) If Ω is of class $\mathcal{C}^2$, then

$$\|H - U\|_{W^{1,2}(\Omega)} \leq C|k - 1|^{\frac{1}{2}} \|g\|_{L^2(\partial\Omega)}\epsilon^{\frac{3d}{4}}.$$

(v) If Ω and $D \subset\subset \Omega$ are Lipschitz, then

$$\|\nabla H - \nabla U\|_{L^\infty(\overline{D})} \leq C\|u - U\|_{L^2(\partial\Omega)} \leq C|k - 1|\|g\|_{L^2(\partial\Omega)}\epsilon^d,$$

where C depends on $\text{dist}(\partial\Omega, D)$.

Proof.   (i) and (ii) follow from the fact that

$$H - U = (\frac{1}{2}I + \mathcal{K}_\Omega)(u - U) \quad \text{on } \partial\Omega$$

together with

$$||\mathcal{K}_\Omega v||_{L^2(\partial\Omega)} \leq C||v||_{L^2(\partial\Omega)} \quad \text{if } \Omega \text{ is Lipschitz,}$$
$$||\mathcal{K}_\Omega v||_{L^\infty(\partial\Omega)} \leq C'||v||_{L^\infty(\partial\Omega)} \quad \text{if } \Omega \text{ is of class } \mathcal{C}^2,$$

where the constants C and C' depend only on the Lipschitz and $\mathcal{C}^2$ characters of Ω, respectively. (iii) is a direct application of the maximum principle to the harmonic function $H - U$ and (ii).

To prove (iv) we write

$$||\nabla(H - U)||_{L^2(\Omega)}^2 = \int_{\partial\Omega} \frac{\partial}{\partial\nu}(H - U)(H - U)$$
$$\leq ||H - U||_{L^2(\partial\Omega)}||\frac{\partial}{\partial\nu}(H - U)||_{L^2(\partial\Omega)}.$$

Since, by using the fact from Theorem 2.11 that $\mathcal{K}_\Omega : W_1^2(\partial\Omega) \to W_1^2(\partial\Omega)$ is a bounded operator together with Lemma 4.4, we have

$$||\frac{\partial}{\partial\nu}(H-U)||_{L^2(\partial\Omega)} \leq C||H-U||_{W_1^2(\partial\Omega)} \leq C||u-U||_{W_1^2(\partial\Omega)} \leq C\epsilon^{\frac{d}{2}}||g||_{L^2(\partial\Omega)}.$$

Therefore we get

$$||\nabla(H - U)||_{L^2(\Omega)}^2 \leq C|k - 1|||g||_{L^2(\partial\Omega)}^2 \epsilon^{\frac{3d}{2}}.$$

Now, since $\nabla(H - U) = \nabla\mathcal{D}_\Omega(u - U)$, we obtain

$$||\nabla H - \nabla U||_{L^\infty(\overline{D})} \leq \sup_{x \in \overline{D}} \int_{\partial\Omega} |\nabla_x\Gamma(x - y)|^2 \, d\sigma(y) \, ||u - U||_{L^2(\partial\Omega)},$$

and consequently (v) holds, where the constant C depends on $\text{dist}(\partial\Omega, D)$. This finishes the proof of the proposition. $\quad\square$

4.2 Asymptotic Expansion

Define

$$H_n(x) := \sum_{|i|=0}^{n} \frac{1}{i!}(\partial^i H)(z)(x - z)^i.$$

Here we use the multi-index notation $i = (i_1, \ldots, i_d) \in \mathbb{N}^d$. Then we have from (2.42) that

$$\left\|\frac{\partial H}{\partial \nu} - \frac{\partial H_n}{\partial \nu}\right\|_{L^2(\partial D)} \leq \sup_{x \in \partial D} |\nabla H(x) - \nabla H_n(x)||\partial D|^{1/2}$$

$$\leq \|H\|_{C^{n+1}(\overline{D})}|x-z|^n|\partial D|^{1/2}$$

$$\leq C\|g\|_{L^2(\partial \Omega)}\epsilon^n|\partial D|^{1/2} .$$

Note that

$$\text{if } \int_{\partial D} h\,d\sigma = 0 , \text{ then } \int_{\partial D} (\lambda I - \mathcal{K}_D^*)^{-1}h\,d\sigma = 0 . \qquad (4.10)$$

If $\int_{\partial D} h\,d\sigma = 0$, then we have for $x \in \partial \Omega$ that

$$\left|N_D(\lambda I - \mathcal{K}_D^*)^{-1}h(x)\right| = \left|\int_{\partial D}\Big[N(x,y) - N(x,z)\Big](\lambda I - \mathcal{K}_D^*)^{-1}h(y)\,d\sigma(y)\right|$$

$$\leq C\epsilon|\partial D|^{1/2}\,\|h\|_{L^2(\partial D)} .$$

It then follows that

$$\sup_{x \in \partial D}\left|N_D(\lambda I - \mathcal{K}_D^*)^{-1}\left(\frac{\partial H}{\partial \nu}|\partial D - \frac{\partial H_n}{\partial \nu}|\partial D\right)(x)\right|$$

$$\leq C\epsilon|\partial D|^{1/2}\left\|\frac{\partial H}{\partial \nu} - \frac{\partial H_n}{\partial \nu}\right\|_{L^2(\partial D)} \leq C\|g\|_{L^2(\partial \Omega)}\epsilon^{d+n} .$$

Therefore, we have

$$u(x) = U(x) - N_D(\lambda I - \mathcal{K}_D^*)^{-1}\left(\frac{\partial H_n}{\partial \nu}|\partial D\right)(x) + O(\epsilon^{d+n}), \quad x \in \partial \Omega , \quad (4.11)$$

where the $O(\epsilon^{d+n})$ term is dominated by $C\|g\|_{L^2(\partial \Omega)}\epsilon^{d+n}$ for some C depending only on c_0. Note that

$$(\lambda I - \mathcal{K}_D^*)^{-1}\left(\frac{\partial H_n}{\partial \nu}|\partial D\right)(x) = \sum_{|i|=1}^{n}(\partial^i H)(z)(\lambda I - \mathcal{K}_D^*)^{-1}\left(\frac{1}{i!}\nu_x \cdot \nabla(x-z)^i\right)(x) .$$

Since $D = \epsilon B + z$, one can prove by using the change of variables $y = (x - z)/\epsilon$ and the expression of $\mathcal{K}_D^*$ defined by (2.14) that

$$(\lambda I - \mathcal{K}_D^*)^{-1}\left(\frac{1}{i!}\nu_x \cdot \nabla(x-z)^i\right)(x) = \epsilon^{|i|-1}(\lambda I - \mathcal{K}_B^*)^{-1}\left(\frac{1}{i!}\nu_y \cdot \nabla y^i\right)\left(\frac{1}{\epsilon}(x-z)\right) .$$

Put

$$\phi_i(x) := (\lambda I - \mathcal{K}_B^*)^{-1}\left(\nu_y \cdot \nabla y^i\right)(x), \quad x \in \partial B . \qquad (4.12)$$

Then we get

$$N_D(\lambda I - \mathcal{K}_D^*)^{-1}\left(\frac{\partial H_n}{\partial \nu}\bigg|_{\partial D}\right)(x) \tag{4.13}$$

$$= \sum_{|i|=1}^n \frac{1}{i!}(\partial^i H)(z)\epsilon^{|i|-1}\int_{\partial D} N(x,y)\phi_i(\epsilon^{-1}(y-z))\,d\sigma(y)$$

$$= \sum_{|i|=1}^n \frac{1}{i!}(\partial^i H)(z)\epsilon^{|i|+d-2}\int_{\partial B} N(x,\epsilon y + z)\phi_i(y)\,d\sigma(y)\,.$$

We now have from (2.35) and (4.13)

$$N_D(\lambda I - \mathcal{K}_D^*)^{-1}\left(\frac{\partial H_n}{\partial \nu}\bigg|_{\partial D}\right)(x)$$

$$= \sum_{|i|=1}^n \frac{1}{i!}(\partial^i H)(z)\epsilon^{|i|+d-2}\sum_{|j|=0}^{+\infty}\frac{1}{j!}\epsilon^{|j|}\partial_z^j N(x,z)\int_{\partial B} y^j \phi_i(y)d\sigma(y)\,.$$

Observe that since H is a harmonic function in Ω we may compute

$$\sum_{|i|=l}\frac{1}{i!}(\partial^i H)(z)\Delta(y^i) = \Delta_y\left(\sum_{|i|=l}\frac{1}{i!}(\partial^i H)(z)y^i\right) = 0\,,$$

and therefore, by Green's theorem, it follows that

$$\int_{\partial B}\sum_{|i|=l}\frac{1}{i!}(\partial^i H)(z)\nabla(y^i)\cdot\nu_y\,d\sigma(y) = 0\,.$$

Thus, in view of (4.10) and (4.12), the following identity holds:

$$\sum_{|i|=l}\frac{1}{i!}(\partial^i H)(z)\int_{\partial B}\phi_i(y)d\sigma(y) = 0 \quad \forall\, l \geq 1\,.$$

Recalling now from Lemma 2.16 the fact that

$$\epsilon^{d-2}N(x,\epsilon y + z) = \epsilon^{d-2}\sum_{|j|=0}^{n-|i|+1}\frac{1}{j!}\epsilon^{|j|}\partial_z^j N(x,z)y^j + O(\epsilon^{d+n-|i|})$$

for all $i, 1 \leq |i| \leq n$, and the definition of GPT's, we obtain the following pointwise asymptotic formula. For $x \in \partial\Omega$,

$$u(x) = U(x) - \epsilon^{d-2}\sum_{|i|=1}^n\sum_{|j|=1}^{n-|i|+1}\frac{\epsilon^{|i|+|j|}}{i!j!}(\partial^i H)(z)M_{ij}\partial_z^j N(x,z)$$
$$+ O(\epsilon^{d+n})\,. \tag{4.14}$$

Observing that the formula (4.14) still contains $\partial^i H$ factors, we see that the remaining task is to convert (4.14) to a formula given solely by U and its derivatives.

As a simplest case, let us now take $n = 1$ to find the leading-order term in the asymptotic expansion of $u|_{\partial\Omega}$ as $\epsilon \to 0$. According to (v) in Proposition 4.6 we have

$$|\nabla H(z) - \nabla U(z)| \leq C\epsilon^d \|g\|_{L^2(\partial\Omega)} \,,$$

and therefore, we deduce from (4.14) that

$$u(x) = U(x) - \epsilon^d \sum_{|i|=1, |j|=1} (\partial^i U)(z) M_{ij} \partial^j N(x, z) + O(\epsilon^{d+1}) \,, \quad x \in \partial\Omega \,,$$

which is, in view of (3.7), exactly the formula derived in [123] and [73] when D has a $\mathcal{C}^{1,\alpha}$-boundary for some $\alpha > 0$.

We now return to (4.14). Recalling that by Green's theorem $U = -\mathcal{S}_\Omega(g) + \mathcal{D}_\Omega(U|_{\partial\Omega})$ in Ω, substitution of (4.14) into (2.40) immediately yields that, for any $x \in \Omega$,

$$
\begin{aligned}
H(x) &= U(x) \\
&- \epsilon^{d-2} \sum_{|i|=1}^{n} \sum_{|j|=1}^{n-|i|+1} \frac{\epsilon^{|i|+|j|}}{i!j!} (\partial^i H)(z) M_{ij} \mathcal{D}_\Omega(\partial_z^j N(\cdot, z))(x) + O(\epsilon^{d+n}) \,.
\end{aligned}
\tag{4.15}
$$

In (4.15) the remainder $O(\epsilon^{d+n})$ is uniform in the $\mathcal{C}^n$-norm on any compact subset of Ω for any n, and therefore

$$(\partial^l H)(z) + \sum_{|i|=1}^{n} \epsilon^{d-2} \sum_{|j|=1}^{n-|i|+1} \epsilon^{|i|+|j|} (\partial^i H)(z) P_{ijl} = (\partial^l U)(z) + O(\epsilon^{d+n}) \tag{4.16}$$

for all $l \in \mathbb{N}^d$ with $|l| \leq n$, where

$$P_{ijl} = \frac{1}{i!j!} M_{ij} \partial_x^l \mathcal{D}_\Omega(\partial_z^j N(\cdot, z)) \Big|_{x=z} \,.$$

Define the operator

$$\mathcal{P}_\epsilon : (v_l)_{l \in \mathbb{N}^d, |l| \leq n} \mapsto \left(v_l + \epsilon^{d-2} \sum_{|i|=1}^{n} \sum_{|j|=1}^{n-|i|+1} \epsilon^{|i|+|j|} v_i P_{ijl} \right)_{l \in \mathbb{N}^d, |l| \leq n} \,.$$

Observe that

$$\mathcal{P}_\epsilon = I + \epsilon^d \mathcal{R}_1 + \cdots + \epsilon^{n+d-1} \mathcal{R}_{n-1} \,.$$

Defining the matrices $\mathcal{Q}_p, p = 1, \ldots, n-1$, by

$$
\begin{aligned}
(I + \epsilon^d \mathcal{R}_1 + \cdots + \epsilon^{n+d-1} \mathcal{R}_{n-1})^{-1} &= I + \epsilon^d \mathcal{Q}_1 + \cdots + \epsilon^{n+d-1} \mathcal{Q}_{n-1} \\
&+ O(\epsilon^{n+d})
\end{aligned}
\tag{4.17}
$$

for small ϵ, we finally obtain that

$$((\partial^i H)(z))_{i \in \mathbb{N}^d, |i| \le n} = \left(I + \sum_{p=1}^n \epsilon^{d+p-1} \mathcal{Q}_p \right) ((\partial^i U)(z))_{i \in \mathbb{N}^d, |i| \le n} + O(\epsilon^{d+n}) ,$$

(4.18)

which yields the main result of this chapter stated in Theorem 4.1.

We also have a complete asymptotic expansion of the solutions of the Dirichlet problem.

Theorem 4.7 *Suppose that the inclusion consists of a single component, and let v be the solution of (4.1) with the Neumann condition replaced by the Dirichlet condition $v|_{\partial\Omega} = f$. Let V be the solution of $\Delta V = 0$ in Ω with $V|_{\partial\Omega} = f$. The following pointwise asymptotic expansion on $\partial\Omega$ holds for $d = 2, 3$:*

$$\frac{\partial v}{\partial \nu}(x) = \frac{\partial V}{\partial \nu}(x) - \epsilon^{d-2} \sum_{|i|=1}^n \sum_{|j|=1}^{n-|i|+1} \frac{\epsilon^{|i|+|j|}}{i!j!}$$
$$\times \left[\left(\left(I + \sum_{p=1}^{n+2-|i|-|j|-d} \epsilon^{d+p-1} \mathcal{Q}_p \right) (\partial^l V(z)) \right)_i M_{ij} \partial_z^j \frac{\partial}{\partial \nu_x} G(x,z) \right]$$
$$+ O(\epsilon^{d+n}) ,$$

(4.19)

where the remainder $O(\epsilon^{d+n})$ is dominated by $C\epsilon^{d+n} \|f\|_{W_{\frac{1}{2}}^2(\partial\Omega)}$ for some C independent of $x \in \partial\Omega$. Here $G(x,z)$ is the Dirichlet Green's function, M_{ij}, $i, j \in \mathbb{N}^d$, are the GPT's, and $\mathcal{Q}_p$ is the operator defined in (4.17), where $\mathcal{P}_{ijk}$ is defined, in this case, by

$$P_{ijl} = \frac{1}{i!j!} M_{ij} \partial_x^l \mathcal{S}_\Omega \left(\partial_z^j \left(\frac{\partial}{\partial \nu_x} G \right) (\cdot, z) \right) \Big|_{x=z} .$$

Theorem 4.7 can be proved in the exactly same manner as Theorem 4.1. We begin with Theorem 2.20. Then the same arguments give us

$$v(x) = V(x) - \epsilon^{d-2} \sum_{|i|=1}^n \sum_{|j|=1}^{n-|i|+1} \frac{\epsilon^{|i|+|j|}}{i!j!} (\partial^i H)(z) M_{ij} \partial_z^j G(x,z) + O(\epsilon^{d+n}) .$$

From this we can get (4.19) as before.

We conclude this section by making a remark. The following formulae (4.20) and (4.21) are not exactly asymptotic formulae since the function H still depends on ϵ. However, since the formula is simple and useful for solving the inverse problem in later sections, we make a record of them as a theorem.

Theorem 4.8 *We have*

$$u(x) = H(x) + \epsilon^{d-2} \sum_{|i|=1}^{n} \sum_{|j|=1}^{n-|i|+1} \frac{\epsilon^{|i|+|j|}}{i!j!} \partial^i H(z) M_{ij} \partial^j \Gamma(x-z) \qquad (4.20)$$
$$+ O(\epsilon^{d+n}) \,,$$

where $x \in \partial\Omega$ and the $O(\epsilon^{d+n})$ term is dominated by $C\|g\|_{L^2(\partial\Omega)}\epsilon^{d+n}$ for some C depending only on c_0, and H is given in (2.40).
Moreover,

$$H(x) = -\sum_{|i|=1}^{n} \sum_{|j|=1}^{n-|i|+1} \frac{1}{i!j!} \epsilon^{|i|+|j|+d-2} \partial^i H(z) M_{ij} \partial^j \Gamma(x-z) + O(\epsilon^{d+n}) \,, \quad (4.21)$$

for all $x \in \mathbb{R}^d \setminus \overline{\Omega}$.

Proof.   Beginning with the representation formula (2.39), one can show in the same way as in the derivation of (4.11) that

$$u(x) = H(x) + \mathcal{S}_D(\lambda I - \mathcal{K}_D^*)^{-1}\left(\frac{\partial H_n}{\partial \nu}\Big|_{\partial D}\right)(x) + O(\epsilon^{d+n}) \,, \quad x \in \partial\Omega \,,$$

for $x \in \partial\Omega$. Then the rest is parallel to the previous arguments.
The formula (4.21) can be derived using (2.43).   □

4.3 Derivation of the Asymptotic Formula for Closely Spaced Small Inclusions

An asymptotic formula similar to (4.2) was obtained for closely spaced inclusions in [21]. In this section we present the formula and its derivation in brief.

Let D denote a set of m closely spaced inclusions inside Ω:

$$D = \cup_{s=1}^m D_s := \cup_{s=1}^m (\epsilon B_s + z) \,,$$

where $z \in \Omega$, $\epsilon > 0$ is small and B_s for $s = 1,\ldots,m$ is a bounded Lipschitz domain in $\mathbb{R}^d$. We suppose in addition to (H1) and (H2) in Sect. 3.6 that the set D is well-separated from the boundary $\partial\Omega$, i.e., $\mathrm{dist}(D,\partial\Omega) > c_0 > 0$.

Let $g \in L_0^2(\partial\Omega)$. The voltage potential in the presence of the set D of conductivity inclusions is denoted by u. It is the solution to

$$\begin{cases} \nabla\cdot\left(\chi(\Omega \setminus \bigcup_{s=1}^m \overline{D_s}) + \sum_{s=1}^m k_s\chi(D_s)\right)\nabla u = 0 & \text{in } \Omega \,, \\ \dfrac{\partial u}{\partial \nu}\Big|_{\partial\Omega} = g \,, \int_{\partial\Omega} u = 0 \,. \end{cases} \quad (4.22)$$

The background voltage potential is denoted by U as before.
Based on the arguments given in Theorem 2.17, the following theorem was proved in [190].

Theorem 4.9 *The solution u of the problem (4.22) can be represented as*

$$u(x) = H(x) + \sum_{s=1}^{m} \mathcal{S}_{D_s} \psi^{(s)}(x) , \quad x \in \Omega , \tag{4.23}$$

where the harmonic function H is given by

$$H(x) = -\mathcal{S}_{\Omega}(g)(x) + \mathcal{D}_{\Omega}(f)(x), \quad x \in \Omega, \quad f := u|_{\partial\Omega} ,$$

and $\psi^{(s)} \in L_0^2(\partial D_s)$, $s = 1, \cdots , m$, satisfies the integral equation

$$(\lambda_s I - \mathcal{K}_{D_s}^*)\psi^{(s)} - \sum_{l \neq s} \frac{\partial(\mathcal{S}_{D_l}\psi^{(l)})}{\partial\nu^{(s)}}\Big|_{\partial D_s} = \frac{\partial H}{\partial\nu^{(s)}}\Big|_{\partial D_s} \quad \text{on } \partial D_s .$$

Moreover, $\forall\, n \in \mathbb{N}$, there exists a constant $C_n = C(n, \Omega, \text{dist}(D, \partial\Omega))$ independent of $|D|$ and the conductivities $k_s, s = 1, \ldots, m$, such that

$$\|H\|_{C^n(\overline{D})} \leq C_n \|g\|_{L^2(\partial\Omega)} .$$

One can also prove the following theorem.

Theorem 4.10 *The solution u of (4.22) can be represented as*

$$u(x) = U(x) - \sum_{s=1}^{m} N_{D_s} \psi^{(s)}(x) , \quad x \in \partial\Omega ,$$

where $\psi^{(s)}$, $s = 1, \ldots, m$, is defined by (2.41).

Following the arguments presented in Sect. 4.2, we only outline the derivation of an asymptotic expansion of u leaving the details to the reader.

For $x \in \partial\Omega$, by using the change of variables $y = (x - z)/\epsilon$ we may write

$$\sum_{s=1}^{m} N_{D_s} \psi^{(s)}(x) = \epsilon^{d-1} \sum_{s=1}^{m} \int_{\partial B_s} N(x, \epsilon y + z)\psi^{(s)}(\epsilon y + z)\, d\sigma(y) . \tag{4.24}$$

We expand the Neumann function as in (2.35)

$$N(x, \epsilon y + z) = \sum_{|j|=0}^{\infty} \frac{1}{j!}\epsilon^{|j|}\partial_z^j N(x, z)y^j . \tag{4.25}$$

We then use the uniqueness of the solution to the integral equation (3.27) and the expansion of the harmonic function H,

$$H(x) := H(z) + \sum_{|i|=1}^{\infty} \frac{1}{i!}(\partial^i H)(z)(x - z)^i, \quad x \in \overline{D} ,$$

to show that

$$\psi^{(s)}(\epsilon y + z) = \sum_{|i|=1}^{\infty} \frac{\epsilon^{|i|-1}}{i!}(\partial^i H)(z)\varphi_i^{(s)}(y) , \quad y \in \partial B_s , \tag{4.26}$$

where $\varphi_i^{(s)}$ is the solution of (3.27). Substituting (4.25) and (4.26) into (4.24), we obtain

$$\sum_{s=1}^{m} N_{D_s}\psi^{(s)}(x) = \sum_{|i|=1}^{\infty}\sum_{|j|=0}^{\infty} \frac{\epsilon^{|i|+|j|+d-2}}{i!j!}(\partial^i H)(z)\partial_z^j N(x,z)$$

$$\times \sum_{s=1}^{m} \int_{\partial B_s} y^j \varphi_i^{(s)}(y)d\sigma(y) .$$

If $j = 0$, then $\int_{\partial B_s} y^j \varphi_i^{(s)}(y) \, d\sigma(y) = 0$ for $s = 1, \ldots, m$, and hence we get

$$\sum_{s=1}^{m} N_{D_s}\psi^{(s)}(x) = \sum_{|i|=1}^{\infty}\sum_{|j|=1}^{\infty} \frac{\epsilon^{|i|+|j|+d-2}}{i!j!}(\partial^i H)(z)\partial_z^j N(x,z)M_{ij} , \tag{4.27}$$

where M_{ij} is the generalized polarization tensor defined in (3.28).

We now convert the formula (4.27) to the one given solely by U and its derivatives, not H. Using formula (4.16), we can show analogously to (v) in Proposition 4.6 that

$$|\partial^i H(z) - \partial^i U(z)| \leq C\epsilon^d \|g\|_{L^2(\partial\Omega)} \quad \text{for } i \in \mathbb{N}^d ,$$

where C is independent of ϵ and g. We finally have the following theorem.

Theorem 4.11 *The following pointwise asymptotic expansion holds uniformly in $x \in \partial\Omega$ for $d = 2$ or 3:*

$$u(x) = U(x) - \sum_{|i|=1}^{d}\sum_{|j|=1}^{d} \frac{\epsilon^{|i|+|j|+d-2}}{i!j!}(\partial^i U)(z)\partial_z^j N(x,z)M_{ij} + O(\epsilon^{2d}) ,$$

where the remainder $O(\epsilon^{2d})$ is dominated by $C\epsilon^{2d}\|g\|_{L^2(\partial\Omega)}$ for some constant C independent of $x \in \partial\Omega$.

5

Detection of Inclusions

Taking advantage of the smallness of the inclusions, Cedio-Fengya, Moskow, and Vogelius [73] used the leading-order term in the asymptotic expansion of u to find the locations z_s, $s = 1, \ldots, m$, of the inclusions and certain properties of the domains B_s, $s = 1, \ldots, m$ (relative size, orientation). The approach proposed in [73] is based on a least-squares algorithm. Ammari, Moskow, and Vogelius [28] also utilized this leading-order term to design a direct reconstruction method based on variational formulation. The idea in [28] is to form the integral of the "measured boundary data" against harmonic test functions and choose the input current g so as to obtain an expression involving the inverse Fourier transform of distributions supported at the locations z_s, $s = 1, \ldots, m$. Applying a direct Fourier transform to this data then pins down the locations. This approach is similar to the method developed by Calderón [69] in his proof of uniqueness of the linearized conductivity problem and later by Sylvester and Uhlmann in their important work [248] on uniqueness of the three-dimensional inverse conductivity problem. The main disadvantage of this algorithm is the fact that it uses current sources of exponential type. This is one important practical issue, which we do not attempt to address, and which needs to be resolved. A more realistic real-time algorithm for determining the locations of the inclusions has been developed by Kwon, Seo, and Yoon [191]. This fast, stable, and efficient algorithm is based on the observation of the pattern of a simple weighted combination of an input current g of the form $g = a \cdot \nu$ for some constant vector a and the corresponding output voltage. In all of these algorithms, the locations z_s, $s = 1, \ldots, m$, of the inclusions are found with an error $O(\epsilon)$, and little about the domains B_s can be reconstructed. Moreover, to put these algorithms into use, one requires that the size of the inclusions be very small.

In this chapter we apply the accurate asymptotic formula (4.2) for the purpose of identifying the location and certain properties of the shape of the conductivity inclusions. By improving the algorithm of Kwon, Seo, and Yoon [191] we first design two real-time algorithms with good resolution and accuracy. We then describe the least-squares algorithm and the variational algorithm introduced in [28] and review the interesting approach proposed by Brühl, Hanke, and Vogelius [64]. Their method is in the spirit of the linear sampling method of Colton and Kirsch [89]. Furthermore, we give Lipschitz-continuous dependence estimates for the reconstruction problem. These estimates, established by Friedman and Vogelius [123], bound the difference in the location and relative size of two sets of inclusions by the difference in the boundary voltage potentials corresponding to a fixed current distribution. We conclude the chapter by presenting upper and lower bounds on the moments of the unknown inclusions. We refer to [181, 184, 240, 244, 100, 247, 65, 62, 63, 147, 151, 152, 194, 201, 142, 102, 59, 60] for other numerical methods aimed at solving the inverse conductivity problem in different settings.

We conclude this introduction with a comment on the uniqueness question of the inverse conductivity problem with one (or two) measurements. This question is whether a single (or two) Cauchy data $(u|_{\partial\Omega}, g)$ is sufficient to determine the conductivity inclusion D uniquely. It has been studied extensively recently. However, it is still wide open. Even the uniqueness within the classes of ellipses and ellipsoids is not known. The global uniqueness results are only obtained when D is restricted to convex polyhedrons and balls in three-dimensional space and polygons and disks in the plane (see [48, 46, 122, 159, 242, 169, 170]). We refer the reader to Appendix A.4.2 for the unique determination of disks with one measurement.

5.1 Constant Current Projection Algorithm – Reconstruction of Single Inclusion

The constant current projection algorithm has been developed by Ammari and Seo in [29]. To bring out the main ideas of this algorithm we only consider the case where D has one component of the form $\epsilon B + z$. Based on Theorem 4.1 and two more observations we rigorously reconstruct, with good resolution and accuracy, the location, the size and the polarization tensors from the observation in the near field (x near $\partial\Omega$) and the far field (x far from $\partial\Omega$) of the pattern $H(x)$, which is computed directly from the current-voltage pairs.

The mathematical analysis provided in this section indicates that the constant current projection algorithm has good resolution and accuracy.

As we said before, this algorithm makes use of constant current sources. For any unit vector $a \in \mathbb{R}^d$, $d = 2, 3$, let $H[a \cdot \nu]$ denote the function H in (2.40) corresponding to the Neumann data

$$g(y) = \frac{\partial}{\partial \nu}(a \cdot y) = a \cdot \nu_y , \quad y \in \partial\Omega .$$

The expression $D = \epsilon B + z$ requires some care because it can be expressed in infinitely many different ways. For a unique representation, we need to select a canonical domain B which is a representative domain of the set of all

$D = \epsilon B + z$. Assume for simplicity that $k > 1$. Let $\mathcal{T}_\lambda$ be the set of all strictly star shaped domains B satisfying

$$\int_B x \, dx = 0 \;, \quad |M(k, B)| = 1 \;,$$

where $|M|$ is the determinant of the matrix M and $M(k, B)$ is the polarization tensor of Pólya–Szegö associated with the domain B and the conductivity $k = (2\lambda + 1)/(1 - 2\lambda)$. Then, by using the essential fact from Chap. 3 that $M(k, B)$ is a symmetric positive definite matrix and so, its determinant cannot vanish, it is not hard to see that if $\epsilon_1 B_1 + z_1 = \epsilon_2 B_2 + z_2$, where B_1 and B_2 belong to $\mathcal{T}_\lambda$ then $z_1 = z_2, \epsilon_1 = \epsilon_2$, and $B_1 = B_2$. Note that if $0 < k < 1$ then $M(k, B)$ is a symmetric negative definite matrix. Throughout this chapter, we assume that $B \in \mathcal{T}_\lambda$.

The first step for the reconstruction procedure is to compute ϵ and $M(k, B)$ up to an error of order ϵ^d.

Theorem 5.1 (Size estimation) *Let S be a C^2-closed surface (or curve in $\mathbb{R}^2$) enclosing the domain Ω. Then for any vectors a and a^* we have*

$$\int_S \frac{\partial H[a \cdot \nu]}{\partial \nu}(x) \, a^* \cdot x \, d\sigma(x) - \int_S H[a \cdot \nu](x) \, a^* \cdot \nu_x \, d\sigma(x) \tag{5.1}$$
$$= -a^* \cdot (\epsilon^d M(k, B) \cdot a) + O(\epsilon^{2d}) \;.$$

Proof. Let Ω' denote the domain inside S, that is, $\partial \Omega' = S$. Since $S \subset \mathbb{R}^d \backslash \overline{\Omega}$, it follows from (2.43) that for any vector a, $H[a \cdot \nu] = -\mathcal{S}_D \phi$ on S, where

$$\phi = (\lambda I - \mathcal{K}_D^*)^{-1} \left(\frac{\partial H}{\partial \nu} \big|_{\partial D} \right) .$$

Thus the left side of (5.1) is in fact equal to

$$-\int_S \frac{\partial}{\partial \nu} (\mathcal{S}_D \phi(x)) \; a^* \cdot x \, d\sigma(x) + \int_S \mathcal{S}_D \phi(x) \; a^* \cdot \nu_x \, d\sigma(x) \;. \tag{5.2}$$

Using the fact that $\Delta \mathcal{S}_D \phi = 0$ in $\mathbb{R}^d \backslash \partial D$ and the divergence theorem on $\Omega' \backslash \overline{D}$, we can see that the term in (5.2) equals to

$$-\int_{\partial D} \frac{\partial (\mathcal{S}_D \phi(x))}{\partial \nu} \bigg|_+ a^* \cdot x \, d\sigma(x) + \int_{\partial D} \mathcal{S}_D \phi(x) \, a^* \cdot \nu_x \, d\sigma(x) \;.$$

Then by the jump relation (2.12), it equals to

$$-\int_{\partial D} a^* \cdot x \, \phi(x) \, d\sigma(x) \;.$$

Setting $\widetilde{h}(y) = \frac{\partial}{\partial \nu} H[a \cdot \nu](z + \epsilon y)$, we have by a change of variables

$$\int_{\partial D} a^* \cdot y\, \phi(y)\, d\sigma(y) = \epsilon^d \int_{\partial B} a^* \cdot y\, (\lambda I - K_B^*)^{-1}\widetilde{h}(y)\, d\sigma(y) \ .$$

The estimate (v) in Proposition 4.6 provides the expansion

$$\int_{\partial B} a^* \cdot y\, (\lambda I - K_B^*)^{-1}\widetilde{h}(y)\, d\sigma(y) = \int_{\partial B} a^* \cdot y\, (\lambda I - K_B^*)^{-1}(\nu \cdot a)\, d\sigma(y) + O(\epsilon^d) \ ,$$

which leads us to the identity (5.1). $\square$

Now, let us explain how to compute ϵ and $M(k, B)$ up to an error of order ϵ^d using Theorem 5.1.

Let $\mathbb{A}$ be the $d \times d$ matrix defined by $\mathbb{A} = \sqrt{\mathbb{B}\mathbb{B}^T}$, where the pq-component of $\mathbb{B}$ is equal to

$$\int_S \frac{\partial H[e_p \cdot \nu]}{\partial \nu}(x)e_q \cdot x\, d\sigma(x) - \int_S H[e_p \cdot \nu](x)e_q \cdot \nu_x\, d\sigma(x) \ .$$

Here $\mathbb{B}^T$ is the transpose of the matrix $\mathbb{B}$ and $\{e_p\}_{p=1}^d$ is an orthonormal basis of $\mathbb{R}^d$. Define

$$\epsilon^* = \sqrt[d^2]{|\mathbb{A}|}, \quad M^* := \frac{1}{(\epsilon^*)^d}\mathbb{A} \ .$$

According to Theorem 5.1 we immediately see that

$$\epsilon^* = \epsilon(1 + O(\epsilon^d)) \text{ and } M^* = M(k, B) + O(\epsilon^d) \ . \tag{5.3}$$

Note that slightly different size estimations can be obtained by making use of Lemma 3.8, see [71, 72]. It can be shown that

$$\frac{1}{d - 1 + \frac{1}{k}}\frac{\text{Tr}(\mathbb{A})}{k - 1} \leq |D|(1 + O(\epsilon)) \leq \frac{\text{Tr}(\mathbb{A})}{k - 1}\frac{d - 1 + k}{d^2} \ .$$

Because of the normalization $|M(k, B)| = 1$, the knowledge of M does not determine k. Equivalently, it is not possible to determine ϵ and k simultaneously from the knowledge of the lowest-order term in the asymptotic expansion of the pattern H.

Observe that by construction the real matrix M^* is symmetric positive definite. Let $0 < \tau_1 \leq \tau_{d-1} \leq \tau_d$ be the eigenvalues of $M(k, B)$. Using once again the fact that $M(k, B)$ is a symmetric positive definite matrix it follows that there is a constant C depending only on the Lipschitz character of B such that $C < \tau_p < 1/C$ and therefore, for ϵ small enough, the eigenvalues $\{\tau_*^{(p)}\}_{p=1}^d$ of M^* satisfy the same estimates.

Having recovered (approximately) the polarization tensor of Pólya–Szegö $M(k, B)$, we now compute an orthonormal basis of eigenvectors $a_*^{(1)}, \ldots, a_*^{(d)}$ of M^*. We will use these eigenvectors for recovering the location z. Let Σ_p be a line parallel to $a_*^{(p)}$ such that

$$\text{dist}\,(\partial\Omega, \Sigma_p) = O\left(\frac{1}{(\epsilon^*)^{d-1}}\right), \quad p = 1, \ldots, d \ .$$

For any $x \in \Sigma_p$ it is readily seen from (4.21) and (5.3) that for background potentials U_p given by

$$U_p(x) = a_*^{(p)} \cdot x - \frac{1}{|\partial \Omega|} \int_{\partial \Omega} a_*^{(p)} \cdot y \, d\sigma(y) , \quad p = 1, \ldots, d ,$$

(or equivalently for the currents $g = a_*^{(p)} \cdot \nu$) the following asymptotic expansion holds:

$$H[a_*^{(p)} \cdot \nu](x) = -\tau_*^{(p)} \frac{(\epsilon^*)^d}{\omega_d |x - z|^d} (x - z) \cdot a_*^{(p)} + O(\frac{\epsilon^{2d}}{|x - z|^{d-1}}) \qquad (5.4)$$

for all $x \in \Sigma_p$, where $\tau_*^{(p)}$ is the eigenvalue of M^* associated with the eigenvector $a_*^{(p)}$.

In fact, this is the far field expansion of the pattern $H[a_*^{(p)} \cdot \nu]$, from which we find the location z with an error of order $O(\epsilon^d)$. To get some insight, let us neglect the asymptotically small remainder $O(\epsilon^{2d})$ in the asymptotic expansion (5.4).

Our second important observation is that, since M^* is symmetric positive and the set of eigenvalues $a_*^{(1)}, \ldots, a_*^{(d)}$ forms an orthonormal basis of $\mathbb{R}^d$, we will find exactly d points $z_*^p \in \Sigma_p, p = 1, \ldots, d$, so that $H[a_*^{(p)} \cdot \nu](z_*^p) = 0$. Finally, the point $z_* = \sum_{p=1}^d (z_*^p \cdot a_*^{(p)}) a_*^{(p)}$ is very close to z, namely $|z_* - z| = O(\epsilon^d)$.

Theorem 5.2 (Detection of the location) *Let $a_*^{(1)}, \ldots, a_*^{(d)}$ denote the mutually orthonormal eigenvectors of the symmetric matrix M^*. For each $p = 1, \cdots, d$, let $H[a_*^{(p)} \cdot \nu]$ be the function H in (2.40) corresponding to the Neumann data $g = a_*^{(p)} \cdot \nu$ and let Σ_p be a line with the direction $a_*^{(p)}$ so that $\mathrm{dist}\,(\partial \Omega, \Sigma_p) = O(1/(\epsilon^*)^{d-1})$. Then there exist $z_*^p \in \Sigma_p$ so that $H[a_*^{(p)} \cdot \nu](z_*^p) = 0$. Moreover, the point $z_* = \sum_{p=1}^d (z_*^p \cdot a_*^{(p)}) a_*^{(p)}$ satisfies the following estimate*

$$|z_* - z| \leq C \epsilon^d , \qquad (5.5)$$

where the constant C is independent of ϵ and z.

Proof. From (5.4) it follows that there exists a positive constant C, independent of x, z, and ϵ such that

$$H[a_*^{(p)} \cdot \nu](x) \geq -\frac{\epsilon^d}{\omega_d |x - z|^{d-1}} \left(\tau_*^{(p)} \frac{(x - z)}{|x - z|} \cdot a_*^{(p)} + C \epsilon^d \right) \text{ for all } x \in \Sigma_p ,$$

$$H[a_*^{(p)} \cdot \nu](x) \leq -\frac{\epsilon^d}{\omega_d |x - z|^{d-1}} \left(\tau_*^{(p)} \frac{(x - z)}{|x - z|} \cdot a_*^{(p)} - C \epsilon^d \right) \text{ for all } x \in \Sigma_p .$$

For $x \in \Sigma_p$ satisfying

$$\frac{(x-z)}{|x-z|} \cdot a_*^{(p)} < -\frac{C}{\tau_*^{(p)}} \epsilon^d$$

we have $H[a_*^{(p)} \cdot \nu](x) > 0$. On the other hand, for $x \in \Sigma_p$ satisfying

$$\frac{(x-z)}{|x-z|} \cdot a_*^{(p)} > \frac{C}{\tau_*^{(p)}} \epsilon^d \,,$$

we similarly have $H[a_*^{(p)} \cdot \nu](x) < 0$. Therefore, the zero point z_* satisfies

$$|(z_* - z) \cdot a_*^{(p)}| \le C\epsilon^d, \text{ for } p = 1, \ldots, d \,,$$

which implies that (5.5) holds, since $\{a_*^{(p)}\}_{p=1}^d$ forms an orthonormal basis of $\mathbb{R}^d$. □

Finally, to find more geometric features of the domain B and its conductivity k, we use higher-order terms in the asymptotic expansion of H which follows from a combination of the estimate (v) in Proposition 4.6 and the expansion (4.21):

$$H[a \cdot \nu](x) = -\sum_{|i|=1}^{d} \sum_{|j|=1}^{d} \frac{\epsilon^{|i|+|j|+d-2}}{j!} a \partial_z^j \Gamma(x-z) M_{ij} + O(\epsilon^{2d}) \,. \tag{5.6}$$

Since z, ϵ, and the polarization tensor M are now recovered with an error $O(\epsilon^d)$, the reconstruction of the higher-order polarization tensors, M_{ij}, for $|i| = 1$ and $2 \le |j| \le d$, could easily been done by inverting an appropriate linear system arising from (5.6). Then we could determine the conductivity k from the knowledge of M_{ij}, for $1 \le |i|, |j| \le d$.

The main results in this section are summarized in the following reconstruction procedure.

[Constant current projection algorithm]

For any unit vector a let $H[a \cdot \nu]$ be the function H in (2.40) corresponding to the Neumann data $g(y) = a \cdot \nu_y, y \in \partial\Omega$. Let $\{e_p\}_{p=1}^d$ denote the standard orthonormal basis of $\mathbb{R}^d$. Let S be a C^2-closed surface (or curve in $\mathbb{R}^2$) enclosing the domain Ω.

Step 1: Compute $H[e_p \cdot \nu](x)$ for $x \in S$ to calculate the matrix $\mathbb{A} = \sqrt{\mathbb{B}\mathbb{B}^T}$, where the pq-component of $\mathbb{B}$ is equal to

$$\int_S \frac{\partial H[e_p \cdot \nu]}{\partial\nu}(x) e_q \cdot x \, d\sigma(x) - \int_S H[e_p \cdot \nu](x) e_q \cdot \nu_x \, d\sigma(x)$$

and $\mathbb{B}^T$ is the transpose of the matrix $\mathbb{B}$. Then

$$\epsilon^* = \sqrt[d^2]{|\mathbb{A}|} = \epsilon(1 + O(\epsilon^d)) \text{ and } M^* := \frac{1}{(\epsilon^*)^d} \mathbb{A} = M(k, B) + O(\epsilon^d) \,.$$

Here $|\mathbb{A}|$ denotes the determinant of $\mathbb{A}$.

Step 2: Compute an orthonormal basis $\{a_*^{(p)}\}_{p=1}^d$ of eigenvalues of the symmetric positive definite matrix M^*.

Step 3: Consider Σ_p to be a line with the direction $a_*^{(p)}$ so that dist $(\partial\Omega, \Sigma_p) = O(1/(\epsilon^*)^{d-1})$ and $z_*^p \in \Sigma_p$ so that $H[a_*^{(p)} \cdot \nu](z_*^p) = 0$. Then the point $z_* = \sum_{p=1}^d (z_*^p \cdot a_*^{(p)}) a_*^{(p)}$ satisfies the estimate $|z_* - z| = O(\epsilon^d)$.

Step 4: Recover the higher-order polarization tensors M_{ij}, for $|i| = 1$ and $2 \le |j| \le d$, by solving an appropriate linear system arising from (5.6) and then determine the conductivity k from the knowledge of M_{ij}, for $1 \le |i|, |j| \le d$.

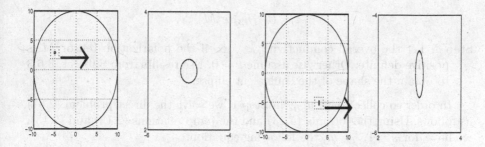

Fig. 5.1. Detection of the location and the polarization tensor of a small inclusion by the constant current projection algorithm.

5.2 Quadratic Algorithm – Detection of Closely Spaced Inclusions

Recall that the constant current projection algorithm uses only linear solutions. In this section, we design an other algorithm using quadratic solutions. We apply this algorithm for the purpose of reconstructing the first-order polarization tensor and the center of closely spaced small inclusions from a finite number of boundary measurements. As before, the algorithm is based on the asymptotic expansion formula (4.19). For $g \in L_0^2(\partial\Omega)$, define the harmonic function $H[g](x)$, $x \in \mathbb{R}^d \setminus \overline{\Omega}$, by

$$H[g](x) := -\mathcal{S}_\Omega(g)(x) + \mathcal{D}_\Omega(u|_{\partial\Omega})(x), \quad x \in \mathbb{R}^d \setminus \overline{\Omega}, \qquad (5.7)$$

where u is the solution of (4.22). Then by substituting (4.19) into (5.7) and using a simple formula $\mathcal{D}_\Omega(N(\cdot - z))(x) = \Gamma(x - z)$ for $z \in \Omega$ and $x \in \mathbb{R}^d \setminus \overline{\Omega}$, we get

$$H[g](x) = -\sum_{|i|=1}^d \sum_{|j|=1}^d \frac{\epsilon^{|i|+|j|+d-2}}{i!j!} (\partial^i U)(z)\partial_z^j \Gamma(x - z) M_{ij} + O(\epsilon^{2d}) . \qquad (5.8)$$

Assume for the sake of simplicity that $d = 2$. The reconstruction procedure is the following.

[Quadratic algorithm]

Step 1: For $g_p = \partial x_p / \partial \nu$, $p = 1, 2$, measure $u|_{\partial\Omega}$.

Step 2: Compute the first-order polarization tensor $\epsilon^2 M = \epsilon^2 (m_{pq})^d_{p,q=1}$ for D by

$$\epsilon^2 m_{pq} = \lim_{t \to \infty} 2\pi t H[g_p](t e_q) .$$

Step 3: Compute $h_p = \lim_{t \to \infty} 2\pi t H[g_3](t e_p)$ for $g_3 = \frac{\partial(x_1 x_2)}{\partial \nu}$, $p = 1, 2$. Then the center is estimated by solving

$$z = (h_1, h_2)(\epsilon^2 M)^{-1} .$$

Step 4: Let the overall conductivity $\bar{k} = \infty$ if the polarization tensor M is positive definite. Otherwise assume $\bar{k} = 0$. Use results from Subsect. 3.6.3 to obtain the shape of the equivalent ellipse.

In order to collect data $u|_{\partial\Omega}$ in Step 1, we solve the direct problem (2.37) as follows. Using the formula (4.23) and the jump relations (2.12) and (2.13), we have, for $s = 1, \ldots, m$, the following equation:

$$\begin{cases} u = \dfrac{u}{2} + \mathcal{K}_\Omega u - \mathcal{S}_\Omega g + \displaystyle\sum_{s=1}^m \mathcal{S}_{D_s} \psi^{(s)} & \text{on } \partial\Omega , \\[2mm] (\lambda_s I - \mathcal{K}^*_{D_s}) \psi^{(s)} - \displaystyle\sum_{l \neq s} \dfrac{\partial(\mathcal{S}_{D_l} \psi^{(l)})}{\partial \nu^{(s)}} \Big|_{\partial D_s} = \dfrac{\partial H}{\partial \nu^{(s)}} \Big|_{\partial D_s} & \text{on } \partial D_s . \end{cases}$$

We solve the integral equation using the collocation method and obtain $u|_{\partial\Omega}$ on $\partial\Omega$ for given data g.

A few words are required for the Step 4. In order to find the overall conductivity, it is necessary to know the individual conductivity k_s and the size of B_s, $s = 1, \ldots, m$, which seems impossible. Thus we assume *a priori* that $\bar{k}$ is either ∞ or 0 depending upon the sign of detected polarization tensor. Therefore it is natural that the quadratic algorithm gives better information when the conductivity contrast between the background and inclusions is high. We illustrate in Fig. 5.2 the viability of this algorithm. Rigorous justification of the validity of this algorithm follows from the arguments we just went through for the constant current projection algorithm.

We conclude this section with a comment on stability. In general, the measured voltage potential contains the unavoidable observation noise, so that we have to answer the stability question. Fortunately, the constant current projection and quadratic algorithms are totally based on the observation of the pattern of H; thus if

$$H^{meas}[g](x) := -\mathcal{S}_\Omega(g)(x) + \mathcal{D}_\Omega(u^{meas}) \quad \text{for } x \in \mathbb{R}^d \setminus \overline{\Omega} ,$$

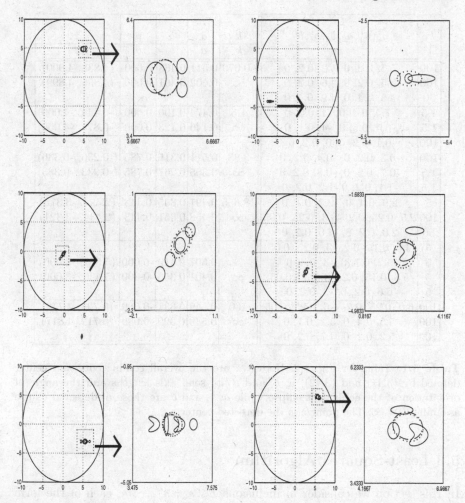

Fig. 5.2. Reconstruction of closely spaced small inclusions. The dashed line is the equivalent ellipse and the dash-dot line is the detected ellipse. The numerical values are given in Table 5.1.

where u^{meas} is the measured voltage on the boundary, then we have the following stability estimate

$$\left| H^{meas}[g](x) - H[g](x) \right| \leq \left| \int_{\partial\Omega} \frac{\partial\Gamma}{\partial\nu_y}(x-y)\left(u^{meas} - u\right)(y)\, d\sigma(y) \right|$$

$$\leq C \left\| u^{meas} - u \right\|_{L^2(\partial\Omega)},$$

where C is a constant depending only on the distance of x from $\partial\Omega$. Thus we have to conclude that the constant current projection and quadratic algorithms are not sensitive to the observation noise.

k_i	$a_0^i, a_1^i, a_2^i, b_0^i, b_1^i, b_2^i$	$\bar{k}$	$\bar{a}$	$\bar{b}$	$\bar{\theta}$	$\bar{z}$
		k	a	b	θ	z
100	5.5, 0.2, 0, 5.2, 0.2, 0	60.079	0.511	0.468	0.000	(4.838, 4.900)
100	5.5, 0.2, 0, 4.6, 0.2, 0	∞	0.502	0.461	0.000	(4.856, 4.899)
50	4.5, 0.4, 0, 4.9, 0.4, 0					
1.5	-7.4, 0.2, 0, -4, 0.2, 0	1.5	0.474	0.190	0.000	(-6.844, -4.000)
1.5	-6.4, 0.5, 0, -4, 0.1, 0	∞	0.146	0.123	0.000	(-6.875, -4.000)
100	0.1, 0.2, 0, 0, 0.2, 0					
100	-0.3, 0.2, 0, -0.4, 0.2, 0	3.88	0.511	0.315	0.785	(-0.236, -0.336)
1.5	-0.7, 0.2, 0, -0.8, 0.2, 0	∞	0.355	0.267	0.785	(-0.233, -0.333)
1.5	-1.1, 0.2, 0,-1.2, 0.2, 0					
5	2.9, 0.4, 0, -2.7, 0.1, 0	18.655	0.491	0.365	0.443	(2.494, -3.375)
100	2.5, 0.25, 0.2, -3.3, 0.25, 0.05	∞	0.458	0.351	0.443	(2.434, -3.321)
50	2.0, 0.2, 0, -4.0, 0.2, 0					
5	4.5, 0.15, 0.2, -3, 0.25, 0.05					
5	5.2, 0.1, 0, -3, 0.4, 0	5	0.507	0.419	-0.000	(5.502, -3.000)
5	5.8, 0.15, 0.2, -3, 0.25,0.05	∞	0.401	0.353	-0.000	(5.436, -3.000)
5	6.6, 0.2, 0, -3, 0.2, 0					
100	6.0, 0.25, 0.2, 4.6, 0.25, 0.05	100	0.549	0.331	-0.089	(5.728, 4.772)
100	5.5, 0.4, 0, 5.2, 0.1, 0	∞	0.540	0.329	-0.089	(5.712, 4.817)
100	5.2, 0.2, 0, 4.7, 0.2, 0					

Table 5.1. Table for Fig. 5.2. Here $\bar{k}, \bar{z}$ are the overall conductivity and center defined by (3.41) and (3.42). $\bar{a}$, $\bar{b}$, and $\bar{\theta}$ are semi axis lengths and the angle of orientation of the equivalent ellipse while a, b, and θ are those of detected ellipse assuming $k = \infty$. The point z is the detected center.

5.3 Least-Squares Algorithm

In this section we consider m inclusions $D_s, s = 1, \ldots, m$, each of the form $D_s = \epsilon_s B_s + z_s$ where each $B_s \in \mathcal{T}_{\lambda_s}$. Here $\lambda_s = (k_s+1)/(2(k_s-1))$. Let S be a $\mathcal{C}^2$-closed surface (or curve in $\mathbb{R}^2$) enclosing the domain Ω. The least-squares algorithm is based on the minimization of a discrete L^2-norm of the residual

$$H[g](x) + \sum_{s=1}^{m} \sum_{|i|=1}^{d} \sum_{|j|=1}^{d} \frac{\epsilon_s^{|i|+|j|+d-2}}{i!j!} (\partial^i U)(z_s)\partial_z^j \Gamma(x - z_s)M_{ij}^s$$

on S. Here $M_{ij}^s = M_{ij}(k_s, B_s)$. We select L equidistant points, $x_1, \ldots, x_L$, on S and we seek the unknown parameters of the inclusions D_s as the solution to the nonlinear least-squares problem

$$\min \sum_{l=1}^{L} \left| H[g](x_l) + \sum_{s=1}^{m} \sum_{|i|=1}^{d} \sum_{|j|=1}^{d} \frac{\epsilon_s^{|i|+|j|+d-2}}{i!j!} (\partial^i U)(z_s)\partial_z^j \Gamma(x_l - z_s)M_{ij}^s \right|^2 .$$

We minimize over $\{m, z_s, \epsilon_s, k_s, B_s\}$ when all the parameters are unknown; however, there may be considerable non-uniqueness of the minimizer in this general case. If the inclusions are assumed to be of the form $z_s + \epsilon_s Q_s B$, for a common known domain B, but unknown locations, z_s, and rotations Q_s then if ϵ_s and k_s are known, the least-squares algorithm can be applied to successfully determine the number m of inclusions, the locations z_s and the rotations Q_s, as demonstrated by numerical examples in [73].

5.4 Variational Algorithm

This algorithm is based on the original idea of Calderón [69], which was, by the way of a low amplitude perturbation formula, to reduce the reconstruction problem to the calculations of an inverse Fourier transform. It may require quite a number of boundary measurements, but if these are readily available, then the approach is rapid and simple to implement.

For arbitrary $\eta \in \mathbb{R}^d$, one assumes that one is in possession of the boundary data for the voltage potential, u, whose corresponding background potential is given by $U(y) = e^{i(\eta + i\eta^\perp) \cdot y}$ (boundary current $g_\eta(y) = i(\eta + i\eta^\perp) \cdot \nu_y e^{i(\eta + i\eta^\perp) \cdot y}$ on $\partial\Omega$), where $\eta^\perp \in \mathbb{R}^d$ is orthogonal to η with $|\eta| = |\eta^\perp|$.

If S is a $\mathcal{C}^2$-closed surface (or curve in $\mathbb{R}^2$) enclosing the domain Ω then, analogously to (5.1), one can easily prove that

$$
\begin{aligned}
\mathcal{E}(\eta) &:= \int_S \frac{\partial}{\partial\nu} H[g_\eta](y)\, e^{i(\eta - i\eta^\perp) \cdot y}\, d\sigma(y) \\
&\quad - i \int_S H[g_\eta](y)\, \nu_y \cdot (\eta - i\eta^\perp)\, e^{i(\eta - i\eta^\perp) \cdot y}\, d\sigma(y) \\
&= \sum_{s=1}^m \epsilon_s^d (\eta + i\eta^\perp) \cdot M^s \cdot (\eta - i\eta^\perp) e^{2i\eta \cdot z_s} + O(\epsilon^{d+1})\,,
\end{aligned}
\tag{5.9}
$$

where $\epsilon = \sup_s \epsilon_s$ and $M^s = M(k_s, B_s)$.

Recall that the function $e^{2i\eta \cdot z_s}$ (up to a multiplicative constant) is exactly the Fourier transform of the Dirac function δ_{-2z_s} (a point mass located at $-2z_s$). Multiplication by powers of η in the Fourier space corresponds to differentiation of the Dirac function. The function $\mathcal{E}(\eta)$ is therefore (approximately) the Fourier transform of a linear combination of derivatives of point masses, or

$$
\check{\mathcal{E}} \simeq \sum_{s=1}^m \epsilon_s^d L_s \delta_{-2z_s}\,,
$$

where L_s is a second-order constant coefficient, differential operator whose coefficients depend on the polarization tensor of Pólya–Szegö M^s, and $\check{\mathcal{E}}$ represents the inverse Fourier transform of $\mathcal{E}(\eta)$.

The variational algorithm consists then of sampling the values of $\mathcal{E}(\eta)$ at some discrete set of points and then calculating the corresponding discrete inverse Fourier transform. After a re-scaling (by $-1/2$) the support of this inverse Fourier transform yields the location of the inclusions. Once the locations are known, one may calculate the polarization tensors of Pólya–Szegö by solving the appropriate linear system arising from (5.9).

To arrive at some idea of the number of the sampling points needed for an accurate discrete Fourier inversion of $\mathcal{E}(\eta)$ we remind the reader of the main assertion of the so-called Shannon's sampling theorem [96]: A function f is completely specified (by a very explicit formula) by the sampled values $\{f(c_0 + n/h)\}_{n=-\infty}^{+\infty}$ if and only if the support of the Fourier transform of f is contained inside a square of side h. For the variational algorithm this suggests two things: (1) if the inclusions are contained inside a square of side h, then we need to sample $\mathcal{E}(\eta)$ at a uniform, infinite, rectangular grid of mesh-size $1/h$ to obtain an accurate reconstruction, (2) if we only sample the points in this grid for which the absolute values of the coordinates are less than K, then the resulting discrete inverse Fourier transform will recover the location of the inclusions with a resolution of $\delta = 1/2K$. In summary: we need (conservatively) of the order h^d/δ^d sampled values of $\mathcal{E}(\eta)$ to reconstruct, with a resolution δ, a collection of inclusions that lie inside a square of side h. The reader is referred to [261] for a review of the fundamental mechanism behind the FFT method for inverting the quantity in (5.9).

The following numerical examples from [28] clearly demonstrate the viability of the variational approach.

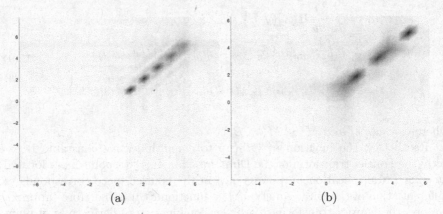

(a) (b)

Fig. 5.3. Five inclusions - (a): 30 x 30 sample points, (b): 20 x 20 sample points.

We take the domain Ω to be the square $[-10, 10] \times [-10, 10]$ and we insert five inclusions in the shape of balls, with the sth-ball positioned at the point (s, s). We take each M^s to be $10 \times I$ and $\epsilon = 0.1$. We sample $\mathcal{E}(\eta)$ on the square $[-3, 3] \times [-3, 3]$ with a uniform 30×30 grid in (a) (900 sample points) and 20×20 grid in (b) (400 sample points). We are thus following the recipe from above, with $h = 5$ and $K = 3$. We should expect a recovery of all the locations of the inclusions, with a resolution $\delta = 1/6$. The discrete inverse Fourier transform yields the grey-level (intensity) plot shown in Figs. 5.3. In Fig. 5.3 (a) we see that the five balls are still visible.

In order to simulate errors in the boundary measurements, as well as in the different approximations, we add on the order of 10% of random noise to the values of $\mathcal{E}(\eta)$. We see from the following figures that the reconstruction is quite stable.

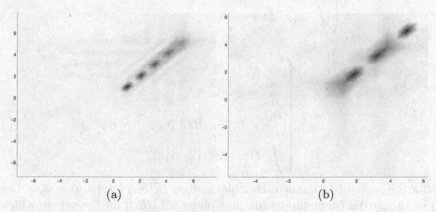

(a)   (b)

Fig. 5.4. Five inclusions with 10% noise- (a): 30 x 30 sample points, (b): 20 x 20 sample points.

5.5 Linear Sampling Method

We now describe the interesting approach proposed by Brühl, Hanke, and Vogelius in their recent paper [64]. This approach is related to the linear sampling method of Colton and Kirsch [89] (see also [177] and [62]) and allows one to reconstruct small inclusions by taking measurements only on some portion of $\partial\Omega$. It has also some similarities to a MUltiple Signal Classification (MUSIC)-type algorithm developed by Devaney [99, 193] for estimating the locations of a number of pointlike scatterers. We refer to Cheney [77] and Kirsch [178] for detailed discussions of the connection between MUSIC algorithm and the linear sampling method. See also Sect. 13.2.2.

Let $\partial\omega$ be a subset of $\partial\Omega$ with positive measure and define $L_0^2(\partial\omega)$ to be those functions in $L^2(\partial\omega)$ with zero integral over $\partial\omega$. Let $D = \cup_{s=1}^m (\epsilon B_s + z_s)$ be a collection of small inclusions with conductivities $0 < k_s \neq 1 < +\infty, s = 1,\dots,m$, and satisfying

$$|z_s - z_{s'}| \geq 2c_0 > 0 \quad \forall\, s \neq s' \quad \text{and} \quad \text{dist}(z_s, \partial\Omega) \geq 2c_0 > 0 \quad \forall\, s. \quad (5.10)$$

For a function $g \in L_0^2(\partial\omega)$ we can solve the problems

$$
\begin{cases}
\nabla \cdot \left(1 + (k-1)\chi(D)\right)\nabla u = 0 & \text{in } \Omega , \\[2mm]
\dfrac{\partial u}{\partial \nu} = g & \text{on } \partial\omega , \\[2mm]
\dfrac{\partial u}{\partial \nu} = 0 & \text{on } \partial\Omega \setminus \partial\omega , \\[2mm]
\displaystyle\int_{\partial\omega} u(x)\, d\sigma(x) = 0 ,
\end{cases}
\tag{5.11}
$$

and

$$
\begin{cases}
\Delta U = 0 & \text{in } \Omega , \\[2mm]
\dfrac{\partial U}{\partial \nu} = g & \text{on } \partial\omega , \\[2mm]
\dfrac{\partial U}{\partial \nu} = 0 & \text{on } \partial\Omega \setminus \partial\omega , \\[2mm]
\displaystyle\int_{\partial\omega} U(x)\, d\sigma(x) = 0 .
\end{cases}
\tag{5.12}
$$

Define the partial Neumann-to-Dirichlet map on $L_0^2(\partial\omega)$ by $\Lambda_D(g) = u|_{\partial\omega}$. Let Λ_0 be the partial Neumann-to-Dirichlet map on $L_0^2(\partial\omega)$ for the case in which no conductivity inclusions are present. We seek to use $\Lambda_D - \Lambda_0$ to determine D. In this connection we shall first establish the following.

Lemma 5.3 *The operator* $\Lambda_D - \Lambda_0 : L_0^2(\partial\omega) \to L_0^2(\partial\omega)$ *is compact, self-adjoint, positive (respectively negative) semi-definite, if* $0 < k_s < 1$ *(respectively* $+\infty > k_s > 1$*) for all* $s = 1, \ldots, m$.

Proof.   Let $g \in L_0^2(\partial\omega)$, and let u and U denote the solutions of (5.11) and (5.12). An easy application of Green's theorem gives

$$
u(x) = \mathcal{D}_\Omega(u|_{\partial\Omega}) + (k-1)\int_D \nabla u(y) \cdot \nabla \Gamma(x-y)\, dy - \int_{\partial\omega} g(y)\Gamma(x-y)\, d\sigma(y)
$$

and

$$
U(x) = \mathcal{D}_\Omega(U|_{\partial\Omega}) - \int_{\partial\omega} g(y)\Gamma(x-y)\, d\sigma(y), \quad x \in \Omega .
$$

Subtracting these two equations and letting x goes to $\partial\Omega$ yields

$$
(\tfrac{1}{2}I - \mathcal{K}_\Omega)(u - U) = (k-1)\int_D \nabla u(y) \cdot \nabla \Gamma(x-y)\, dy \quad \text{for any } x \in \partial\Omega .
$$

By using (2.13) and (2.52) together with the fact that

$$
\int_{\partial\omega} (u - U)(y)\, d\sigma(y) = 0 ,
$$

the above equation implies that

$$(\Lambda_D - \Lambda_0)g(x) = (1 - k) \int_D \nabla u(y) \cdot \nabla \Gamma(x - y)\, dy + C \quad \text{for } x \in \partial\omega \,,$$

where the constant C is given by

$$C = (k - 1) \int_{\partial\omega} \int_D \nabla u(y) \cdot \nabla \Gamma(x - y)\, dy\, d\sigma(x) \,.$$

From the smoothness of $\int_D \nabla u(y) \cdot \nabla \Gamma(x-y)\, dy|_{\partial\omega}$ we conclude that $\Lambda_D - \Lambda_0 :$ $L_0^2(\partial\omega) \to L_0^2(\partial\omega)$ is compact. Moreover, by using (2.58) and (2.59), we can prove that

$$\|u - U\|_{L^2(\partial\omega)} \le C\|\nabla u\|_{L^2(D)} \le C'\|g\|_{L^2(\partial\omega)} \,,$$

for some positive constant C' independent of g which indicates that $\Lambda_D - \Lambda_0$ is bounded. Thus, to prove that $\Lambda_D - \Lambda_0$ is self-adjoint it suffices to show that it is symmetric. Consider $h \in L_0^2(\partial\omega)$ and v and V to be the solutions of (5.11) and (5.12) corresponding to the Neumann data h. Using integration by parts we can establish the following identity

$$\int_{\partial\omega} (\Lambda_D - \Lambda_0)(g)h = -\int_\Omega \nabla(u - U) \cdot \nabla(v - V) + \sum_{s=1}^m (1 - k_s) \int_{\epsilon B_s + z_s} \nabla u \cdot \nabla v$$

$$= \int_\Omega \left(1 + \sum_{s=1}^m (k_s - 1)\chi(\epsilon B_s + z_s)\right) \nabla(u - U) \cdot \nabla(v - V)$$

$$+ \sum_{s=1}^m (1 - k_s) \int_{\epsilon B_s + z_s} \nabla U \cdot \nabla V \,.$$

This gives, as desired, that $\Lambda_D - \Lambda_0 : L_0^2(\partial\omega) \to L_0^2(\partial\omega)$ is self-adjoint, positive (respectively negative) semi-definite, if $0 < k_s < 1$ (respectively $+\infty > k_s > 1$) for all $s = 1, \dots, m$. $\square$

Next, let $\widetilde{N}$ be the solution to

$$\begin{cases} \Delta_x \widetilde{N}(x, z) = -\delta_z & \text{in } \Omega, \\[2mm] \dfrac{\partial \widetilde{N}}{\partial \nu_x}\Big|_{\partial\omega} = -\dfrac{1}{|\partial\omega|}\,, \\[2mm] \dfrac{\partial \widetilde{N}}{\partial \nu_x}\Big|_{\partial\Omega \setminus \partial\omega} = 0\,, \\[2mm] \displaystyle\int_{\partial\omega} \widetilde{N}(x, z)\, d\sigma(x) = 0 & \text{for } z \in \Omega \,. \end{cases} \tag{5.13}$$

Similarly to (4.4), we can prove without any new difficulties that

$$(\Lambda_D - \Lambda_0)(g)(x) = -\epsilon^d \sum_{s=1}^m \partial U(z_s) M(k_s, B_s) \partial_z \widetilde{N}(x, z_s) + O(\epsilon^{d+1}) \,,$$

uniformly on $\partial\omega$, where the remainder $O(\epsilon^{d+1})$ is bounded by $C\epsilon^{d+1}$ in the operator norm of $\mathcal{L}(L_0^2(\partial\omega), L_0^2(\partial\omega))$ and U is the background solution, that is, the solution of (5.12). Here $\mathcal{L}(L_0^2(\partial\omega), L_0^2(\partial\omega))$ is the set of linear bounded operators on $L_0^2(\partial\omega)$. Define the operator $T : L_0^2(\partial\omega) \to L_0^2(\partial\omega)$ by

$$T(g) = -\sum_{s=1}^m \partial U(z_s) M(k_s, B_s) \partial_z \widetilde{N}(\cdot, z_s) \, .$$

Since U depends linearly on g this operator is linear. Corresponding to Lemma 5.3 the following result can be obtained.

Lemma 5.4 *The operator* $T : L_0^2(\partial\omega) \to L_0^2(\partial\omega)$ *is compact, self-adjoint, positive (respectively negative) semi-definite, if* $0 < k_s < 1$ *(respectively* $+\infty > k_s > 1$*) for all* $s = 1, \ldots, m$.

Proof.   We first observe that T is a finite-dimensional operator and hence, it is compact. Moreover, to prove that T is self-adjoint it suffices to show that it is symmetric. Let g and h be in $L_0^2(\partial\omega)$ and denote by U and V the background solutions corresponding to g and h. We have

$$\int_{\partial\omega} T(g)h = -\sum_{s=1}^m \partial U(z_s) M(k_s, B_s) \int_{\partial\omega} \partial_z \widetilde{N}(x, z_s) \frac{\partial V}{\partial\nu}(x) \, d\sigma(x)$$

$$= -\sum_{s=1}^m \partial U(z_s) M(k_s, B_s) \int_{\partial\Omega} \partial_z \widetilde{N}(x, z_s) \frac{\partial V}{\partial\nu}(x) \, d\sigma(x) \, .$$

But since $\partial_x \widetilde{N} = -\partial_z \widetilde{N}$ we have $\Delta_x \partial_z \widetilde{N} = \partial_x \delta_{x=z}$ and therefore

$$\int_{\partial\Omega} \partial_z \widetilde{N}(x, z_s) \frac{\partial V}{\partial\nu}(x) \, d\sigma(x) = \partial V(z_s) \, .$$

Consequently,

$$\int_{\partial\omega} T(g)h = -\sum_{s=1}^m \partial U(z_s) M(k_s, B_s) \partial V(z_s) \, .$$

From the symmetry and the positive definiteness of the matrices $M(k_s, B_s)$ established in Theorem 3.6 we infer that T is self-adjoint, positive (respectively negative) semi-definite, if $0 < k_s < 1$ (respectively $+\infty > k_s > 1$) for all $s = 1, \ldots, m$.   □

Introduce now the linear operator $\mathcal{G} : L_0^2(\partial\omega) \to \mathbb{R}^{d \times m}$ defined by

$$\mathcal{G}g = (\partial U(z_1), \ldots, \partial U(z_m)) \, .$$

Endowing $\mathbb{R}^{d \times m}$ with the standard Euclidean inner product,

$$\langle a, b \rangle = \sum_{s=1}^m a_s \cdot b_s \text{ for } a = (a_1, \ldots, a_m), b = (b_1, \ldots, b_m), a_s, b_s \in \mathbb{R}^d \, ,$$

we then obtain

$$\langle \mathcal{G}g, a \rangle = \sum_{s=1}^{m} a_s \cdot \partial U(z_s) = \int_{\partial\omega} \left(\sum_{s=1}^{m} a_s \cdot \partial \widetilde{N}(x, z_s) \right) g(x) \, d\sigma(x) \,,$$

for arbitrary $a = (a_1, \ldots, a_m) \in \mathbb{R}^{d \times m}$.

Therefore, the adjoint $\mathcal{G}^* : \mathbb{R}^{d \times m} \to L_0^2(\partial\omega)$ is given by $\mathcal{G}^* a = \sum_{s=1}^{m} a_s \cdot \partial \widetilde{N}(\cdot, z_s)$.

A characterization of the range of the operator T is obtained in the following lemma due to Brühl, Hanke, and Vogelius [64].

Lemma 5.5 *(i) $\mathcal{G}^*$ is injective;*
(ii) $\mathcal{G}$ is surjective;
(iii) $T = \mathcal{G}^* \mathcal{M} \mathcal{G}$, where

$$\mathcal{M} a = \left(M(k_1, B_1) a_1, \ldots, M(k_m, B_m) a_m \right), a = (a_1, \ldots, a_m) \in \mathbb{R}^{d \times m} \,;$$

(iv) $Range(T) = span \left\{ e_p \cdot \partial \widetilde{N}(\cdot, z_s), p = 1, \ldots, d; s = 1, \ldots, m \right\}$, where

$\{e_p\}_{p=1}^{d}$ *is an orthonormal basis of $\mathbb{R}^d$.*

Proof.   Suppose that $\mathcal{G}^* a = 0$ then the function $w(x) = \sum_{s=1}^{m} a_s \cdot \widetilde{N}(x, z_s)$ solves the Cauchy problem $\Delta w = 0$ in $\Omega \setminus \cup_{s=1}^{m} \{z_s\}, w = \partial w / \partial \nu = 0$ on $\partial\omega$, and from the uniqueness of a solution to this problem we deduce that $w \equiv 0$. The dipole singularity of $\widetilde{N}(x, z_s)$ at z_s implies that

$$a_s \cdot \frac{(x - z_s)}{|x - z_s|^d} \to 0 \quad \text{as } x \to z_s, s = 1, \ldots, m \,.$$

This proves that $a_s = 0$, and thus assertion (i) holds. (ii) follows from (i) and the well-known relation between the ranges and the null spaces of adjoint finite-dimensional operators: $Range(\mathcal{G}) = Ker(\mathcal{G}^*)^\perp$. Using the above formula for $\mathcal{G}$ and $\mathcal{G}^*$ it is easy to see that (iii) holds. Now according to (iii), we write $Range(T) = Range(\mathcal{G}^* \mathcal{M} \mathcal{G}) = Range(\mathcal{G}^*)$, since $\mathcal{M}$ and $\mathcal{G}$ are surjective. This yields (iv). $\square$

Now we present the main tool for the identification of the locations z_s. The following theorem is also due to Brühl, Hanke, and Vogelius [64].

Theorem 5.1. *Let $e \in \mathbb{R}^d \setminus \{0\}$. A point $z \in \Omega$ belongs to the set $\{z_s : s = 1, \ldots, m\}$ if and only if $e \cdot \partial_z \widetilde{N}(\cdot, z)\Big|_{\partial\omega} \in Range\ (T)$.*

Proof.   Assume that $g_{z,e} = e \cdot \partial_z \widetilde{N}(\cdot, z)\Big|_{\partial\omega} \in Range\ (T)$. As a consequence of (iv), $g_{z,e}$ may be represented as

$$g_{z,e}(x) = \sum_{s=1}^{m} a_s \cdot \partial_z \widetilde{N}(x,z) \quad \text{for } x \in \partial\omega .$$

But then by the uniqueness of a solution to the Cauchy problem it follows that

$$\sum_{s=1}^{m} a_s \cdot \partial_z \widetilde{N}(x,z) = e \cdot \partial_z \widetilde{N}(x,z) \quad \text{for all } x \in \Omega \setminus (\cup_{s=1}^{m}\{z_s\} \cup \{z\}) .$$

This is only possible if $z \in \{z_s : s = 1,\dots,m\}$, and so we have established the necessity of this condition. The sufficiency follows immediately from (iv) in Lemma 5.5. $\square$

The finite-dimensional self-adjoint operator T can be decomposed as

$$T = \sum_{p=1}^{dm} \lambda_p v_p v_p^*, \quad ||v_p||_{L^2(\partial\omega)} = 1 ,$$

say with $|\lambda_1| \geq |\lambda_2| \geq \dots \geq |\lambda_{dm}| > 0$. Let $P_p : L_0^2(\partial\omega) \to \text{span } \{v_1,\dots,v_p\}$, $p = 1,\dots,dm$, be the orthogonal projector $P_p = \sum_{q=1}^{p} v_q v_q^*$. From Theorem 5.1 it follows that

$$z \in \{z_s : s = 1,\dots,m\} \text{ iff } (I - P_{dm})(e \cdot \partial_z \widetilde{N}(\cdot,z)|_{\partial\omega}) = 0 ,$$

or equivalently, if we define the angle $\theta(z) \in [0,\pi/2)$ by

$$\cot\theta(z) = \frac{||P_{dm}(e \cdot \partial_z \widetilde{N}(\cdot,z)|_{\partial\omega})||_{L^2(\partial\omega)}}{||(I - P_{dm})(e \cdot \partial_z \widetilde{N}(\cdot,z)|_{\partial\omega})||_{L^2(\partial\omega)}} ,$$

then we have

$$z \in \{z_s : s = 1,\dots,m\} \text{ iff } \cot\theta(z) = +\infty .$$

On the other hand, since $\Lambda_D - \Lambda_0$ is self-adjoint and compact operator on $L_0^2(\partial\omega)$ it admits, by the spectral theorem, the following spectral decomposition

$$\Lambda_D - \Lambda_0 = \sum_{p=1}^{+\infty} \lambda_p^\epsilon v_p^\epsilon (v_p^\epsilon)^* , \quad ||v_p^\epsilon||_{L^2(\partial\omega)} = 1 ,$$

with $|\lambda_1^\epsilon| \geq |\lambda_2^\epsilon| \geq \dots \geq |\lambda_{dm}^\epsilon| \geq \dots \geq 0$. Let $P_p^\epsilon : L_0^2(\partial\omega) \to \text{span } \{v_1^\epsilon,\dots,v_p^\epsilon\}$, $p = 1,2,\dots$, be the orthogonal projector $P_p^\epsilon = \sum_{q=1}^{p} v_q^\epsilon (v_q^\epsilon)^*$. Standard arguments from perturbation theory for linear operators [173] give (after appropriate enumeration of $\lambda_p^\epsilon, p = 1,\dots,dm$)

$$\lambda_p^\epsilon = \epsilon^d \lambda_p + O(\epsilon^{d+1}) \quad \text{for } p = 1,2,\dots , \tag{5.14}$$

where we have set $\lambda_p = 0$ for $p > dm$, and

$$P_p^\epsilon = P_{dm} + O(\epsilon) \quad \text{for } p \geq dm, \tag{5.15}$$

provided that one makes appropriate choices of eigenvectors v_p^ϵ and $v_p, p = 1, \ldots, dm$.

Now in view of (5.14) the number m of inclusions may be estimated by looking for a gap in the set of eigenvalues of $\Lambda_D - \Lambda_0$. In order to recover the locations $z_s, s = 1, \ldots, m$, one can estimate, using (5.15), the $\cot \theta(z)$ by

$$\cot \theta_p(z) = \frac{\|P_p^\epsilon(e \cdot \partial_z \widetilde{N}(\cdot, z))|_{\partial\omega})\|_{L^2(\partial\omega)}}{\|(I - P_p^\epsilon)(e \cdot \partial_z \widetilde{N}(\cdot, z)|_{\partial\omega})\|_{L^2(\partial\omega)}}.$$

If one plots $\cot \theta_{dm}(z)$ as a function of z, we may see large values for z which are close to the positions z_s. The viability of this direct approach has been documented by several numerical examples in [64]. In particular, its good ability to efficiently locate a high number of inclusions has been clearly demonstrated.

When comparing the different methods that have been designed for imaging small inclusions, it is fair to point out that the variational method and the sampling linear approach use "many boundary measurements". In contrast, the constant current projection algorithm, the quadratic algorithm, and the least-squares algorithm only rely on "single measurements" and not surprisingly, they are more limited in their abilities to effectively locate a higher number of small inclusions.

5.6 Lipschitz-Continuous Dependence and Moments Estimations

5.6.1 Lipschitz-Continuous Dependence

We now prove a Lipschitz-continuous dependence of the location and relative size of two sets of inclusions on the difference in the boundary voltage potentials corresponding to a fixed current distribution. This explains the practical success of various numerical algorithms to detect the location and size of unknown small inclusions.

Consider two arbitrary collections of inclusions

$$D = \cup_{s=1}^m (\epsilon \rho_s B + z_s) \text{ and } D' = \cup_{s=1}^{m'} (\epsilon \rho_s' B + z_s'),$$

both satisfying (5.10). The parameter ϵ determines the common length scale of the inclusions and the parameters $\rho_s, 0 < c_0 \leq \rho_s \leq C_0$, for some constant C_0, determine their relative size. We suppose that all the inclusions have the same known conductivity $0 < k \neq 1 < +\infty$. Let u and u' denote the corresponding voltage potentials (with fixed boundary current $g \in L_0^2(\partial\Omega)$). It is crucial to assume that $\nabla U(x) \neq 0, \forall x \in \Omega$, where U is the background solution. Introduce $H[g] = -\mathcal{S}_\Omega g + \mathcal{D}_\Omega u$ and $H'[g] = -\mathcal{S}_\Omega g + \mathcal{D}_\Omega u'$.

By iterating the asymptotic formula (5.8) we arrive at the following expansions:

$$H[g](x) = -\sum_{s=1}^{m}\sum_{|i|=1}^{d}\sum_{|j|=1}^{d}\frac{(\epsilon\rho_s)^{|i|+|j|+d-2}}{i!j!}(\partial^i U)(z_s)\partial_z^j\Gamma(x-z_s)M_{ij}(k,B)$$
$$+ O(\epsilon^{2d}),$$

$$H'[g](x) = -\sum_{s=1}^{m'}\sum_{|i|=1}^{d}\sum_{|j|=1}^{d}\frac{(\epsilon\rho_s')^{|i|+|j|+d-2}}{i!j!}(\partial^i U)(z_s')\partial_z^j\Gamma(x-z_s')M_{ij}(k,B)$$
$$+ O(\epsilon^{2d}) .$$

(5.16)

The following theorem, due to Friedman and Vogelius [123], shows that for small ϵ the locations of the inclusions, z_s, and their relative size, ρ_s, depend Lipschitz-continuous on $\epsilon^{-d}||H[g] - H'[g]||_{L^\infty(S)}$ for any C^2-closed surface (or curve in $\mathbb{R}^2$) S enclosing the domain Ω.

Theorem 5.6 *Let S be a C^2-closed surface (or curve in $\mathbb{R}^2$) enclosing the domain Ω. There exist constants $0 < \epsilon_0, \delta_0$, and C such that if $\epsilon < \epsilon_0$ and $\epsilon^{-d}||H[g] - H'[g]||_{L^\infty(S)} < \delta_0$ then*

(i) $m = m'$, and, after appropriate reordering,

(ii) $|z_s - z_s'| + |\rho_s - \rho_s'| \le C\left(\epsilon^{-d}||H[g] - H'[g]||_{L^\infty(S)} + \epsilon\right)$.

The constants ϵ_0, δ_0 and C depend on $c_0, C_0, \Omega, S, B, k$ but are otherwise independent of the two sets of inclusions.

Proof. From (5.16) we get

$$\epsilon^{-d}\left(H[g](x) - H'[g](x)\right) = \left[\sum_{s=1}^{m'}(\rho_s')^d(\partial U)(z_s')\partial_z\Gamma(x-z_s')M(k,B)\right.$$
$$\left. -\sum_{s=1}^{m}(\rho_s)^d(\partial U)(z_s)\partial_z\Gamma(x-z_s)M(k,B)\right] + O(\epsilon) ,$$

for all $x \in S$. Suppose now the assertion $m = m'$ is not true. Then there exists a function of the form $F(x) = \sum_{s=1}^{m'}\partial_z\Gamma(x-z_s')\cdot\alpha_s' - \sum_{s=1}^{m}\partial_z\Gamma(x-z_s)\cdot\alpha_s$, with $\alpha_s' \ne 0$ and $\alpha_s \ne 0$, such that $F(x) = 0$ for all $x \in S$. To see that α_s' as well as α_s are not zero we use the fact that ∇U never vanishes and that the polarization tensor $M(k, B)$ is invertible. Let Ω' denote the region outside S. From the uniqueness of a solution to $\Delta F = 0$ in $\Omega', F = 0$ on $S, F(x) = O(|x|^{-d+1})$ as $|x| \to +\infty$, it follows that $\partial F/\partial\nu = 0$ on S. But F is also harmonic in $\mathbb{R}^d \setminus (\{z_s\} \cup \{z_s'\})$. From the uniqueness of a solution to the Cauchy problem for the Laplacian we then conclude that $F \equiv 0$ in $\mathbb{R}^d$. This contradicts the fact that $m \ne m'$.

When it comes to proving (ii) suppose for simplicity that $U(x) = x_1 - (1/|\partial|\Omega|) \int_{\partial\Omega} U$ (corresponding to the boundary current $g = \nu_1$). Then

$$\epsilon^{-d}\left(H[g](x) - H'[g](x)\right) = \sum_{s=1}^{m} \left[(\rho'_s)^d \partial_z \Gamma(x - z'_s)\right.$$
$$\left. - (\rho_s)^d \partial_z \Gamma(x - z_s)\right](M(k,B))_1 + O(\epsilon) ,$$

(5.17)

for all $x \in S$, where $(M(k,B))_1$ is the first column of the matrix $M(k,B)$. A simple calculation shows that, for some $\overline{\rho_s}$ and $\overline{z_s}$,

$$\sum_{s=1}^{m} \left[(\rho'_s)^d \partial_z \Gamma(x - z'_s) - (\rho_s)^d \partial_z \Gamma(x - z_s)\right](M(k,B))_1$$

$$= \sum_{s=1}^{m} \left[d(\rho'_s - \rho_s)(\overline{\rho_s})^{d-1} \partial_z \Gamma(x - z'_s) + \rho_s^d(z'_s - z_s) \cdot \partial_z^2 \Gamma(x - \overline{z_s})\right](M(k,B))_1$$

$$= \sum_{s=1}^{m} \left(|z_s - z'_s| + |\rho_s - \rho'_s|\right)$$

$$\times \sum_{s=1}^{m} \left[d \partial\rho_s (\overline{\rho_s})^{d-1} \partial_z \Gamma(x - z'_s) + \rho_s^d \partial z_s \cdot \partial_z^2 \Gamma(x - \overline{z_s})\right](M(k,B))_1 ,$$

where

$$\partial\rho_s = \frac{(\rho'_s - \rho_s)}{\sum_{s=1}^{m} \left(|z_s - z'_s| + |\rho_s - \rho'_s|\right)}$$

and

$$\partial z_s = \frac{(z'_s - z_s)}{\sum_{s=1}^{m} \left(|z_s - z'_s| + |\rho_s - \rho'_s|\right)} .$$

Suppose the estimate (ii) is not true. Then there exist perturbations $\partial\rho_s$ and ∂z_s with $\sum_{s=1}^{m} |\partial\rho_s| + |\partial z_s| = 1$, points z_s $(= z'_s = \overline{z_s})$, and parameters $\rho_s(= \rho'_s = \overline{\rho_s})$ so that

$$G(x) = \sum_{s=1}^{m} \left[d \partial\rho_s (\rho_s)^{d-1} \partial_z \Gamma(x - z_s) + \rho_s^d \partial z_s \cdot \partial_z^2 \Gamma(x - z_s)\right](M(k,B))_1 = 0$$

for all $x \in S$. Just as was the case with F, the function G has a vanishing normal derivative on S and it is harmonic except at the points $\{z_s\}$ and $\{z'_s\}$. Therefore, by the unique continuation property of harmonic functions, $G \equiv 0$ and thus, $\partial\rho_s = \partial z_s = 0, s = 1, \ldots m$. This, however, would be a contradiction to the fact that $\sum_{s=1}^{m} |\partial\rho_s| + |\partial z_s| = 1$. We therefore conclude that the desired Lipschitz-continuous dependence estimate holds. $\square$

The factor ϵ^{-d} in front of $\|H[g] - H'[g]\|_{L^\infty(S)}$ is best possible; it follows immediately from (5.16) that even $|z_s - z'_s|$ and $|\rho_s - \rho'_s|$ are of order 1 then

$||H[g] - H'[g]||_{L^\infty(S)}$ is of order ϵ^d. The use of the L^∞-norm of $H[g] - H'[g]$ on S is not essential; in fact other norms, such as the L^1-norm, can be used.

In the two-dimensional case, the results in Theorem 5.6 have some similarity to the results about the location of poles for meromorphic functions, found in [204]. The idea is quite simple. Suppose $d = 2$ then (5.17) reads

$$\epsilon^{-2}\left(H[g](x) - H'[g](x)\right) = -\frac{1}{2\pi}\sum_{s=1}^{m'}(\rho'_s)^2\frac{(x - z'_s)}{|x - z'_s|^2}(M(k,B))_1$$

$$+\frac{1}{2\pi}\sum_{s=1}^{m}(\rho_s)^2\frac{(x - z'_s)}{|x - z'_s|^2}(M(k,B))_1 + O(\epsilon) .$$

Identifying $\mathbb{R}^2$ with $\mathbb{C}$ yields

$$\epsilon^{-2}\left(H[g](x) - H'[g](x)\right) = \Re\left(\sum_{s=1}^{m'}\frac{\alpha'_s}{x - z'_s} - \sum_{s=1}^{m}\frac{\alpha_s}{x - z_s}\right) + O(\epsilon) ,$$

for all $x \in S$, where the constants $\alpha_s = -(1/2\pi)\rho_s^2((M(k,B))_{11}+i(M(k,B))_{12})$ and $\alpha'_s = -(1/2\pi)(\rho'_s)^2((M(k,B))_{11}+i(M(k,B))_{12})$ are of order 1. Therefore,

$$\sum_{s=1}^{m'}\frac{\alpha'_s}{x - z'_s} - \sum_{s=1}^{m}\frac{\alpha_s}{x - z_s} = \epsilon^{-2}\left(H[g](x) - H'[g](x)\right)$$

$$+i\epsilon^{-2}\int_a^x\frac{\partial}{\partial\nu}\left(H[g](y) - H'[g](y)\right)d\sigma(y) + O(\epsilon) ,$$

for some complex constant $a \in S$. Here

$$\int_a^x\frac{\partial}{\partial\nu}\left(H[g](y) - H'[g](y)\right)d\sigma(y)$$

is the harmonic conjugate to $H[g](x) - H'[g](x)$ and satisfies

$$||\int_a^x\frac{\partial}{\partial\nu}\left(H[g](y) - H'[g](y)\right)d\sigma(y)||_{L^\infty(S)} \leq C||H[g] - H'[g]||_{L^\infty(S)} ,$$

for some constant C independent of ϵ. Since the poles $\{z_s\}$ and $\{z'_s\}$ are well separated and the pole residues $\{\rho_s\}$ and $\{\rho'_s\}$ are well bounded from zero then it follows from Theorem 1 in [204] that (i) and (ii) in Theorem 5.6 hold.

5.6.2 Moments Estimations

Consider a collection of inclusions

$$D = \cup_{s=1}^{m}(\epsilon B_s + z_s) ,$$

satisfying (5.10). Our goal is to obtain upper and lower bounds on the moments of the unknown inclusions B_s. Observe that $\sum_{|i|=1}^{d} \partial^i U(z_s) x^i$ is a harmonic polynomial. Let S be a C^2-closed surface (or curve in $\mathbb{R}^2$) enclosing the domain Ω. From

$$H[g](x) = -\sum_{s=1}^{m} \sum_{|i|=1}^{d} \sum_{|j|=1}^{d} \frac{\epsilon^{|i|+|j|+d-2}}{i!j!} (\partial^i U)(z_s) \partial_z^j \Gamma(x - z_s) M_{ij}^s + O(\epsilon^{2d}) \text{ on } S \,,$$

where $g = \partial U/\partial \nu$, we compute by Green's formula

$$\int_S \left(\frac{\partial}{\partial \nu} H[g] \, U - H[g] \, g \right) d\sigma$$

$$= -\sum_{s=1}^{m} \sum_{|i|=1}^{d} \sum_{|j|=1}^{d} \frac{\epsilon^{|i|+|j|+d-2}}{i!j!} \partial^i U(z_s) M_{ij}^s \partial^j U(z_s) + O(\epsilon^{2d}) \,.$$

Then, as a direct consequence of Theorem 3.9, the following moments estimations hold. Note that, in general, they are only meaningful if the conductivities $\{k_s\}_{s=1}^{m}$ and the locations $\{z_s\}_{s=1}^{m}$ are known.

Theorem 5.7 *We have*

$$\epsilon^{2-d} \left| \int_S \left(\frac{\partial}{\partial \nu} H[g] \, U - H[g] \, g \right) d\sigma \right|$$

$$\approx \sum_{s=1}^{m} \frac{|k_s - 1|}{k_s + 1} \int_{B_s} \left| \nabla \left(\sum_{|i|=1}^{d} \frac{\epsilon^{|i|}}{i!} \partial^i U(z_s) x^i \right) \right|^2 dx \,.$$

Part II

Detection of Small Elastic Inclusions

Suppose that an elastic medium occupies a bounded domain Ω in $\mathbb{R}^d$ with a connected Lipschitz boundary $\partial\Omega$. Let C_{ijkl}, $i,j,k,l = 1,\ldots,d$, be the elasticity tensor of Ω, which is piecewise constant. Then by the generalized Hooke's law, the displacement vector $\mathbf{u}$ caused by the traction $\mathbf{g}$ applied on the boundary $\partial\Omega$ is the solution to the following transmission problem associated to the system of elastostatics with the traction boundary condition:

$$
\begin{cases}
\displaystyle\sum_{j,k,l=1}^{d} \frac{\partial}{\partial x_j}\left(C_{ijkl}\frac{\partial u_k}{\partial x_l}\right) = 0 \quad \text{in } \Omega, \quad i = 1,\ldots,d\,, \\
\dfrac{\partial \mathbf{u}}{\partial \nu}\bigg|_{\partial\Omega} = \mathbf{g}\,.
\end{cases}
$$

If Ω is an isotropic elastic material and contains isotropic inclusions $D = \cup_{s=1}^{m} D_s$, the elasticity tensor C_{ijkl} takes the following form

$$
C_{ijkl} := \left(\lambda\,\chi(\Omega \setminus D) + \sum_{s=1}^{m}\lambda_s\,\chi(D_s)\right)\delta_{ij}\delta_{kl}
$$

$$
+ \left(\mu\,\chi(\Omega \setminus D) + \sum_{s=1}^{m}\mu_s\,\chi(D_s)\right)(\delta_{ik}\delta_{jl} + \delta_{il}\delta_{jk})\,,
$$

where $\chi(D)$ is the characteristic function of D, and (λ,μ) and (λ_s,μ_s) are pairs of Lamé constants of $\Omega \setminus D$ and D_s, respectively, for $s = 1,\ldots,m$.

The problem we consider in this part is to detect unknown inclusions D_s, $s = 1,\ldots,m$, by means of a finite number of pairs of traction-displacement $(\mathbf{g}, \mathbf{u}|_{\partial\Omega})$ measured on $\partial\Omega$. In particular, we are interested in finding diametrically small inclusions D_s, $s = 1,\ldots,m$. So we assume that each D_s takes the form $D_s = \epsilon B_s + z_s$, $s = 1,\ldots,m$, where ϵ is a small number and denotes the common magnitude of the inclusions, B_s is a bounded Lipschitz domain, and z_s indicates the location of D_s.

Most of the existing algorithms to solve inverse problems for the Lamé system are iterative and are then based on regularization techniques. See, for example, [67, 118, 57, 144, 214]. Inspired by Part I, we design efficient and robust direct (non-iterative) algorithms to reconstruct the location and certain features of the elastic inclusions D_s.

One medical problem for which knowledge of internal elastic properties would be useful is breast tumor detection [50, 137]. The elastic properties are very different for cancerous and normal tissues. A variety of nonclinical applications of our mathematical model is also possible. These include earth imaging [141] and nondestructive evaluation of materials [109].

Inclusions of small size are believed to be the starting point of crack development in elastic bodies.

A striking example is provided by an aircraft accident due to a fatigue failure initiated by an inclusion in an engine component. We quote paragraphs from the website www.irc.bham.ac.uk/theme1/plasma/production.htm, since they describe very well one of the motivations of the present part.

"Imperfection in the metallic structure can lead to a significant reduction in the performance of a given item, but worse still can be 'inclusion' or 'defect' (small particles of other materials trapped in the metal). Metallic items normally ultimately fail by cracking and inclusions can act as the starting points for cracks - the larger the inclusion, the larger the crack and the quicker it will grow. In aerospace applications, inclusions as small as 1-hundredth of a millimeter are important. To put this in perspective an inclusion of about 20 millionth of a gramme can lead to failure in a component a meter long.

Since the early 1980's, a number of near air disasters have occurred caused by engine problem traceable to the presence of inclusions. On 19 July 1989, United Airlines Flight 232, a wide-bodied DC-10, crashed at Sioux City, Iowa, ultimately resulting in 112 deaths (Randall, 1991). This crash was a direct consequence of a fatigue failure initiated by the presence of a 'hard alpha' inclusion in a titanium alloy engine component. Ensuring the safe performance of such components is therefore of paramount importance. However, it is not just the aerospace industry which requires predictable long life from significantly stressed components - in both the medical and offshore industries, the effects of component failure could be disastrous."

The main aim of this part is to develop a method to detect the size and location of an inclusion in an elastic body in a mathematically rigorous way. We find a complete asymptotic formula of solutions of the linear elastic system in terms of the size of the inclusion. This formula describes the perturbation of the solution caused by the presence of an anomaly (inclusion) of small size. Based on this asymptotic expansion we derive formulae to find the location and the order of magnitude of the elastic inclusion with high accuracy. The formulae are explicit and can be easily implemented numerically. The general approach we will take is parallel to that in Part I except for some technical difficulties due to the fact that we are dealing with a system, not a single equation, and the equations inside and outside the inclusion are different.

In the course of deriving the asymptotic formula, we introduce the concepts of elastic moment tensors (EMT) and prove some of their basic properties. These concepts are defined in a way analogous to the GPT's. The first-order EMT was introduced by Maz'ya and Nazarov [202].

The main body of this part is from [23] and [166].

Transmission Problem for Elastostatics

In this chapter we review some well-known results on the solvability and layer potentials for the Lamé system, mostly from [93] and [111], and prove a representation formula for solutions of the Lamé system which will be our main tool in later chapters.

6.1 Layer Potentials for the Lamé System

Let D be a bounded Lipschitz domain in $\mathbb{R}^d$, $d = 2, 3$, and (λ, μ) be the Lamé constants for D satisfying

$$\mu > 0 \quad \text{and} \quad d\lambda + 2\mu > 0 .$$

See Kupradze [189]. The elastostatic system corresponding to the Lamé constants λ, μ is defined by

$$\mathcal{L}_{\lambda,\mu} \mathbf{u} := \mu \Delta \mathbf{u} + (\lambda + \mu) \nabla \nabla \cdot \mathbf{u} .$$

The corresponding conormal derivative $\partial \mathbf{u} / \partial \nu$ on ∂D is defined to be

$$\frac{\partial \mathbf{u}}{\partial \nu} := \lambda (\nabla \cdot \mathbf{u}) N + \mu (\nabla \mathbf{u} + \nabla \mathbf{u}^T) N \quad \text{on } \partial D , \tag{6.1}$$

where N is the outward unit normal to ∂D and the superscript T denotes the transpose of a matrix. Let us note a simple, but important relation.

Lemma 6.1 *If $\mathbf{u} \in W^{1,2}(D)$ and $\mathcal{L}_{\lambda,\mu} \mathbf{u} = 0$ in D, then for all $\mathbf{v} \in W^{1,2}(D)$,*

$$\int_{\partial D} \mathbf{v} \cdot \frac{\partial \mathbf{u}}{\partial \nu} \, d\sigma = \int_D \lambda (\nabla \cdot \mathbf{u})(\nabla \cdot \mathbf{v}) + \frac{\mu}{2} (\nabla \mathbf{u} + \nabla \mathbf{u}^T) \cdot (\nabla \mathbf{v} + \nabla \mathbf{v}^T) \, dx , \tag{6.2}$$

where for $d \times d$ matrices $a = (a_{ij})$ and $b = (b_{ij})$, $a \cdot b = \sum_{ij} a_{ij} b_{ij}$.

Proof. By the definition (6.1) of the conormal derivative, we get

$$\int_{\partial D} \mathbf{v} \cdot \frac{\partial \mathbf{u}}{\partial \nu} \, d\sigma = \int_{\partial D} \lambda (\nabla \cdot \mathbf{u}) \mathbf{v} \cdot N + \mu \mathbf{v} \cdot (\nabla \mathbf{u} + \nabla \mathbf{u}^T) N \, d\sigma$$

$$= \int_D \lambda \nabla \cdot ((\nabla \cdot \mathbf{u}) \mathbf{v}) + \mu \nabla \cdot ((\nabla \mathbf{u} + \nabla \mathbf{u}^T) \mathbf{v}) \, dx \,.$$

Since

$$\nabla \cdot \left((\nabla \mathbf{u} + \nabla \mathbf{u}^T) \mathbf{v} \right) = \nabla (\nabla \cdot \mathbf{u}) \cdot \mathbf{v} + \Delta \mathbf{u} \cdot \mathbf{v} + \frac{1}{2} (\nabla \mathbf{u} + \nabla \mathbf{u}^T) \cdot (\nabla \mathbf{v} + \nabla \mathbf{v}^T) \,,$$

we obtain (6.2) and the proof is complete. □

We give now a fundamental solution to the Lamé system $\mathcal{L}_{\lambda,\mu}$ in $\mathbb{R}^d$.

Lemma 6.2 *A fundamental solution $\mathbf{\Gamma} = (\Gamma_{ij})_{i,j=1}^d$ to the Lamé system $\mathcal{L}_{\lambda,\mu}$ is given by*

$$\Gamma_{ij}(x) := \begin{cases} -\dfrac{A}{4\pi} \dfrac{\delta_{ij}}{|x|} - \dfrac{B}{4\pi} \dfrac{x_i x_j}{|x|^3} & \text{if } d = 3 \,, \\ \dfrac{A}{2\pi} \delta_{ij} \log |x| - \dfrac{B}{2\pi} \dfrac{x_i x_j}{|x|^2} & \text{if } d = 2 \,, \end{cases} \qquad x \neq 0,$$

where

$$A = \frac{1}{2} \left(\frac{1}{\mu} + \frac{1}{2\mu + \lambda} \right) \quad and \quad B = \frac{1}{2} \left(\frac{1}{\mu} - \frac{1}{2\mu + \lambda} \right) . \tag{6.3}$$

The function $\mathbf{\Gamma}$ is known as the Kelvin matrix of fundamental solutions.

Proof. We seek a solution $\mathbf{\Gamma} = (\Gamma_{ij})_{i,j=1}^d$ of

$$\mu \Delta \mathbf{\Gamma} + (\lambda + \mu) \nabla \nabla \cdot \mathbf{\Gamma} = \delta_0 I_d \quad \text{in } \mathbb{R}^d \,, \tag{6.4}$$

where I_d is the $d \times d$ identity matrix and δ_0 is the Dirac function at 0. Taking the divergence of (6.4), we have

$$(\lambda + 2\mu) \Delta (\nabla \cdot \mathbf{\Gamma}) = \nabla \delta_0 \,.$$

Thus by Lemma 2.2

$$\nabla \cdot \mathbf{\Gamma} = \frac{1}{\lambda + 2\mu} \nabla \Gamma \,,$$

where Γ is given by (2.4). Inserting this into (6.4) gives

$$\mu \Delta \mathbf{\Gamma} = \delta_0 I_d - \frac{\lambda + \mu}{\lambda + 2\mu} \nabla \nabla \Gamma \,.$$

Hence it follows that

$$
\Gamma_{ij}(x) := \begin{cases} -\dfrac{A}{4\pi}\dfrac{\delta_{ij}}{|x|} - \dfrac{B}{4\pi}\dfrac{x_i x_j}{|x|^3} & \text{if } d = 3 \,, \\[2mm] \dfrac{A}{2\pi}\delta_{ij}\log|x| - \dfrac{B}{2\pi}\dfrac{x_i x_j}{|x|^2} & \text{if } d = 2 \,, \end{cases} \qquad x \neq 0,
$$

modulo constants, where A and B are given by (6.3). $\square$

The single and double layer potentials of the density function φ on D associated with the Lamé parameters (λ, μ) are defined by

$$
\mathcal{S}_D\varphi(x) := \int_{\partial D} \boldsymbol{\Gamma}(x-y)\varphi(y)\,d\sigma(y) \,, \quad x \in \mathbb{R}^d \,, \tag{6.5}
$$

$$
\mathcal{D}_D\varphi(x) := \int_{\partial D} \frac{\partial}{\partial \nu_y}\boldsymbol{\Gamma}(x-y)\varphi(y)\,d\sigma(y) \,, \quad x \in \mathbb{R}^d \setminus \partial D \,, \tag{6.6}
$$

where $\partial/\partial\nu$ denotes the conormal derivative defined in (6.1). Thus, for $m = 1, \ldots, d$,

$$
(\mathcal{D}_D\varphi(x))_m = \int_{\partial D} \lambda\frac{\partial \Gamma_{mi}}{\partial y_i}(x-y)\varphi(y)\cdot N(y)
$$
$$
+ \mu\Big(\frac{\partial \Gamma_{mi}}{\partial y_j} + \frac{\partial \Gamma_{mj}}{\partial y_i}\Big)(x-y)N_i(y)\varphi_j(y)\,d\sigma(y) \,.
$$

Here we used the Einstein convention for summation notation. As an immediate consequence of (6.2) we obtain the following lemma which can be proved in the same way as the Green's representation (2.7) of harmonic functions.

Lemma 6.3 *If* $\mathbf{u} \in W^{1,2}(D)$ *and* $\mathcal{L}_{\lambda,\mu}\mathbf{u} = 0$ *in* D, *then*

$$
\mathbf{u}(x) = \mathcal{D}_D(\mathbf{u}|_{\partial D})(x) - \mathcal{S}_D\left(\frac{\partial\mathbf{u}}{\partial\nu}\Big|_{\partial D}\right)(x) \,, \quad x \in D \,, \tag{6.7}
$$

and

$$
\mathcal{D}_D(\mathbf{u}|_{\partial D})(x) - \mathcal{S}_D\left(\frac{\partial\mathbf{u}}{\partial\nu}\Big|_{\partial D}\right)(x) = 0 \,, \quad x \in \mathbb{R}^d \setminus \overline{D} \,. \tag{6.8}
$$

As before, let $\mathbf{u}|_+$ and $\mathbf{u}|_-$ denote the limits from outside D and inside D, respectively.

The following theorems are due to Dahlberg, Kenig, and Verchota [93].

Theorem 6.4 (Jump formula, [93]) *Let* D *be a bounded Lipschitz domain in* $\mathbb{R}^d, d = 2$ *or* 3. *For* $\varphi \in L^2(\partial D)$

$$
\mathcal{D}_D\varphi|_\pm = (\mp\frac{1}{2}I + \mathcal{K}_D)\varphi \quad \text{a.e. on } \partial D \,, \tag{6.9}
$$

$$
\frac{\partial}{\partial\nu}\mathcal{S}_D\varphi\Big|_\pm = (\pm\frac{1}{2}I + \mathcal{K}_D^*)\varphi \quad \text{a.e. on } \partial D \,, \tag{6.10}
$$

where $\mathcal{K}_D$ *is defined by*

$$\mathcal{K}_D\varphi(x) := p.v. \int_{\partial D} \frac{\partial}{\partial \nu_y}\mathbf{\Gamma}(x-y)\varphi(y)\,d\sigma(y) \quad a.e.\ x \in \partial D\ ,$$

and $\mathcal{K}_D^*$ is the adjoint operator of $\mathcal{K}_D$ on $L^2(\partial D)$, that is,

$$\mathcal{K}_D^*\varphi(x) := p.v. \int_{\partial D} \frac{\partial}{\partial \nu_x}\mathbf{\Gamma}(x-y)\varphi(y)\,d\sigma(y) \quad a.e.\ x \in \partial D\ .$$

It must be emphasized that, in contrast to the corresponding singular integral operators defined in (2.14) and (2.15) that arise when studying Laplace's equation, the singular integral operators $\mathcal{K}_D$ and $\mathcal{K}_D^*$ are not compact, even on bounded $\mathcal{C}^\infty$-domains [93].

Let Ψ be the vector space of all linear solutions of the equation $\mathcal{L}_{\lambda,\mu}\mathbf{u} = 0$ and $\partial\mathbf{u}/\partial\nu = 0$ on ∂D, or alternatively,

$$\Psi := \left\{ \boldsymbol{\psi} : \partial_i\psi_j + \partial_j\psi_i = 0, \quad 1 \le i,j \le d \right\}.$$

Observe now that the space Ψ is defined independently of the Lamé constants λ, μ and its dimension is 3 if $d = 2$ and 6 if $d = 3$. Define

$$L_\Psi^2(\partial D) := \left\{ \mathbf{f} \in L^2(\partial D) : \int_{\partial D} \mathbf{f} \cdot \boldsymbol{\psi}\,d\sigma = 0 \text{ for all } \boldsymbol{\psi} \in \Psi \right\}.$$

In particular, since Ψ contains constant functions, we get

$$\int_{\partial D} \mathbf{f}\,d\sigma = 0$$

for any $\mathbf{f} \in L_\Psi^2(\partial D)$. The following fact, which immediately follows from (6.2), is useful in later sections.

If $\mathbf{u} \in W^{1,\frac{3}{2}}(D)$ satisfies $\mathcal{L}_{\lambda,\mu}\mathbf{u} = 0$ in D, then $\left.\dfrac{\partial\mathbf{u}}{\partial\nu}\right|_{\partial D} \in L_\Psi^2(\partial D)\ .$ (6.11)

One of fundamental results in the theory of linear elasticity using layer potentials is the following invertibility result.

Theorem 6.5 ([93]) *The operator $\mathcal{K}_D$ is bounded on $L^2(\partial D)$, and $-(1/2)\,I + \mathcal{K}_D^*$ and $(1/2)\,I + \mathcal{K}_D^*$ are invertible on $L_\Psi^2(\partial D)$ and $L^2(\partial D)$, respectively.*

As a consequence of (6.10) and Theorem 6.5, we obtain the following result.

Corollary 6.6 ([93]) *For a given $\mathbf{g} \in L_\Psi^2(\partial D)$, the function $\mathbf{u} \in W^{1,2}(D)$ defined by*

$$\mathbf{u}(x) := \mathcal{S}_D(-\frac{1}{2}I + \mathcal{K}_D^*)^{-1}\mathbf{g} \tag{6.12}$$

is a solution to the problem

$$\begin{cases} \mathcal{L}_{\lambda,\mu}\mathbf{u} = 0 & in\ D\ , \\ \dfrac{\partial\mathbf{u}}{\partial\nu}|_{\partial D} = \mathbf{g}\ , \quad (\mathbf{u}|_{\partial D} \in L_\Psi^2(\partial D))\ . \end{cases} \tag{6.13}$$

If $\psi \in \Psi$ and $x \in \mathbb{R}^d \setminus \overline{D}$, it follows from (6.2) that $\mathcal{D}_D \psi(x) = 0$. Hence we obtain from (6.9) that $(-(1/2) I + \mathcal{K}_D)\psi = 0$. Since the dimension of the orthogonal complement of the range of the operator $-(1/2) I + \mathcal{K}_D^*$ is less than 3 if $d = 2$ and 6 if $d = 3$, which is the dimension of the space Ψ, we have the following corollary.

Corollary 6.7 *The null space of $-(1/2) I + \mathcal{K}_D$ on $L^2(\partial D)$ is Ψ.*

The following formulation of Korn's inequality will be of use to us. See Nečas [223] and Ciarlet [85] (Theorem 6.3.4).

Lemma 6.8 *Let D be a bounded Lipschitz domain in $\mathbb{R}^d$. Consider functions $\mathbf{u} \in W^{1,2}(D)$ so that*

$$\int_D \left(\mathbf{u} \cdot \psi + \nabla \mathbf{u} \cdot \nabla \psi \right) = 0 \quad \text{for all } \psi \in \Psi .$$

Then there is a constant C depending only on the Lipschitz character of D so that

$$\int_D \left(|\mathbf{u}|^2 + |\nabla \mathbf{u}|^2 \right) dx \leq C \int_D |\nabla \mathbf{u} + \nabla \mathbf{u}^T|^2 \, dx . \tag{6.14}$$

6.2 Kelvin Matrix Under Unitary Transforms

Unlike the fundamental solution to the Laplacian, the fundamental solution Γ to $\mathcal{L}_{\lambda,\mu}$ is not invariant under unitary transforms. In this section we find formulae for Γ and the single layer potential under unitary transforms.

Lemma 6.9 *Let R be a unitary transform on $\mathbb{R}^d$. Then,*

(i) $\mathcal{L}_{\lambda,\mu}(R^{-1}(\mathbf{u} \circ R)) = R^{-1}(\mathcal{L}_{\lambda,\mu}\mathbf{u}) \circ R$,

(ii) $(\frac{\partial \mathbf{u}}{\partial \nu}) \circ R = R \frac{\partial}{\partial \nu} \left(R^{-1}(\mathbf{u} \circ R) \right)$.

Proof. Since, for a vector $\mathbf{u}$ and a scalar function f,

$$(\nabla \cdot \mathbf{u}) \circ R = \nabla \cdot \left(R^{-1}(\mathbf{u} \circ R) \right) ,$$

$$R^{-1}(\nabla f) \circ R = \nabla (f \circ R) ,$$

we have that

$$\mathcal{L}_{\lambda,\mu}\left(R^{-1}(\mathbf{u} \circ R) \right) = \mu \Delta (R^{-1}(\mathbf{u} \circ R)) + (\lambda + \mu)\nabla\nabla \cdot (R^{-1}(\mathbf{u} \circ R))$$

$$= \mu R^{-1}(\Delta(\mathbf{u} \circ R)) + (\lambda + \mu)\nabla((\nabla \cdot \mathbf{u}) \circ R))$$

$$= \mu R^{-1}((\Delta\mathbf{u}) \circ R)) + (\lambda + \mu)R^{-1}(\nabla\nabla \cdot \mathbf{u}) \circ R$$

$$= R^{-1}(\mathcal{L}_{\lambda,\mu}\mathbf{u}) \circ R ,$$

which proves (i).

To prove (ii), note first that if N_1 and N_2 are the unit normals to ∂D and $\partial R(D)$, then $N_2(R(x)) = R(N_1(x))$ for $x \in \partial D$. Therefore, we have

$$
\left(\frac{\partial \mathbf{u}}{\partial \nu}\right) \circ R = \lambda(\nabla \cdot \mathbf{u} \circ R)N \circ R + \mu((\nabla \mathbf{u}) \circ R + (\nabla \mathbf{u})^T \circ R)N \circ R
$$

$$
= \lambda \nabla \cdot (R^{-1}(\mathbf{u} \circ R))RN + \mu R(R^{-1}(\nabla \mathbf{u}) \circ R + R^{-1}(\nabla \mathbf{u})^T \circ R)RN
$$

$$
= \lambda \nabla \cdot (R^{-1}(\mathbf{u} \circ R))RN + \mu R(\nabla(\mathbf{u} \circ R)R + (\nabla(\mathbf{u} \circ R)^T R)N
$$

$$
= \lambda \nabla \cdot (R^{-1}(\mathbf{u} \circ R))RN + \mu R(\nabla(R^{-1}(\mathbf{u} \circ R)) + (\nabla(R^{-1}(\mathbf{u} \circ R))^T)N
$$

$$
= R\frac{\partial}{\partial \nu}(R^{-1}(\mathbf{u} \circ R)) \,,
$$

and thus (ii) is proved. □

Lemma 6.10 *Suppose that $\mathcal{L}_{\lambda,\mu}\mathbf{u} = 0$ in $\mathbb{R}^d$. If, in addition, $\mathbf{u}$ is bounded for $d = 2$ and behaves like $O(|x|^{-1})$ as $|x| \to \infty$ for $d = 3$, and $\nabla \mathbf{u} = O(|x|^{1-d})$ as $|x| \to \infty$, then $\mathbf{u} = constant$ if $d = 2$ and $\mathbf{u} = 0$ if $d = 3$.*

Proof.   Let B_r be a ball of radius r centered at 0. Then by (6.7),

$$
\mathbf{u}(x) = \mathcal{D}_{B_r}(\mathbf{u}|_{\partial B_r})(x) - \mathcal{S}_{B_r}\left(\frac{\partial \mathbf{u}}{\partial \nu}\Big|_{\partial B_r}\right)(x) \,, \quad x \in B_r \,.
$$

By (6.11), $\partial \mathbf{u}/\partial \nu \in L^2_{\Psi}(\partial B_r)$, which, in particular, shows that

$$
\int_{\partial B_r} \frac{\partial \mathbf{u}}{\partial \nu}\, d\sigma = 0 \,.
$$

Thus we have

$$
\nabla \mathbf{u}(x) = O(\frac{1}{r}) \quad \text{as } r \to +\infty \,,
$$

provided that x is in a bounded set. This immediately implies that $\mathbf{u}$ is constant and ends the proof. □

Lemma 6.11 (Rotation Formula)

$$
\Gamma(R(x)) = R\Gamma(x)R^{-1}, \quad x \in \mathbb{R}^d. \tag{6.15}
$$

Proof.   It follows from Lemma 6.9 (i) that

$$
\mathcal{L}_{\lambda,\mu}(R^{-1}(\Gamma \circ R))(x) = R^{-1}(\mathcal{L}_{\lambda,\mu}\Gamma)(R(x)) = \delta_0(R(x))R^{-1} = \delta_0(x)R^{-1} \,.
$$

Consequently,

$$
\mathcal{L}_{\lambda,\mu}(R^{-1}(\Gamma \circ R) - \Gamma R^{-1}) = 0 \quad \text{in } \mathbb{R}^d \,.
$$

Observe that $R^{-1}(\mathbf{\Gamma} \circ R) - \mathbf{\Gamma} R^{-1}$ is bounded if $d = 2$ and behaves like $O(|x|^{-1})$ as $|x| \to \infty$ if $d = 3$. Moreover,

$$\nabla\big(R^{-1}(\mathbf{\Gamma} \circ R) - \mathbf{\Gamma} R^{-1}\big)(x) = O(|x|^{1-d}) \quad \text{as } |x| \to \infty .$$

It then follows from Lemma 6.10 that

$$R^{-1}(\mathbf{\Gamma} \circ R) - \mathbf{\Gamma} R^{-1} = \text{constant},$$

which is obviously zero. $\square$

As a consequence of (6.15) we obtain the following rotation formula for the single layer potential.

Lemma 6.12 *Let $\widehat{D}$ be a bounded domain in $\mathbb{R}^d$ and $D = R(\widehat{D})$. Then for any vector potential $\varphi \in L^2(\partial D)$, we have*

$$(\mathcal{S}_D \varphi)(R(x)) = R \mathcal{S}_{\widehat{D}}(R^{-1}(\varphi \circ R))(x) , \tag{6.16}$$

$$\frac{\partial(\mathcal{S}_D \varphi)}{\partial \nu}(R(x)) = R \frac{\partial}{\partial \nu} \mathcal{S}_{\widehat{D}}(R^{-1}(\varphi \circ R))(x) . \tag{6.17}$$

Proof. Using (6.15) we compute

$$\begin{aligned}
(\mathcal{S}_D \varphi)(R(x)) &= \int_{\partial D} \mathbf{\Gamma}(R(x) - y)\varphi(y)\, d\sigma(y) \\
&= \int_{\partial \widehat{D}} \mathbf{\Gamma}(R(x) - R(y))\varphi(R(y))\, d\sigma(y) \\
&= R \int_{\partial \widehat{D}} \mathbf{\Gamma}(x - y)R^{-1}\varphi(R(y))\, d\sigma(y) \\
&= R \mathcal{S}_{\widehat{D}}(R^{-1}(\varphi \circ R))(x) ,
\end{aligned}$$

which proves (6.16).

Applying Lemma 6.9 (ii), we arrive at

$$\frac{\partial}{\partial \nu}(\mathcal{S}_D \varphi)(R(x)) = R \frac{\partial}{\partial \nu}\Big(R^{-1}(\mathcal{S}_D \varphi) \circ R \Big)(x) .$$

Then (6.17) follows from (6.16) and the above identity. This completes the proof. $\square$

6.3 Transmission Problem

We suppose that the elastic medium Ω contains a single inclusion D which is also a bounded Lipschitz domain. Let the constants (λ, μ) denote the background Lamé coefficients, that are the elastic parameters in the absence of any inclusions. Suppose that D has the pair of Lamé constants $(\widetilde{\lambda}, \widetilde{\mu})$ which is different from that of the background elastic body, (λ, μ). It is always assumed that

$$\mu > 0, \quad d\lambda + 2\mu > 0, \quad \widetilde{\mu} > 0 \quad \text{and} \quad d\widetilde{\lambda} + 2\widetilde{\mu} > 0. \tag{6.18}$$

We also assume that

$$(\lambda - \widetilde{\lambda})(\mu - \widetilde{\mu}) \geq 0, \ \left((\lambda - \widetilde{\lambda})^2 + (\mu - \widetilde{\mu})^2 \neq 0\right).$$

We consider the transmission problem

$$\begin{cases} \displaystyle\sum_{j,k,l=1}^{d} \frac{\partial}{\partial x_j} \left(C_{ijkl} \frac{\partial u_k}{\partial x_l} \right) = 0 \quad \text{in } \Omega, \quad i = 1, \ldots, d, \\ \displaystyle \frac{\partial \mathbf{u}}{\partial \nu}\bigg|_{\partial\Omega} = \mathbf{g}, \end{cases} \tag{6.19}$$

where the elasticity tensor is given by

$$\begin{aligned} C_{ijkl} := & \left(\lambda \chi(\Omega \setminus D) + \widetilde{\lambda} \chi(D) \right) \delta_{ij} \delta_{kl} \\ & + \left(\mu \chi(\Omega \setminus D) + \widetilde{\mu} \chi(D) \right) (\delta_{ik}\delta_{jl} + \delta_{il}\delta_{jk}). \end{aligned} \tag{6.20}$$

In order to ensure existence and uniqueness of the solution to (6.19), we assume that $\mathbf{g} \in L^2_\Psi(\partial\Omega)$ and seek for a solution $\mathbf{u} \in W^{1,2}(\Omega)$ such that $\mathbf{u}|_{\partial\Omega} \in L^2_\Psi(\partial\Omega)$. The problem (6.19) is understood in a weak sense, namely, for any $\boldsymbol{\varphi} \in W^{1,2}(\Omega)$ the following equality holds:

$$\sum_{i,j,k,l=1}^{d} \int_\Omega C_{ijkl} \frac{\partial u_k}{\partial x_l} \frac{\partial \varphi_i}{\partial x_j} dx = \int_{\partial\Omega} \mathbf{g} \cdot \boldsymbol{\varphi} \, d\sigma.$$

Let $\mathcal{L}_{\widetilde{\lambda}, \widetilde{\mu}}$ and $\partial/\partial\widetilde{\nu}$ be the Lamé system and the conormal derivative associated with $(\widetilde{\lambda}, \widetilde{\mu})$, respectively. Then for any $\boldsymbol{\varphi} \in \mathcal{C}_0^\infty(\Omega)$, we compute

$$\begin{aligned} 0 = & \sum_{i,j,k,l=1}^{d} \int_\Omega C_{ijkl} \frac{\partial u_k}{\partial x_l} \frac{\partial \varphi_i}{\partial x_j} \, dx \\ = & \int_{\Omega \setminus \overline{D}} \lambda (\nabla \cdot \mathbf{u})(\nabla \cdot \boldsymbol{\varphi}) + \frac{\mu}{2} (\nabla\mathbf{u} + \nabla\mathbf{u}^T) \cdot (\nabla\boldsymbol{\varphi} + \nabla\boldsymbol{\varphi}^T) \, dx \\ & + \int_D \widetilde{\lambda}(\nabla \cdot \mathbf{u})(\nabla \cdot \boldsymbol{\varphi}) + \frac{\widetilde{\mu}}{2} (\nabla\mathbf{u} + \nabla\mathbf{u}^T) \cdot (\nabla\boldsymbol{\varphi} + \nabla\boldsymbol{\varphi}^T) \, dx \\ = & -\int_{\Omega \setminus \overline{D}} \mathcal{L}_{\lambda,\mu}\mathbf{u} \cdot \boldsymbol{\varphi} \, dx - \int_{\partial D} \frac{\partial\mathbf{u}}{\partial\nu} \cdot \boldsymbol{\varphi} \, d\sigma - \int_D \mathcal{L}_{\widetilde{\lambda},\widetilde{\mu}}\mathbf{u} \cdot \boldsymbol{\varphi} \, dx + \int_{\partial D} \frac{\partial\mathbf{u}}{\partial\widetilde{\nu}} \cdot \boldsymbol{\varphi} \, d\sigma, \end{aligned}$$

where the last equality follows from (6.2). Thus (6.19) is equivalent to the following problem:

$$\begin{cases} \mathcal{L}_{\lambda,\mu}\mathbf{u} = 0 \quad \text{in } \Omega \setminus \overline{D}\,, \\ \mathcal{L}_{\widetilde{\lambda},\widetilde{\mu}}\mathbf{u} = 0 \quad \text{in } D\,, \\ \mathbf{u}\big|_{-} = \mathbf{u}\big|_{+} \quad \text{on } \partial D\,, \\ \dfrac{\partial \mathbf{u}}{\partial \widetilde{\nu}}\bigg|_{-} = \dfrac{\partial \mathbf{u}}{\partial \nu}\bigg|_{+} \quad \text{on } \partial D\,, \\ \dfrac{\partial \mathbf{u}}{\partial \nu}\big|_{\partial \Omega} = \mathbf{g}\,, \quad \left(\mathbf{u}|_{\partial \Omega} \in L^2_{\Psi}(\partial \Omega)\right). \end{cases} \tag{6.21}$$

We denote by $\mathcal{S}_D$ and $\widetilde{\mathcal{S}}_D$ the single layer potentials on ∂D corresponding to the Lamé constants (λ, μ) and $(\widetilde{\lambda}, \widetilde{\mu})$, respectively.

We use the following solvability theorem due to Escauriaza and Seo (Theorem 4, [111]).

Theorem 6.13 *Suppose that* $(\lambda - \widetilde{\lambda})(\mu - \widetilde{\mu}) \geq 0$ *and* $0 < \widetilde{\lambda}, \widetilde{\mu} < \infty$. *For any given* $(\mathbf{F}, \mathbf{G}) \in W^2_1(\partial D) \times L^2(\partial D)$, *there exists a unique pair* $(\mathbf{f}, \mathbf{g}) \in L^2(\partial D) \times L^2(\partial D)$ *such that*

$$\begin{cases} \widetilde{\mathcal{S}}_D\mathbf{f}\big|_{-} - \mathcal{S}_D\mathbf{g}\big|_{+} = \mathbf{F} \quad \text{on } \partial D\,, \\ \dfrac{\partial}{\partial \widetilde{\nu}}\widetilde{\mathcal{S}}_D\mathbf{f}\bigg|_{-} - \dfrac{\partial}{\partial \nu}\mathcal{S}_D\mathbf{g}\bigg|_{+} = \mathbf{G} \quad \text{on } \partial D\,, \end{cases} \tag{6.22}$$

and there exists a constant C *depending only on* λ, μ, $\widetilde{\lambda}$, $\widetilde{\mu}$, *and the Lipschitz character of* D *such that*

$$\|\mathbf{f}\|_{L^2(\partial D)} + \|\mathbf{g}\|_{L^2(\partial D)} \leq C\left(\|\mathbf{F}\|_{W^2_1(\partial D)} + \|\mathbf{G}\|_{L^2(\partial D)}\right). \tag{6.23}$$

Moreover, if $\mathbf{G} \in L^2_{\Psi}(\partial D)$, *then* $\mathbf{g} \in L^2_{\Psi}(\partial D)$.

Proof. The unique solvability of the integral equation (6.22) was proved in [111]. By (6.11), $\partial \widetilde{\mathcal{S}}_D\mathbf{f}/\partial \widetilde{\nu}|_{-} \in L^2_{\Psi}(\partial D)$. Thus if $\mathbf{G} \in L^2_{\Psi}(\partial D)$, then $\partial \mathcal{S}_D\mathbf{g}/\partial \nu|_{+} \in L^2_{\Psi}(\partial D)$. Since

$$\mathbf{g} = \dfrac{\partial}{\partial \nu}\mathcal{S}_D\mathbf{g}|_{+} - \dfrac{\partial}{\partial \nu}\mathcal{S}_D\mathbf{g}|_{-}\,,$$

by (6.10) and $\partial \mathcal{S}_D\mathbf{g}/\partial \nu|_{-} \in L^2_{\Psi}(\partial D)$, we conclude that $\mathbf{g} \in L^2_{\Psi}(\partial D)$. $\square$

Lemma 6.14 *Let* $\varphi \in \Psi$. *If the pair* $(\mathbf{f}, \mathbf{g}) \in L^2(\partial D) \times L^2_{\Psi}(\partial D)$ *is the solution of*

$$\begin{cases} \widetilde{\mathcal{S}}_D\mathbf{f}\big|_{-} - \mathcal{S}_D\mathbf{g}\big|_{+} = \varphi|_{\partial D}\,, \\ \dfrac{\partial}{\partial \widetilde{\nu}}\widetilde{\mathcal{S}}_D\mathbf{f}\bigg|_{-} - \dfrac{\partial}{\partial \nu}\mathcal{S}_D\mathbf{g}\bigg|_{+} = 0\,, \end{cases} \tag{6.24}$$

then $\mathbf{g} = 0$.

Proof. Define **u** by

$$\mathbf{u}(x) := \begin{cases} \mathcal{S}_D \mathbf{g}(x) , & x \in \mathbb{R}^d \setminus \overline{D} , \\ \widetilde{\mathcal{S}}_D \mathbf{f}(x) - \boldsymbol{\varphi}(x) , & x \in D . \end{cases}$$

Since $\mathbf{g} \in L^2_{\Psi}(\partial D)$, then $\int_{\partial D} \mathbf{g} \, d\sigma = 0$, and hence

$$\mathcal{S}_D \mathbf{g}(x) = O(|x|^{1-d}) \quad \text{as } |x| \to \infty .$$

Therefore **u** is the unique solution of

$$\begin{cases} \mathcal{L}_{\lambda,\mu} \mathbf{u} = 0 & \text{in } \mathbb{R}^d \setminus \overline{D} , \\ \mathcal{L}_{\widetilde{\lambda},\widetilde{\mu}} \mathbf{u} = 0 & \text{in } D , \\ \mathbf{u}|_+ = \mathbf{u}|_- & \text{on } \partial D , \\ \dfrac{\partial \mathbf{u}}{\partial \nu}\Big|_+ = \dfrac{\partial \mathbf{u}}{\partial \widetilde{\nu}}\Big|_- & \text{on } \partial D , \\ \mathbf{u}(x) = O(|x|^{1-d}) & \text{as } |x| \to \infty . \end{cases} \tag{6.25}$$

Since the trivial solution is the unique solution to (6.25), we see that $\mathcal{S}_D \mathbf{g}(x) = 0$ for $x \in \mathbb{R}^d \setminus \overline{D}$. It then follows that $\mathcal{L}_{\lambda,\mu} \mathcal{S}_D \mathbf{g}(x) = 0$ for $x \in D$ and $\mathcal{S}_D \mathbf{g}(x) = 0$ for $x \in \partial D$. Thus $\mathcal{S}_D \mathbf{g}(x) = 0$ for $x \in D$. Using the fact that

$$\mathbf{g} = \frac{\partial(\mathcal{S}_D \mathbf{g})}{\partial \nu}\Big|_+ - \frac{\partial(\mathcal{S}_D \mathbf{g})}{\partial \nu}\Big|_- ,$$

we conclude that $\mathbf{g} = 0$. $\square$

We now prove a representation theorem for the solution of the transmission problem (6.21) which will be the main ingredient in deriving the asymptotic expansions in Chap. 8.

Theorem 6.15 *There exists a unique pair $(\boldsymbol{\varphi}, \boldsymbol{\psi}) \in L^2(\partial D) \times L^2_{\Psi}(\partial D)$ such that the solution **u** of (6.21) is represented by*

$$\mathbf{u}(x) = \begin{cases} \mathbf{H}(x) + \mathcal{S}_D \boldsymbol{\psi}(x) , & x \in \Omega \setminus \overline{D} , \\ \widetilde{\mathcal{S}}_D \boldsymbol{\varphi}(x) , & x \in D , \end{cases} \tag{6.26}$$

*where **H** is defined by*

$$\mathbf{H}(x) = \mathcal{D}_\Omega(\mathbf{u}|_{\partial\Omega})(x) - \mathcal{S}_\Omega(\mathbf{g})(x) , \quad x \in \mathbb{R}^d \setminus \partial\Omega . \tag{6.27}$$

In fact, the pair $(\boldsymbol{\varphi}, \boldsymbol{\psi})$ is the unique solution in $L^2(\partial D) \times L^2_{\Psi}(\partial D)$ of

$$\begin{cases} \widetilde{\mathcal{S}}_D \boldsymbol{\varphi}\Big|_- - \mathcal{S}_D \boldsymbol{\psi}\Big|_+ = \mathbf{H}|_{\partial D} & \text{on } \partial D , \\ \dfrac{\partial}{\partial \widetilde{\nu}} \widetilde{\mathcal{S}}_D \boldsymbol{\varphi}\Big|_- - \dfrac{\partial}{\partial \nu} \mathcal{S}_D \boldsymbol{\psi}\Big|_+ = \dfrac{\partial \mathbf{H}}{\partial \nu}\Big|_{\partial D} & \text{on } \partial D . \end{cases} \tag{6.28}$$

There exists C such that

$$\|\boldsymbol{\varphi}\|_{L^2(\partial D)} + \|\boldsymbol{\psi}\|_{L^2(\partial D)} \leq C \|\mathbf{H}\|_{W_1^2(\partial D)} . \tag{6.29}$$

For each integer n there exists C_n depending only on c_0 and λ, μ (not on $\widetilde{\lambda}, \widetilde{\mu}$) such that

$$\|\mathbf{H}\|_{C^n(\overline{D})} \leq C_n \|\mathbf{g}\|_{L^2(\partial \Omega)} . \tag{6.30}$$

Moreover,

$$\mathbf{H}(x) = -\mathcal{S}_D \boldsymbol{\psi}(x) , \quad x \in \mathbb{R}^d \setminus \overline{\Omega} . \tag{6.31}$$

Proof.   Let $\boldsymbol{\varphi}$ and $\boldsymbol{\psi}$ be the unique solutions of (6.28). Then clearly $\mathbf{u}$ defined by (6.26) satisfies the transmission condition (the third and fourth conditions in (6.21)). By (6.11), $\partial \mathbf{H}/\partial \nu|_{\partial D} \in L_{\Psi}^2(\partial D)$. Thus by Theorem 6.13, $\boldsymbol{\psi} \in L_{\Psi}^2(\partial D)$.

We now prove that $\partial \mathbf{u}/\partial \nu|_{\partial \Omega} = \mathbf{g}$. To this end we consider the following two phases transmission problem:

$$\begin{cases} \mathcal{L}_{\lambda,\mu}\mathbf{u} = 0 & \text{in } (\Omega \setminus \overline{D}) \cup (\mathbb{R}^d \setminus \overline{\Omega}) , \\ \mathcal{L}_{\widetilde{\lambda},\widetilde{\mu}}\mathbf{u} = 0 & \text{in } D , \\ \mathbf{u}|_- = \mathbf{u}|_+ \text{ and } \dfrac{\partial \mathbf{u}}{\partial \widetilde{\nu}}\Big|_- = \dfrac{\partial \mathbf{u}}{\partial \nu}\Big|_+ & \text{on } \partial D , \\ \mathbf{u}|_- - \mathbf{u}|_+ = \mathbf{f} \text{ and } \dfrac{\partial \mathbf{u}}{\partial \widetilde{\nu}}\Big|_- - \dfrac{\partial \mathbf{u}}{\partial \nu}\Big|_+ = \mathbf{g} & \text{on } \partial \Omega , \\ |x||\mathbf{u}(x)| + |x|^2|\nabla \mathbf{u}(x)| \leq C & \text{as } |x| \to \infty , \end{cases} \tag{6.32}$$

where $\mathbf{f} = \mathbf{u}|_{\partial \Omega}$. If $\mathbf{v} \in W^{1,2}(\Omega)$ is the solution of (6.21), then $\mathbf{U}_1$, defined by

$$\mathbf{U}_1(x) := \begin{cases} \mathbf{v}(x) , & x \in \Omega , \\ \mathbf{0} , & x \in \mathbb{R}^d \setminus \Omega , \end{cases}$$

is a solution of (6.32). On the other hand, it can be easily seen from the jump relations of the layer potentials, (6.9) and (6.10), that $\mathbf{U}_2$ defined by

$$\mathbf{U}_2(x) = \begin{cases} -\mathcal{S}_\Omega(\mathbf{g})(x) + \mathcal{D}_\Omega(\mathbf{u}|_{\partial \Omega})(x) + \mathcal{S}_D \boldsymbol{\psi}(x) , & x \in \mathbb{R}^d \setminus (\overline{D} \cup \partial \Omega) , \\ \widetilde{\mathcal{S}}_D \boldsymbol{\varphi}(x) , & x \in D , \end{cases}$$

is also a solution of (6.32). Thus $\mathbf{U}_1 - \mathbf{U}_2$ is a solution of (6.32) with $\mathbf{f} = 0$ and $\mathbf{g} = 0$. Moreover, $\mathbf{U}_1 - \mathbf{U}_2 \in W^{1,2}(\mathbb{R}^d)$ and therefore, $\mathbf{U}_1 - \mathbf{U}_2 = 0$, which implies, in particular, that $\partial \mathbf{u}/\partial \nu|_{\partial \Omega} = \mathbf{g}$. Indeed, $\mathbf{U}_2(x) = 0$ for $x \in \mathbb{R}^d \setminus \overline{\Omega}$ and hence (6.31) has been verified.

Now it remains to prove (6.30). Let

$$\Omega' := \left\{ x \in \Omega : \text{ dist}(x, \partial \Omega) > c_0 \right\}$$

so that D is compactly contained in Ω'. Then by an identity of Rellich type available for general constant coefficient systems (see Lemma 1.14 (i) of [93]) there exists a constant C such that

$$\|\nabla \mathbf{u}\|_{L^2(\partial\Omega)} \leq C\left(\|\mathbf{g}\|_{L^2(\partial\Omega)} + \|\nabla \mathbf{u}\|_{L^2(\Omega\setminus\Omega')}\right) . \tag{6.33}$$

It then follows from the Korn's inequality (6.14) and the divergence theorem that

$$\|\nabla \mathbf{u}\|_{L^2(\Omega\setminus\Omega')} \leq C\|\nabla \mathbf{u} + \nabla \mathbf{u}^T\|_{L^2(\Omega\setminus\overline{D})}$$
$$\leq C \int_\Omega \left(\lambda\chi(\Omega\setminus\overline{D}) + \tilde\lambda\chi(D)\right) |\nabla \cdot \mathbf{u}|^2$$
$$+ \frac{1}{2}\left(\mu\chi(\Omega\setminus\overline{D}) + \tilde\mu\chi(D)\right) |\nabla \mathbf{u} + \nabla \mathbf{u}^T|^2 dx$$
$$\leq C \int_{\partial\Omega} \mathbf{u} \cdot \frac{\partial \mathbf{u}}{\partial \nu} d\sigma ,$$

where the constants, generically denoted by C, do not depend on $\tilde\lambda, \tilde\mu$. Further, by (6.33) and the Poincaré inequality (2.1),

$$\|\mathbf{u}\|_{L^2(\partial\Omega)} \leq C\|\nabla \mathbf{u}\|_{L^2(\partial\Omega)}$$
$$\leq C\left(\|\mathbf{g}\|_{L^2(\partial\Omega)} + \|\mathbf{g}\|_{L^2(\partial\Omega)}\|\mathbf{u}\|_{L^2(\partial\Omega)}\right) ,$$

and hence

$$\|\mathbf{u}\|_{L^2(\partial\Omega)} \leq C\|\mathbf{g}\|_{L^2(\partial\Omega)} . \tag{6.34}$$

Clearly the desired estimate (6.30) immediately follows from the definition of $\mathbf{H}$ and (6.34) and the proof is complete. $\square$

We now derive a representation for $\mathbf{u}$ in terms of the background solution. Let $\mathbf{N}(x,y)$ be the Neumann function for $\mathcal{L}_{\lambda,\mu}$ in Ω corresponding to a Dirac mass at y. That is, $\mathbf{N}$ is the solution to

$$\begin{cases} \mathcal{L}_{\lambda,\mu}\mathbf{N}(x,y) = -\delta_y(x)I_d & \text{in } \Omega, \\ \left.\dfrac{\partial \mathbf{N}}{\partial \nu}\right|_{\partial\Omega} = -\dfrac{1}{|\partial\Omega|}I_d , \\ \mathbf{N}(\cdot,y) \in L^2_\Psi(\partial\Omega) & \text{for each } y \in \Omega , \end{cases} \tag{6.35}$$

where the differentiations act on the x-variables, and I_d is the $d \times d$ identity matrix.

For $\mathbf{g} \in L^2_\Psi(\partial\Omega)$, define

$$\mathbf{U}(x) := \int_{\partial\Omega} \mathbf{N}(x,y)\mathbf{g}(y) d\sigma(y) , \quad x \in \Omega . \tag{6.36}$$

Then $\mathbf{U}$ is the solution to (6.13) with D replaced by Ω. On the other hand, by (6.12), the solution to (6.13) is given by

$$\mathbf{U}(x) := \mathcal{S}_\Omega \left(-\frac{1}{2}I + \mathcal{K}_\Omega^* \right)^{-1} \mathbf{g}(x) .$$

Thus we have

$$\int_{\partial\Omega} \mathbf{N}(x,y)\mathbf{g}(y)\, d\sigma(y) = \int_{\partial\Omega} \mathbf{\Gamma}(x-y)(-\frac{1}{2}I + \mathcal{K}_\Omega^*)^{-1}\mathbf{g}(y)\, d\sigma(y) ,$$

or equivalently,

$$\int_{\partial\Omega} \mathbf{N}(x,y)(-\frac{1}{2}I + \mathcal{K}_\Omega^*)\mathbf{g}(y)\, d\sigma(y) = \int_{\partial\Omega} \mathbf{\Gamma}(x-y)\mathbf{g}(y)\, d\sigma(y) , \quad x \in \Omega ,$$

for any $\mathbf{g} \in L_\Psi^2(\partial\Omega)$. Consequently, it follows that, for any simply connected Lipschitz domain D compactly contained in Ω and for any $\mathbf{g} \in L_\Psi^2(\partial D)$, the following identity holds:

$$\int_{\partial D} (-\frac{1}{2}I + \mathcal{K}_\Omega)(\mathbf{N}_y)(x)\mathbf{g}(y)\, d\sigma(y) = \int_{\partial D} \mathbf{\Gamma}_y(x)\mathbf{g}(y)\, d\sigma(y) ,$$

for all $x \in \partial\Omega$. The following lemma has been proved.

Lemma 6.16 *For $y \in \Omega$ and $x \subset \partial\Omega$, let $\mathbf{\Gamma}_y(x) := \mathbf{\Gamma}(x-y)$ and $\mathbf{N}_y(x) := \mathbf{N}(x,y)$. Then*

$$\left(-\frac{1}{2}I + \mathcal{K}_\Omega \right)(\mathbf{N}_y)(x) = \mathbf{\Gamma}_y(x) \quad modulo \ \Psi . \tag{6.37}$$

We fix one more notation. Let

$$N_D\mathbf{f}(x) := \int_{\partial D} \mathbf{N}(x,y)\mathbf{f}(y)\, d\sigma(y) , \quad x \in \overline{\Omega} .$$

Theorem 6.17 *Let $\mathbf{u}$ be the solution to (6.21) and $\mathbf{U}$ the background solution, i.e., the solution to (6.13). Then the following holds:*

$$\mathbf{u}(x) = \mathbf{U}(x) - N_D\psi(x), \quad x \in \partial\Omega , \tag{6.38}$$

where ψ is defined by (6.28).

Proof. By substituting (6.26) into the equation (6.27), we obtain

$$\mathbf{H}(x) = -\mathcal{S}_\Omega(\mathbf{g})(x) + \mathcal{D}_\Omega \left(\mathbf{H}|_{\partial\Omega} + (\mathcal{S}_D\psi)|_{\partial\Omega} \right)(x) , \quad x \in \Omega .$$

It then follows from (6.9) that

$$(\frac{1}{2}I - \mathcal{K}_\Omega)(\mathbf{H}|_{\partial\Omega}) = -(\mathcal{S}_\Omega\mathbf{g})|_{\partial\Omega} + (\frac{1}{2}I + \mathcal{K}_\Omega)((\mathcal{S}_D\psi)|_{\partial\Omega}) \quad \text{on } \partial\Omega . \quad (6.39)$$

Since $\mathbf{U}(x) = -\mathcal{S}_\Omega(\mathbf{g})(x) + \mathcal{D}_\Omega(\mathbf{U}|_{\partial\Omega})(x)$ for all $x \in \Omega$, we have

$$(\frac{1}{2}I - \mathcal{K}_\Omega)(\mathbf{U}|_{\partial\Omega}) = -(\mathcal{S}_\Omega\mathbf{g})|_{\partial\Omega} . \quad (6.40)$$

By Theorem 6.13, it follows from (6.37) that

$$(-\frac{1}{2}I + \mathcal{K}_\Omega)((N_D\psi)|_{\partial\Omega})(x) = (\mathcal{S}_D\psi)(x) , \quad x \in \partial\Omega , \quad (6.41)$$

since $\psi \in L^2_\Psi(\partial D)$. From (6.39), (6.40), and (6.41), we conclude that

$$(\frac{1}{2}I - \mathcal{K}_\Omega)\Big(\mathbf{H}|_{\partial\Omega} - \mathbf{U}|_{\partial\Omega} + (\frac{1}{2}I + \mathcal{K}_\Omega)((N_D\psi)|_{\partial\Omega})\Big) = 0 \quad \text{on } \partial\Omega ,$$

and hence, by Corollary 6.7, we obtain that

$$\mathbf{H}|_{\partial\Omega} - \mathbf{U}|_{\partial\Omega} + (\frac{1}{2}I + \mathcal{K}_\Omega)((N_D\psi)|_{\partial\Omega}) \in \Psi .$$

Note that

$$(\frac{1}{2}I + \mathcal{K}_\Omega)((N_D\psi)|_{\partial\Omega}) = (N_D\psi)|_{\partial\Omega} + (\mathcal{S}_D\psi)|_{\partial\Omega} ,$$

which comes from (6.37). Thus we see from (6.26) that

$$\mathbf{u}|_{\partial\Omega} = \mathbf{U}|_{\partial\Omega} - (N_D\psi)|_{\partial\Omega} \quad \text{modulo } \Psi . \quad (6.42)$$

Since all the functions entering in (6.42) belong to $L^2_\Psi(\partial\Omega)$, we have (6.38). This completes the proof.   □

We have a similar representation for solutions of the Dirichlet problem. Let $\mathbf{G}(x, y)$ be the Green's function for the Dirichlet problem, i.e., the solution to

$$\begin{cases} \mathcal{L}_{\lambda,\mu}\mathbf{G}(x, y) = -\delta_y(x)I_d & \text{in } \Omega , \\ \mathbf{G}(x, y) = 0 , & x \in \partial\Omega \text{ for each } y \in \partial\Omega . \end{cases}$$

Then, the function $\mathbf{V}$, for $\mathbf{f} \in W^2_{\frac{1}{2}}(\partial\Omega)$, defined by

$$\mathbf{V}(x) := -\int_{\partial\Omega} \frac{\partial}{\partial\nu_y}\mathbf{G}(x, y)\mathbf{f}(y)\, d\sigma(y) ,$$

is the solution of the problem

$$\begin{cases} \mathcal{L}_{\lambda,\mu}\mathbf{V} = 0 & \text{in } \Omega , \\ \mathbf{V}|_{\partial\Omega} = \mathbf{f} . \end{cases}$$

We have the following theorem.

Theorem 6.18 *We have*

$$(\frac{1}{2}I + \mathcal{K}_\Omega^*)^{-1}(\frac{\partial}{\partial \nu}\mathbf{\Gamma}_z)(x) = \frac{\partial}{\partial \nu}\mathbf{G}_z(x) , \quad x \in \partial\Omega , \ z \in \Omega .$$

Moreover, let $\mathbf{u}$ *be the solution of (6.21) with the Neumann condition on* $\partial\Omega$ *replaced by the Dirichlet condition* $\mathbf{u}|_{\partial\Omega} = \mathbf{f} \in W_{\frac{1}{2}}^2(\partial\Omega)$. *Then the following identity holds:*

$$\frac{\partial \mathbf{u}}{\partial \nu}(x) = \frac{\partial \mathbf{V}}{\partial \nu}(x) - G_D\psi(x) , \quad x \in \partial\Omega ,$$

where ψ *is defined in (6.26) and*

$$G_D\psi(x) := \int_{\partial D} \frac{\partial}{\partial \nu}\mathbf{G}(x,y)\psi(y) \, d\sigma(y) .$$

Theorem 6.18 can be proved in the same way as Theorem 6.17. In fact, it is simpler because of the solvability of the Dirichlet problem, or equivalently, the invertibility of $(1/2)\, I + \mathcal{K}_\Omega^*$ on $L^2(\partial\Omega)$. So we omit the proof.

6.4 Complex Representation of Displacement Vectors

This section is devoted to a representation of the solution to (6.19) by a pair of holomorphic functions in the two-dimensional case. The results of this section will be used to compute the elastic moment tensors in Chap. 7.

The following theorem is from [216]. We include a proof of the theorem for the readers' sake.

Theorem 6.19 *Suppose that* Ω *is a simply connected domain in* $\mathbb{R}^2$ *(bounded or unbounded) with the Lamé constants* λ, μ *and let* $\mathbf{u} = (u, v) \in W^{1,\frac{3}{2}}(\Omega)$ *be a solution of* $\mathcal{L}_{\lambda,\mu}\mathbf{u} = 0$ *in* Ω. *Then there are holomorphic functions* φ *and* ψ *in* Ω *such that*

$$2\mu(u + iv)(z) = \kappa\varphi(z) - z\overline{\varphi'(z)} - \overline{\psi(z)} , \quad \kappa = \frac{\lambda + 3\mu}{\lambda + \mu} , \quad z = x + iy . \quad (6.43)$$

Moreover, the conormal derivative $\partial\mathbf{u}/\partial\nu$ *is represented as*

$$\left(\left(\frac{\partial \mathbf{u}}{\partial \nu}\right)_1 + i\left(\frac{\partial \mathbf{u}}{\partial \nu}\right)_2\right) d\sigma = -i\partial\left[\varphi(z) + z\overline{\varphi'(z)} + \overline{\psi(z)}\right] , \quad (6.44)$$

where $d\sigma$ *is the line element of* $\partial\Omega$ *and* $\partial = (\partial/\partial x)\, dx + (\partial/\partial y)\, dy$. *Here* $\partial\Omega$ *is positively oriented.*

Proof. Let $\theta := \nabla \cdot \mathbf{u}$ and

$$X := \lambda\theta + 2\mu\frac{\partial u}{\partial x}, \quad Y := \lambda\theta + 2\mu\frac{\partial v}{\partial y}, \quad Z := \mu\left(\frac{\partial v}{\partial x} + \frac{\partial u}{\partial y}\right)^1.$$

Then we can see by elementary calculation that the equation $\mathcal{L}_{\lambda,\mu}\mathbf{u} = 0$ is equivalent to

$$\frac{\partial X}{\partial x} + \frac{\partial Z}{\partial y} = 0 \quad \text{and} \quad \frac{\partial Z}{\partial x} + \frac{\partial Y}{\partial y} = 0. \tag{6.45}$$

Thus there are two functions A and B such that

$$\nabla B = (-Z, X) \quad \text{and} \quad \nabla A = (Y, -Z).$$

In particular, $\partial A/\partial y = \partial B/\partial x$ and hence there is a function[2] U such that $\nabla U = (A, B)$. Thus,

$$X = \frac{\partial^2 U}{\partial y^2}, \quad Y = \frac{\partial^2 U}{\partial x^2}, \quad Z = -\frac{\partial^2 U}{\partial x \partial y}. \tag{6.46}$$

By taking the x-derivative of the first component of $\mathcal{L}_{\lambda,\mu}\mathbf{u}$ and the y-derivative of the second, we can see that

$$\frac{\partial}{\partial x}(\Delta u) + \frac{\partial}{\partial y}(\Delta v) = 0.$$

It then follows that

$$\Delta(X + Y) = 2(\lambda + \mu)\left[\frac{\partial}{\partial x}(\Delta u) + \frac{\partial}{\partial y}(\Delta v)\right] = 0.$$

Thus U is biharmonic, namely, $\Delta\Delta U = 0$. In short, we proved that there is a biharmonic function U such that

$$\lambda\theta + 2\mu\frac{\partial u}{\partial x} = \frac{\partial^2 U}{\partial y^2}, \lambda\theta + 2\mu\frac{\partial v}{\partial y} = \frac{\partial^2 U}{\partial x^2}, \mu\left(\frac{\partial v}{\partial x} + \frac{\partial u}{\partial y}\right) = -\frac{\partial^2 U}{\partial x \partial y}. \tag{6.47}$$

We claim that there exist two holomorphic functions in Ω, φ and f, such that

$$2U(z) = \bar{z}\varphi(z) + z\overline{\varphi(z)} + f(z) + \overline{f(z)}, \quad z \in \Omega. \tag{6.48}$$

In fact, let $P := \Delta U$. Then P is harmonic in Ω. Let Q be a harmonic conjugate of P so that $P + iQ$ is holomorphic in Ω. Such a function exists since Ω is simply connected. Let $\varphi = p + iq$ be a holomorphic function in Ω so that $4\varphi'(z) = P(z) + iQ(z)$. Then,

$$\frac{\partial p}{\partial x} = \frac{1}{4}P, \quad \frac{\partial p}{\partial y} = -\frac{1}{4}Q. \tag{6.49}$$

Then it is easy to see that

---

[1] These notations are slightly different from those of [216].
[2] This function U is called the stress function or the Airy function.

$$\Delta(U - \Re(\bar{z}\varphi)) = P - \Re\varphi'(z) = 0 .$$

Therefore, there exists a function f holomorphic in Ω such that

$$U - \Re(\bar{z}\varphi) = \Re f(z) .$$

Thus we get (6.48).

Adding the first two equations in (6.47), we get $2(\lambda + \mu)\theta = \Delta U = P$. It then follows from the first equation in (6.47) and (6.49) that

$$2\mu\frac{\partial u}{\partial x} = -\frac{\partial^2 U}{\partial x^2} + \frac{2(\lambda + 2\mu)}{\lambda + \mu}\frac{\partial p}{\partial x} .$$

Likewise, we obtain

$$2\mu\frac{\partial v}{\partial y} = -\frac{\partial^2 U}{\partial y^2} + \frac{2(\lambda + 2\mu)}{\lambda + \mu}\frac{\partial q}{\partial y} ,$$

and therefore

$$2\mu u = -\frac{\partial U}{\partial x} + \frac{2(\lambda + 2\mu)}{\lambda + \mu}p + f_1(y) ,$$

$$2\mu v = -\frac{\partial U}{\partial y} + \frac{2(\lambda + 2\mu)}{\lambda + \mu}q + f_2(x) .$$

Substitute these equations into the third equation in (6.47). Then by the Cauchy–Riemann equation $\partial p/\partial y = -\partial q/\partial x$, we get

$$f_1'(y) + f_2'(x) = 0 ,$$

which implies that

$$f_1(y) = ay + b , \quad f_2(x) = -ax + c ,$$

for some constants a, b, c. Thus we obtain

$$2\mu(u + iv)(x, y) = -\frac{\partial U}{\partial x} - i\frac{\partial U}{\partial y} + \frac{2(\lambda + 2\mu)}{\lambda + \mu}(p + iq) + a(y - ix) + b + ic .$$

It then follows from (6.48) that

$$2\mu(u + iv)(x, y) = \frac{\lambda + 3\mu}{\lambda + \mu}\varphi(z) - z\overline{\varphi(z)} + \overline{\psi'(z)} - aiz + b + ic ,$$

where $\psi(z) = f'(z)$. By adding constants to φ and ψ to define new φ and ψ, we get (6.43).

To prove (6.44) we first observe that

$$\frac{\partial \mathbf{u}}{\partial \nu} = \left(XN_1 + ZN_2, ZN_1 + YN_2\right) ,$$

where $N = (N_1, N_2)$. Since $(-N_2, N_1)$ is positively oriented tangential vector field on $\partial\Omega$, we get

$$-N_2 ds = dx, \quad N_1 ds = dy .\tag{6.50}$$

It then follows from (6.46) that

$$\left(\left(\frac{\partial \mathbf{u}}{\partial \nu} \right)_1 + i \left(\frac{\partial \mathbf{u}}{\partial \nu} \right)_2 \right) d\sigma = \left(\frac{\partial^2 U}{\partial y^2} \, dy + \frac{\partial^2 U}{\partial x \partial y} \, dx \right) - i \left(\frac{\partial^2 U}{\partial x \partial y} \, dy + \frac{\partial^2 U}{\partial x^2} \, dx \right)$$

$$= \partial \left(\frac{\partial U}{\partial y} - i \frac{\partial U}{\partial x} \right) .$$

Now (6.44) follows from (6.48). This completes the proof. $\square$

We now prove that a similar theorem holds for the solution to the problem (6.21).

Theorem 6.20 *Suppose $d = 2$. Let $\mathbf{u} = (u, v)$ be the solution of (6.21) and let $\mathbf{u}_e := \mathbf{u}|_{\mathbb{C} \setminus D}$ and $\mathbf{u}_i := \mathbf{u}|_D$. Then there are functions φ_e and ψ_e holomorphic in $\Omega \setminus \overline{D}$ and φ_i and ψ_i holomorphic in D such that*

$$2\mu(u_e + iv_e)(z) = \kappa\varphi_e(z) - z\overline{\varphi'_e(z)} - \overline{\psi_e(z)} , \quad z \in \mathbb{C} \setminus \overline{D} ,\tag{6.51}$$

$$2\tilde{\mu}(u_i + iv_i)(z) = \tilde{\kappa}\varphi_i(z) - z\overline{\varphi'_i(z)} - \overline{\psi_i(z)} , \quad z \in D ,\tag{6.52}$$

where

$$\kappa = \frac{\lambda + 3\mu}{\lambda + \mu} , \qquad \tilde{\kappa} = \frac{\tilde{\lambda} + 3\tilde{\mu}}{\tilde{\lambda} + \tilde{\mu}} .$$

Moreover, the following holds on ∂D:

$$\frac{1}{2\mu} \left(\kappa\varphi_e(z) - z\overline{\varphi'_e(z)} - \overline{\psi_e(z)} \right) = \frac{1}{2\tilde{\mu}} \left(\tilde{\kappa}\varphi_i(z) - z\overline{\varphi'_i(z)} - \overline{\psi_i(z)} \right) ,\tag{6.53}$$

$$\varphi_e(z) + z\overline{\varphi'_e(z)} + \overline{\psi_e(z)} = \varphi_i(z) + z\overline{\varphi'_i(z)} + \overline{\psi_i(z)} + c ,\tag{6.54}$$

where c is a constant.

Proof. By Theorem 6.15, there exists a unique pair $(\boldsymbol{\varphi}, \boldsymbol{\psi}) \in L^2(\partial D) \times L^2_{\Psi}(\partial D)$ such that

$$\mathbf{u}_e(x) = \mathbf{H}(x) + \mathcal{S}_D \boldsymbol{\psi}(x) , \quad x \in \Omega \setminus \overline{D} ,$$

$$\mathbf{u}_i(x) = \widetilde{\mathcal{S}}_D \boldsymbol{\varphi}(x) , \quad x \in D .$$

Since $\mathcal{L}_{\lambda,\mu}\mathbf{H} = 0$ in Ω and $\mathcal{L}_{\tilde{\lambda},\tilde{\mu}}\widetilde{\mathcal{S}}_D \boldsymbol{\varphi} = 0$ in D, by Theorem 6.19, $\mathbf{H}$ and $\widetilde{\mathcal{S}}_D \boldsymbol{\varphi}$ have the desired representation by holomorphic functions. So, in order to prove (6.51), it suffices to show that there are functions f and g holomorphic in $\Omega \setminus \overline{D}$ such that

$$2\mu\left[(\mathcal{S}_D\psi)_1 + i(\mathcal{S}_D\psi)_2\right](z) = \kappa f(z) - z\overline{f'(z)} - \overline{g(z)}\,, \quad z \in \Omega \setminus \overline{D}\,. \quad (6.55)$$

Observe that for $i = 1, 2$,

$$(\mathcal{S}_D\psi)_i(x) = \frac{A}{2\pi}\int_{\partial D}\log|x - y|\psi_i(y)\,d\sigma(y)$$

$$-\frac{B}{2\pi}\sum_{j=1}^{2}\int_{\partial D}\frac{(x_i - y_i)(x_j - y_j)}{|x - y|^2}\psi_j(y)\,d\sigma(y)\,.$$

Hence

$$\left[(\mathcal{S}_D\psi)_1 + i(\mathcal{S}_D\psi)_2\right](x) = \frac{A}{2\pi}\int_{\partial D}\log|x - y|\left[\psi_1(y) + i\psi_2(y)\right]d\sigma(y)$$

$$-\frac{B}{2\pi}\int_{\partial D}\frac{(x_1 - y_1) + i(x_2 - y_2)}{|x - y|^2}\left[(x_1 - y_1)\psi_1(y) + (x_2 - y_2)\psi_2(y)\right]d\sigma(y)\,.$$

Let $z = x_1 + ix_2$, $\zeta = y_1 + iy_2$, and $\psi = \psi_1 + i\psi_2$. Then

$$\left[(\mathcal{S}_D\psi)_1 + i(\mathcal{S}_D\psi)_2\right](z) = \frac{A}{2\pi}\int_{\partial D}\log|z - \zeta|\psi(\zeta)\,d\sigma(\zeta)$$

$$-\frac{B}{2\pi}\int_{\partial D}\frac{z - \zeta}{|z - \zeta|^2}\left[(z - \zeta)\overline{\psi(\zeta)} + \overline{(z - \zeta)}\psi(\zeta)\right]d\sigma(\zeta)$$

$$= \frac{A}{4\pi}\int_{\partial D}\log(z - \zeta)\psi(\zeta)\,d\sigma(\zeta) - \frac{B}{4\pi}z\int_{\partial D}\frac{\overline{\psi(\zeta)}}{z - \zeta}\,d\sigma(\zeta)$$

$$+ \frac{A}{4\pi}\int_{\partial D}\overline{\log(z - \zeta)}\psi(\zeta)\,d\sigma(\zeta) + \frac{B}{4\pi}\int_{\partial D}\frac{\overline{\zeta\psi(\zeta)}}{z - \zeta}\,d\sigma(\zeta) - \frac{B}{4\pi}\int_{\partial D}\psi(\zeta)\,d\sigma(\zeta)\,.$$

Observe that

$$\frac{A}{B} = \frac{\lambda + 3\mu}{\lambda + \mu}\,.$$

Then (6.55) follows with f defined by

$$f(z) := \frac{B}{8\mu\pi}\int_{\partial D}\log(z - \zeta)\psi(\zeta)\,d\sigma(\zeta)\,, \quad (6.56)$$

and g defined in an obvious way. It should be noted that f defined by (6.56) is holomorphic outside D. This is because $\psi \in L^2_\Psi(\partial D)$ which implies that $\int_{\partial D}\psi\,d\sigma = 0$.

The equation (6.53) is identical to the third equation in (6.21). By the fourth equation in (6.21) and (6.44), we get

$$\partial\left[\varphi_e(z) + z\overline{\varphi'_e(z)} + \overline{\psi_e(z)}\right] = \partial\left[\varphi_i(z) + z\overline{\varphi'_i(z)} + \overline{\psi_i(z)}\right]\,,$$

from which (6.54) follows immediately. This finishes the proof. □

7

Elastic Moment Tensor

In this chapter, we introduce the notion of an elastic moment tensor (EMT's) as was defined in [23] and investigate some important properties of the first-order EMT such as symmetry and positive-definiteness. We also obtain estimation of its eigenvalues and compute EMT's associated with ellipses, elliptic holes, and hard inclusions of elliptic shape.

7.1 Asymptotic Expansion in Free Space

As in the electrostatic case, the elastic moment tensors describe the perturbation of the displacement vector due to the presence of elastic inclusions. To see this let us consider a transmission problem in the free space.

Let B be a bounded Lipschitz domain in $\mathbb{R}^d$, $d = 2, 3$. Consider the following transmission problem

$$\begin{cases} \displaystyle\sum_{j,k,l=1}^{d} \frac{\partial}{\partial x_j}\left(C_{ijkl}\frac{\partial u_k}{\partial x_l}\right) = 0 \quad \text{in } \mathbb{R}^d, \quad i = 1,\dots,d, \\ \mathbf{u}(x) - \mathbf{H}(x) = O(|x|^{1-d}) \quad \text{as } |x| \to \infty, \end{cases} \tag{7.1}$$

where

$$C_{ijkl} = \left(\lambda\,\chi(\mathbb{R}^d\setminus B) + \widetilde{\lambda}\,\chi(B)\right)\delta_{ij}\delta_{kl} + \left(\mu\,\chi(\mathbb{R}^d\setminus B) + \widetilde{\mu}\,\chi(B)\right)(\delta_{ik}\delta_{jl} + \delta_{il}\delta_{jk}),$$

and $\mathbf{H}$ is a vector-valued function which satisfies $\mathcal{L}_{\lambda,\mu}\mathbf{H} = 0$ in $\mathbb{R}^d$. In a similar way to the proof of Theorem 6.15, we can show that the solution $\mathbf{u}$ to (7.1) is represented as

$$\mathbf{u}(x) = \begin{cases} \mathbf{H}(x) + \mathcal{S}_B\psi(x), & x \in \mathbb{R}^d\setminus\overline{B}, \\ \widetilde{\mathcal{S}}_B\varphi(x), & x \in B, \end{cases} \tag{7.2}$$

for a unique pair $(\varphi, \psi) \in L^2(\partial B) \times L^2_{\widetilde{\Psi}}(\partial B)$ which satisfies

$$\begin{cases} \widetilde{\mathcal{S}}_B\varphi\big|_- - \mathcal{S}_B\psi\big|_+ = \mathbf{H}|_{\partial B} & \text{on } \partial B\,, \\[2mm] \dfrac{\partial}{\partial\widetilde{\nu}}\widetilde{\mathcal{S}}_B\varphi\bigg|_- - \dfrac{\partial}{\partial\nu}\mathcal{S}_B\psi\bigg|_+ = \dfrac{\partial\mathbf{H}}{\partial\nu}\bigg|_{\partial B} & \text{on } \partial B\,. \end{cases} \tag{7.3}$$

Suppose that the origin $0 \in B$ and expand $\mathbf{H}$ in terms of Taylor series to write

$$\mathbf{H}(x) = \left(\sum_{\alpha\in\mathbb{N}^d}\frac{1}{\alpha!}\partial^\alpha H_1(0)x^\alpha,\ldots,\sum_{\alpha\in\mathbb{N}^d}\frac{1}{\alpha!}\partial^\alpha H_d(0)x^\alpha\right)$$

$$= \sum_{j=1}^{d}\sum_{\alpha\in\mathbb{N}^d}\frac{1}{\alpha!}\partial^\alpha H_j(0)\,x^\alpha\mathbf{e}_j\,,$$

where $\{\mathbf{e}_j\}_{j=1}^d$ is the standard basis for $\mathbb{R}^d$. This series converges uniformly and absolutely on any compact set. For multi-index $\alpha\in\mathbb{N}^d$ and $j=1,\ldots,d$, let $\mathbf{f}_\alpha^j$ and $\mathbf{g}_\alpha^j$ in $L^2(\partial B)$ be the solution of

$$\begin{cases} \widetilde{\mathcal{S}}_B\mathbf{f}_\alpha^j\big|_- - \mathcal{S}_B\mathbf{g}_\alpha^j\big|_+ = x^\alpha\mathbf{e}_j|_{\partial B}\,, \\[2mm] \dfrac{\partial}{\partial\widetilde{\nu}}\widetilde{\mathcal{S}}_B\mathbf{f}_\alpha^j\bigg|_- - \dfrac{\partial}{\partial\nu}\mathcal{S}_B\mathbf{g}_\alpha^j\bigg|_+ = \dfrac{\partial(x^\alpha\mathbf{e}_j)}{\partial\nu}|_{\partial B}\,. \end{cases} \tag{7.4}$$

Then, by linearity, we get

$$\psi = \sum_{j=1}^{d}\sum_{\alpha\in\mathbb{N}^d}\frac{1}{\alpha!}\partial^\alpha H_j(0)\,\mathbf{g}_\alpha^j\,. \tag{7.5}$$

By a Taylor expansion, we have

$$\mathbf{\Gamma}(x-y) = \sum_{\beta\in\mathbb{N}^d}\frac{1}{\beta!}\partial^\beta\mathbf{\Gamma}(x)y^\beta, \quad y \text{ in a compact set,}\quad |x|\to\infty\,. \tag{7.6}$$

Combining (7.2), (7.5), and (7.6) yields the expansion

$$\mathbf{u}(x) = \mathbf{H}(x) + \sum_{j=1}^{d}\sum_{\alpha\in\mathbb{N}^d}\sum_{\beta\in\mathbb{N}^d}\frac{1}{\alpha!\beta!}\partial^\alpha H_j(0)\partial^\beta\mathbf{\Gamma}(x)\int_{\partial B}y^\beta\mathbf{g}_\alpha^j(y)\,d\sigma(y)\,, \tag{7.7}$$

which is valid for all x with $|x| > R$ where R is such that $B \subset B_R(0)$.

Definition 7.1 (Elastic moment tensors) *For multi-index $\alpha\in\mathbb{N}^d$ and $j=1,\ldots,d$, let $\mathbf{f}_\alpha^j$ and $\mathbf{g}_\alpha^j$ in $L^2(\partial B)$ be the solution of (7.4). For $\beta\in\mathbb{N}^d$, the elastic moment tensor (EMT) $M_{\alpha\beta}^j$, $j=1,\ldots,d$, associated with the domain B and Lamé parameters (λ,μ) for the background and $(\widetilde{\lambda},\widetilde{\mu})$ for B is defined by*

$$M_{\alpha\beta}^j = (m_{\alpha\beta1}^j,\ldots,m_{\alpha\beta d}^j) = \int_{\partial B}y^\beta\mathbf{g}_\alpha^j(y)\,d\sigma(y)\,.$$

We note the analogy of the EMT with the polarization tensor studied in Chap. 3. For the cavities in elastic body, Maz'ya and Nazarov introduced the notion of Pólya-Szegö tensor in relation to the asymptotic expansion for energy due to existence of a small hole or cavity [202]. See also [212] and [196]. This tensor is exactly the one defined by (7.4) when $|\alpha| = |\beta| = 1$ and B is a cavity.

Theorem 7.2 *Let* $\mathbf{u}$ *be the solution of (7.1). Then for all* x *with* $|x| > R$ *where* $B \subset B_R(0)$, $\mathbf{u}$ *has the expansion*

$$\mathbf{u}(x) = \mathbf{H}(x) + \sum_{j=1}^{d} \sum_{|\alpha| \geq 1} \sum_{|\beta| \geq 1} \frac{1}{\alpha! \beta!} \partial^\alpha H_j(0) \partial^\beta \Gamma(x) M_{\alpha\beta}^j . \qquad (7.8)$$

Proof. We first show that if $\alpha = 0$, then $\mathbf{g}_0^j = 0$ for $j = 1, \ldots, d$. For that, recall that $(\mathbf{f}_0^j, \mathbf{g}_0^j)$ is the unique solution to

$$\begin{cases} \widetilde{\mathcal{S}}_B \mathbf{f}_0^j |_- - \mathcal{S}_B \mathbf{g}_0^j |_+ = \mathbf{e}_j |_{\partial B} , \\ \dfrac{\partial}{\partial \widetilde{\nu}} \widetilde{\mathcal{S}}_B \mathbf{f}_0^j \bigg|_- - \dfrac{\partial}{\partial \nu} \mathcal{S}_B \mathbf{g}_0^j \bigg|_+ = 0 . \end{cases} \qquad (7.9)$$

Thus by Lemma 6.14, $\mathbf{g}_0^j = 0$. Note that $\sum_{j=1}^{d} \sum_{|\alpha|=l} \frac{1}{\alpha!} \partial^\alpha H_j(0) \mathbf{g}_\alpha^j$ is the solution of the integral equation (7.4) when the right-hand side is given by the function $\mathbf{u} := \sum_{j=1}^{d} \sum_{|\alpha|=l} \frac{1}{\alpha!} \partial^\alpha H_j(0) x^\alpha \mathbf{e}_j$. Moreover this function is a solution of $\mathcal{L}_{\lambda,\mu} \mathbf{u} = 0$ in B and therefore, $\partial \mathbf{u}/\partial \nu|_{\partial B} \in L_\Psi^2(\partial B)$. Hence by Theorem 6.13 we obtain that $\sum_{j=1}^{d} \sum_{|\alpha|=l} \frac{1}{\alpha!} \partial^\alpha H_j(z) \mathbf{g}_\alpha^j \in L_\Psi^2(\partial B)$. In particular, we have

$$\sum_{j=1}^{d} \sum_{|\alpha|=l} \frac{1}{\alpha!} \partial^\alpha H_j(z) \int_{\partial B} \mathbf{g}_\alpha^j(y) \, d\sigma(y) = 0 \quad \forall\, l .$$

Now (7.8) follows from (7.7). This completes the proof. $\square$

The asymptotic expansion formula (7.8) shows that the perturbation of the displacement vector in $\mathbb{R}^d$ due to the presence of an inclusion B are completely described by the EMT's $M_{\alpha\beta}^j$.

When $|\alpha| = |\beta| = 1$, we make a slight change of notations: When $\alpha = \mathbf{e}_i$ and $\beta = \mathbf{e}_p$ $(i, p = 1, \ldots, d)$, put

$$m_{pq}^{ij} := m_{\alpha\beta q}^j , \quad q, j = 1, \ldots, d .$$

So, if we set $\mathbf{f}_i^j := \mathbf{f}_\alpha^j$ and $\mathbf{g}_i^j := \mathbf{g}_\alpha^j$, then

$$\begin{cases} \widetilde{\mathcal{S}}_B \mathbf{f}_i^j |_- - \mathcal{S}_B \mathbf{g}_i^j |_+ = x_i \mathbf{e}_j |_{\partial B} , \\ \dfrac{\partial}{\partial \widetilde{\nu}} \widetilde{\mathcal{S}}_B \mathbf{f}_i^j \bigg|_- - \dfrac{\partial}{\partial \nu} \mathcal{S}_B \mathbf{g}_i^j \bigg|_+ = \dfrac{\partial(x_i \mathbf{e}_j)}{\partial \nu} |_{\partial B} , \end{cases} \qquad (7.10)$$

and

$$m_{pq}^{ij} = \int_{\partial B} x_p \mathbf{e}_q \cdot \mathbf{g}_i^j \, d\sigma \ . \tag{7.11}$$

Lemma 7.3 *Suppose that* $0 < \widetilde{\lambda}, \widetilde{\mu} < \infty$. *For* $p, q, i, j = 1, \ldots, d$,

$$m_{pq}^{ij} = \int_{\partial B} \left[-\frac{\partial(x_p \mathbf{e}_q)}{\partial \nu} + \frac{\partial(x_p \mathbf{e}_q)}{\partial \widetilde{\nu}} \right] \cdot \mathbf{u}|_- \, d\sigma \ , \tag{7.12}$$

where $\mathbf{u}$ *is the unique solution of the transmission problem*

$$\begin{cases} \mathcal{L}_{\lambda,\mu} \mathbf{u} = 0 & in \ \mathbb{R}^d \setminus \overline{B} \ , \\ \mathcal{L}_{\widetilde{\lambda},\widetilde{\mu}} \mathbf{u} = 0 & in \ B \ , \\ \mathbf{u}|_+ - \mathbf{u}|_- = 0 & on \ \partial B \ , \\ \dfrac{\partial \mathbf{u}}{\partial \nu}\bigg|_+ - \dfrac{\partial \mathbf{u}}{\partial \widetilde{\nu}}\bigg|_- = 0 & on \ \partial B \ , \\ \mathbf{u}(x) - x_i \mathbf{e}_j = O(|x|^{1-d}) & as \ |x| \to \infty \ . \end{cases} \tag{7.13}$$

Proof. Note first that $\mathbf{u}$ defined by

$$\mathbf{u}(x) := \begin{cases} \mathcal{S}_B \mathbf{g}_i^j(x) + x_i \mathbf{e}_j, & x \in \mathbb{R}^d \setminus \overline{B} \ , \\ \widetilde{\mathcal{S}}_B \mathbf{f}_i^j(x) \ , & x \in B \ , \end{cases}$$

is the solution of (7.13). Using (6.10) and (7.10) we compute

$$\begin{aligned} m_{pq}^{ij} &= \int_{\partial B} x_p \mathbf{e}_q \cdot \mathbf{g}_i^j \, d\sigma \\ &= \int_{\partial B} x_p \mathbf{e}_q \cdot \left[\frac{\partial}{\partial \nu} \mathcal{S}_B \mathbf{g}_i^j \Big|_+ - \frac{\partial}{\partial \nu} \mathcal{S}_B \mathbf{g}_i^j \Big|_- \right] d\sigma \\ &= -\int_{\partial B} x_p \mathbf{e}_q \cdot \frac{\partial(x_i \mathbf{e}_j)}{\partial \nu} \, d\sigma - \int_{\partial B} x_p \mathbf{e}_q \cdot \left[\frac{\partial}{\partial \nu} \mathcal{S}_B \mathbf{g}_i^j \Big|_- - \frac{\partial}{\partial \widetilde{\nu}} \widetilde{\mathcal{S}}_B \mathbf{f}_i^j \Big|_- \right] d\sigma \\ &= -\int_{\partial B} \frac{\partial(x_p \mathbf{e}_q)}{\partial \nu} \cdot x_i \mathbf{e}_j \, d\sigma - \int_{\partial B} \left[\frac{\partial(x_p \mathbf{e}_q)}{\partial \nu} \cdot \mathcal{S}_B \mathbf{g}_i^j - \frac{\partial(x_p \mathbf{e}_q)}{\partial \widetilde{\nu}} \cdot \widetilde{\mathcal{S}}_B \mathbf{f}_i^j \right] d\sigma \\ &= \int_{\partial B} \left[-\frac{\partial(x_p \mathbf{e}_q)}{\partial \nu} + \frac{\partial(x_p \mathbf{e}_q)}{\partial \widetilde{\nu}} \right] \cdot \widetilde{\mathcal{S}}_B \mathbf{f}_i^j \, d\sigma \ , \end{aligned}$$

and hence (7.12) is established. $\square$

7.2 Properties of EMT's

In this section we investigate some important properties of the first-order EMT $M = (m_{pq}^{ij})$ such as symmetry and positive-definiteness. These properties of EMT's were first proved in [23]. It is worth mentioning that these properties make M an (anisotropic in general) elasticity tensor. We first define a bilinear form on a domain B corresponding to the Lamé parameters λ, μ by

$$\langle \mathbf{u}, \mathbf{v} \rangle_B^{\lambda, \mu} := \int_B \left[\lambda (\nabla \cdot \mathbf{u})(\nabla \cdot \mathbf{v}) + \frac{\mu}{2}(\nabla \mathbf{u} + \nabla \mathbf{u}^T) \cdot (\nabla \mathbf{v} + \nabla \mathbf{v}^T) \right] dx \ .$$

The corresponding quadratic form is defined by

$$Q_B^{\lambda, \mu}(\mathbf{u}) := \langle \mathbf{u}, \mathbf{u} \rangle_B^{\lambda, \mu}.$$

If $\mathcal{L}_{\lambda, \mu} \mathbf{u} = 0$, then

$$\int_{\partial B} \frac{\partial \mathbf{u}}{\partial \nu} \cdot \mathbf{v} \, d\sigma = \langle \mathbf{u}, \mathbf{v} \rangle_B^{\lambda, \mu} \ .$$

Proposition 7.4 *Suppose that* $\mu \neq \tilde{\mu}$. *Given a nonzero symmetric matrix* $a = (a_{ij})$, *define* $\boldsymbol{\varphi}_a$, $\mathbf{f}_a$, *and* $\mathbf{g}_a$ *by*

$$\boldsymbol{\varphi}_a := (a_{ij})x = \sum_{i,j=1}^d a_{ij}x_j\mathbf{e}_i \ , \quad \mathbf{f}_a := \sum_{i,j=1}^d a_{ij}\mathbf{f}_i^j \ , \quad \mathbf{g}_a := \sum_{i,j=1}^d a_{ij}\mathbf{g}_i^j \ . \quad (7.14)$$

Define $\bar{a}$ *by*

$$\bar{a} := \frac{\tilde{\mu} + \mu}{\tilde{\mu} - \mu}\left[a - \frac{\operatorname{Tr}(a)}{d}I_d\right] + \frac{d(\tilde{\lambda} + \lambda) + 2(\tilde{\mu} + \mu)}{d(\tilde{\lambda} - \lambda) + 2(\tilde{\mu} - \mu)}\frac{\operatorname{Tr}(a)}{d}I_d \ , \quad (7.15)$$

where I_d *is the* $d \times d$ *identity matrix. Then*

$$\langle \bar{a}, Ma \rangle = \langle \tilde{\mathcal{S}}_B \mathbf{f}_a, \tilde{\mathcal{S}}_B \mathbf{f}_a \rangle_B^{\tilde{\lambda}, \tilde{\mu}} + \langle \mathcal{S}_B \mathbf{g}_a, \mathcal{S}_B \mathbf{g}_a \rangle_{\mathbb{R}^d \setminus B}^{\lambda, \mu} + \langle \boldsymbol{\varphi}_a, \boldsymbol{\varphi}_a \rangle_B^{\lambda, \mu} \ . \quad (7.16)$$

Recall that $\langle a, b \rangle = \sum_{ij} a_{ij}b_{ij}$ *for* $d \times d$ *matrices* $a = (a_{ij})$ *and* $b = (b_{ij})$.

Proof. Set, for convenience, $\boldsymbol{\varphi} = \boldsymbol{\varphi}_a$, $\mathbf{f} = \mathbf{f}_a$, and $\mathbf{g} = \mathbf{g}_a$. Then these functions clearly satisfy

$$\begin{cases} \tilde{\mathcal{S}}_B \mathbf{f} - \mathcal{S}_B \mathbf{g} = \boldsymbol{\varphi}|_{\partial B} \ , \\ \left. \dfrac{\partial}{\partial \tilde{\nu}} \tilde{\mathcal{S}}_B \mathbf{f} \right|_- - \left. \dfrac{\partial}{\partial \nu} \mathcal{S}_B \mathbf{g} \right|_+ = \left. \dfrac{\partial \boldsymbol{\varphi}}{\partial \nu} \right|_{\partial B} \ . \end{cases} \quad (7.17)$$

For $j = 1, 2$, define

$$\boldsymbol{\varphi}_1 := \left[(a_{ij}) - \frac{\operatorname{Tr}(a_{ij})}{d}I \right]x \quad \text{and} \quad \boldsymbol{\varphi}_2 := \frac{\operatorname{Tr}(a_{ij})}{d}x \ . \quad (7.18)$$

Then $\boldsymbol{\varphi} = \boldsymbol{\varphi}_1 + \boldsymbol{\varphi}_2$. Define $\mathbf{f}_j$ and $\mathbf{g}_j$, $j = 1, 2$, by

$$\begin{cases} \tilde{\mathcal{S}}_B \mathbf{f}_j - \mathcal{S}_B \mathbf{g}_j = \boldsymbol{\varphi}_j|_{\partial B} \ , \\ \left. \dfrac{\partial}{\partial \tilde{\nu}} \tilde{\mathcal{S}}_B \mathbf{f}_j \right|_- - \left. \dfrac{\partial}{\partial \nu} \mathcal{S}_B \mathbf{g}_j \right|_+ = \left. \dfrac{\partial \boldsymbol{\varphi}_j}{\partial \nu} \right|_{\partial B} \ . \end{cases} \quad (7.19)$$

It is clear that $\mathbf{f} = \mathbf{f}_1 + \mathbf{f}_2$ and $\mathbf{g} = \mathbf{g}_1 + \mathbf{g}_2$. We now claim that

$$\langle \bar{a}, Ma \rangle = \frac{\widetilde{\mu} + \mu}{\widetilde{\mu} - \mu} \int_{\partial B} \varphi_1 \cdot \mathbf{g}\, d\sigma + \frac{d(\widetilde{\lambda} + \lambda) + 2(\widetilde{\mu} + \mu)}{d(\widetilde{\lambda} - \lambda) + 2(\widetilde{\mu} - \mu)} \int_{\partial B} \varphi_2 \cdot \mathbf{g}\, d\sigma \ . \quad (7.20)$$

In fact, we have

$$\langle \bar{a}, Ma \rangle = \sum_{i,j,p,q=1}^{d} \bar{a}_{pq} m_{pq}^{ij} a_{ij}$$

$$= \int_{pB} \Big(\sum_{pq} \bar{a}_{pq} x_p \mathbf{e}_q\Big) \cdot \Big(\sum_{ij} a_{ij} \mathbf{g}_i^j\Big)\, d\sigma \ .$$

But

$$\sum_{pq} \bar{a}_{pq} x_p \mathbf{e}_q = \frac{\widetilde{\mu} + \mu}{\widetilde{\mu} - \mu} \varphi_1 + \frac{d(\widetilde{\lambda} + \lambda) + 2(\widetilde{\mu} + \mu)}{d(\widetilde{\lambda} - \lambda) + 2(\widetilde{\mu} - \mu)} \varphi_2 \ ,$$

and therefore (7.20) holds.

Next, using the jump relation (6.10) and (7.17), we compute that

$$\int_{\partial B} \varphi_j \cdot \mathbf{g}\, d\sigma = \int_{\partial B} \varphi_j \cdot \Big[\frac{\partial}{\partial \nu} \mathcal{S}_B \mathbf{g}\big|_+ - \frac{\partial}{\partial \nu} \mathcal{S}_B \mathbf{g}\big|_-\Big]\, d\sigma \qquad (7.21)$$

$$= -\int_{\partial B} \varphi_j \cdot \frac{\partial \varphi}{\partial \nu}\, d\sigma - \int_{\partial B} \varphi_j \cdot \Big[\frac{\partial}{\partial \nu} \mathcal{S}_B \mathbf{g}\big|_- - \frac{\partial}{\partial \widetilde{\nu}} \widetilde{\mathcal{S}}_B \mathbf{f}\big|_-\Big]\, d\sigma$$

$$= -\int_{\partial B} \varphi_j \cdot \frac{\partial \varphi}{\partial \nu}\, d\sigma - \int_{\partial B} \Big[\frac{\partial \varphi_j}{\partial \nu} \cdot \mathcal{S}_B \mathbf{g} - \frac{\partial \varphi_j}{\partial \widetilde{\nu}} \cdot \widetilde{\mathcal{S}}_B \mathbf{f}\Big]\, d\sigma$$

$$= \int_{\partial B} \Big[-\frac{\partial \varphi_j}{\partial \nu} + \frac{\partial \varphi_j}{\partial \widetilde{\nu}}\Big] \cdot \widetilde{\mathcal{S}}_B \mathbf{f}\, d\sigma \ .$$

Observe that $\nabla \cdot \varphi_1 = 0$. Put $\alpha := \widetilde{\mu}/\mu$. Then, from the definition of the conormal derivative $\partial/\partial \nu$, we can immediately see that

$$\frac{\partial \varphi_1}{\partial \nu} - \frac{\partial \varphi_1}{\partial \widetilde{\nu}} = (1 - \alpha)\frac{\partial \varphi_1}{\partial \nu} = \frac{1 - \alpha}{\alpha}\frac{\partial \varphi_1}{\partial \widetilde{\nu}} \ . \qquad (7.22)$$

Combining (7.17), (7.19), and (7.21), together with the second relation of (7.22) yields

$$-\frac{\alpha}{1 - \alpha} \int_{\partial B} \varphi_1 \cdot \mathbf{g}\, d\sigma = \int_{\partial B} \frac{\partial \varphi_1}{\partial \widetilde{\nu}} \cdot \widetilde{\mathcal{S}}_B \mathbf{f}\, d\sigma = \int_{\partial B} \varphi_1 \cdot \frac{\partial}{\partial \widetilde{\nu}} \widetilde{\mathcal{S}}_B \mathbf{f}\big|_-\, d\sigma$$

$$= \int_{\partial B} \widetilde{\mathcal{S}}_B \mathbf{f}_1 \cdot \frac{\partial}{\partial \widetilde{\nu}} \widetilde{\mathcal{S}}_B \mathbf{f}\big|_-\, d\sigma - \int_{\partial B} \mathcal{S}_B \mathbf{g}_1 \cdot \frac{\partial}{\partial \nu} \mathcal{S}_B \mathbf{g}\big|_+\, d\sigma - \int_{\partial B} \mathcal{S}_B \mathbf{g}_1 \cdot \frac{\partial \varphi}{\partial \nu}\, d\sigma$$

$$= \langle \widetilde{\mathcal{S}}_B \mathbf{f}_1, \widetilde{\mathcal{S}}_B \mathbf{f} \rangle_B^{\widetilde{\lambda},\widetilde{\mu}} + \langle \mathcal{S}_B \mathbf{g}_1, \mathcal{S}_B \mathbf{g} \rangle_{\mathbb{R}^d \backslash B}^{\lambda,\mu} - \langle \mathcal{S}_B \mathbf{g}_1, \varphi \rangle_B^{\lambda,\mu} \ .$$

On the other hand, it follows from (7.17), (7.21), and the first relation of (7.22) that

$$-\frac{1}{1-\alpha}\int_{\partial B}\varphi_1\cdot\mathbf{g}\,d\sigma = \int_{\partial B}\frac{\partial\varphi_1}{\partial\nu}\cdot\widetilde{S}_B\mathbf{f}\,d\sigma$$

$$= \int_{\partial B}\frac{\partial\varphi_1}{\partial\nu}\cdot S_B\mathbf{g}\,d\sigma + \int_{\partial B}\frac{\partial\varphi_1}{\partial\nu}\cdot\varphi\,d\sigma$$

$$= \langle\varphi_1, S_B\mathbf{g}\rangle_B^{\lambda,\mu} + \langle\varphi_1,\varphi\rangle_B^{\lambda,\mu}.$$

Adding the above two identities we get

$$-\frac{1+\alpha}{1-\alpha}\int_{\partial B}\varphi_1\cdot\mathbf{g}\,d\sigma$$

$$= \langle\widetilde{S}_B\mathbf{f}_1,\widetilde{S}_B\mathbf{f}\rangle_B^{\widetilde{\lambda},\widetilde{\mu}} + \langle S_B\mathbf{g}_1, S_B\mathbf{g}\rangle_{\mathbb{R}^d\setminus B}^{\lambda,\mu} + \langle\varphi_1,\varphi\rangle_B^{\lambda,\mu} \qquad (7.23)$$

$$- \langle S_B\mathbf{g}_1,\varphi\rangle_B^{\lambda,\mu} + \langle\varphi_1, S_B\mathbf{g}\rangle_B^{\lambda,\mu}.$$

Observe that

$$\frac{1+\alpha}{1-\alpha} = \frac{\mu+\widetilde{\mu}}{\mu-\widetilde{\mu}}.$$

Put

$$\beta := \frac{d\lambda+2\mu}{d(\lambda-\widetilde{\lambda})+2(\mu-\widetilde{\mu})} \quad\text{and}\quad \widetilde{\beta} := \frac{d\widetilde{\lambda}+2\widetilde{\mu}}{d(\lambda-\widetilde{\lambda})+2(\mu-\widetilde{\mu})}.$$

It can be easily seen that

$$\frac{\partial\varphi_2}{\partial\nu} - \frac{\partial\varphi_2}{\partial\widetilde{\nu}} = \frac{1}{\beta}\frac{\partial\varphi_2}{\partial\nu} = \frac{1}{\widetilde{\beta}}\frac{\partial\varphi_2}{\partial\widetilde{\nu}}. \qquad (7.24)$$

Following the same lines of derivation of (7.23), we obtain

$$-(\beta+\widetilde{\beta})\int_{\partial B}\varphi_2\cdot\mathbf{g}\,d\sigma$$

$$= \langle\widetilde{S}_B\mathbf{f}_2,\widetilde{S}_B\mathbf{f}\rangle_B^{\widetilde{\lambda},\widetilde{\mu}} + \langle S_B\mathbf{g}_2, S_B\mathbf{g}\rangle_{\mathbb{R}^d\setminus B}^{\lambda,\mu} + \langle\varphi_2,\varphi\rangle_B^{\lambda,\mu} \qquad (7.25)$$

$$- \langle S_B\mathbf{g}_2,\varphi\rangle_B^{\lambda,\mu} + \langle\varphi_2, S_B\mathbf{g}\rangle_B^{\lambda,\mu}.$$

Adding (7.23) and (7.25) yields

$$-\frac{1+\alpha}{1-\alpha}\int_{\partial B}\varphi_1\cdot\mathbf{g}\,d\sigma - (\beta+\widetilde{\beta})\int_{\partial B}\varphi_2\cdot\mathbf{g}\,d\sigma$$

$$= \langle\widetilde{S}_B\mathbf{f},\widetilde{S}_B\mathbf{f}\rangle_B^{\widetilde{\lambda},\widetilde{\mu}} + \langle S_B\mathbf{g}, S_B\mathbf{g}\rangle_{\mathbb{R}^d\setminus B}^{\lambda,\mu} + \langle\varphi,\varphi\rangle_B^{\lambda,\mu} - \langle S_B\mathbf{g},\varphi\rangle_B^{\lambda,\mu} + \langle\varphi, S_B\mathbf{g}\rangle_B^{\lambda,\mu}$$

$$= \langle\widetilde{S}_B\mathbf{f},\widetilde{S}_B\mathbf{f}\rangle_B^{\widetilde{\lambda},\widetilde{\mu}} + \langle S_B\mathbf{g}, S_B\mathbf{g}\rangle_{\mathbb{R}^d\setminus B}^{\lambda,\mu} + \langle\varphi,\varphi\rangle_B^{\lambda,\mu}.$$

Then the final formula (7.16) follows from (7.20), and the proof is complete.

$$\square$$

Theorem 7.5 (Symmetry) *For $p, q, i, j = 1, \ldots, d$, the following holds:*

$$m_{pq}^{ij} = m_{qp}^{ij}, \quad m_{pq}^{ij} = m_{pq}^{ji}, \quad \text{and} \quad m_{pq}^{ij} = m_{ij}^{pq}. \tag{7.26}$$

Proof. By Theorem 6.13 and the definition (7.4) of $\mathbf{g}_i^j$, we have $\mathbf{g}_i^j \in L_\Psi^2(\partial B)$. Since $x_p \mathbf{e}_q - x_q \mathbf{e}_p \in \Psi$, we have

$$\int_{\partial B} (x_p \mathbf{e}_q - x_q \mathbf{e}_p) \cdot \mathbf{g}_i^j \, d\sigma = 0 .$$

The first assertion of (7.26) immediately follows from the above identity.

Since $x_i \mathbf{e}_j - x_j \mathbf{e}_i \in \Psi$, we have $\partial(x_i \mathbf{e}_j - x_j \mathbf{e}_i)/\partial \nu = 0$ on ∂B. Let $\mathbf{g} := \mathbf{g}_i^j - \mathbf{g}_j^i$ and $\mathbf{f} := \mathbf{f}_i^j - \mathbf{f}_j^i$. Then $\mathbf{f}$ and $\mathbf{g}$ satisfies

$$\begin{cases} \widetilde{\mathcal{S}}_B \mathbf{f}\big|_- - \mathcal{S}_B \mathbf{g}\big|_+ = (x_i \mathbf{e}_j - x_j \mathbf{e}_i)\big|_{\partial B} , \\ \dfrac{\partial}{\partial \widetilde{\nu}} \widetilde{\mathcal{S}}_B \mathbf{f}\Big|_- - \dfrac{\partial}{\partial \nu} \mathcal{S}_B \mathbf{g}\Big|_+ = 0 . \end{cases}$$

It then follows from Lemma 6.14 that $\mathbf{g} = 0$ or

$$\mathbf{g}_i^j = \mathbf{g}_j^i .$$

This proves the second assertion of (7.26).

Because of the first and second identities of (7.26), $\langle a, Mb \rangle = \frac{1}{4}\langle a + a^T, M(b + b^T) \rangle$ for any matrices a, b. Therefore, in order to prove the third identity in (7.26), it suffices to show that

$$\langle a, Mb \rangle = \langle b, Ma \rangle , \quad \text{for all symmetric matrices } a, b .$$

Let a, b be two symmetric matrices. Define $\varphi_a, \mathbf{f}_a, \mathbf{g}_a, \varphi_{aj}, \mathbf{f}_{aj}, \mathbf{g}_{aj}, j = 1, 2$, as in (7.14), (7.18), and (7.19). Define $\varphi_b, \mathbf{f}_b, \mathbf{g}_b, \varphi_{bj}, \mathbf{f}_{bj}, \mathbf{g}_{bj}, j = 1, 2$, likewise. Then,

$$\langle a, Mb \rangle = \int_{\partial B} \varphi_a \cdot \mathbf{g}_b \, d\sigma = \int_{\partial B} \varphi_{a1} \cdot \mathbf{g}_b \, d\sigma + \int_{\partial B} \varphi_{a2} \cdot \mathbf{g}_b \, d\sigma .$$

By (7.21), we get for $j = 1, 2$

$$\int_{\partial B} \varphi_{aj} \cdot \mathbf{g}_b \, d\sigma = \int_{\partial B} \left[\frac{\partial \varphi_{aj}}{\partial \nu} - \frac{\partial \varphi_{aj}}{\partial \widetilde{\nu}} \right] \cdot \widetilde{\mathcal{S}}_B \mathbf{f}_b \, d\sigma . \tag{7.27}$$

Let α, β, and $\widetilde{\beta}$ be as before. It then follows from (7.19), the first relation in (7.22), and (7.27) that

$$-\frac{1}{1-\alpha}\int_{\partial B}\boldsymbol{\varphi}_{a1}\cdot\mathbf{g}_b\,d\sigma$$

$$=\int_{\partial B}\frac{\partial\boldsymbol{\varphi}_{a1}}{\partial\nu}\cdot\widetilde{\mathcal{S}}_B\mathbf{f}_b\,d\sigma \tag{7.28}$$

$$=\int_{\partial B}\frac{\partial(\boldsymbol{\varphi}_{a1}+\mathcal{S}_B\mathbf{g}_{a1})}{\partial\nu}\Big|_{-}\cdot(\mathcal{S}_B\mathbf{g}_a+\boldsymbol{\varphi}_a)\,d\sigma$$

$$-\int_{\partial B}\frac{\partial(\mathcal{S}_B\mathbf{g}_{a1})}{\partial\nu}\Big|_{-}\cdot(\mathcal{S}_B\mathbf{g}_a+\boldsymbol{\varphi}_a)\,d\sigma\,.$$

On the other hand, by (7.19), the second relation in (7.22), and (7.27), we have

$$-\frac{\alpha}{1-\alpha}\int_{\partial B}\boldsymbol{\varphi}_{a1}\cdot\mathbf{g}_b\,d\sigma$$

$$=\int_{\partial B}\frac{\partial\boldsymbol{\varphi}_{a1}}{\partial\widetilde{\nu}}\cdot\widetilde{\mathcal{S}}_B\mathbf{f}_b\,d\sigma=\int_{\partial B}\boldsymbol{\varphi}_{a1}\cdot\frac{\partial(\widetilde{\mathcal{S}}_B\mathbf{f}_b)}{\partial\widetilde{\nu}}\Big|_{-}\,d\sigma \tag{7.29}$$

$$=\int_{\partial B}\widetilde{\mathcal{S}}_B\mathbf{f}_{a1}\cdot\frac{\partial(\widetilde{\mathcal{S}}_B\mathbf{f}_b)}{\partial\widetilde{\nu}}\Big|_{-}\,d\sigma$$

$$-\int_{\partial B}\mathcal{S}_B\mathbf{g}_{a1}\cdot\frac{\partial(\mathcal{S}_B\mathbf{g}_b)}{\partial\nu}\Big|_{+}\,d\sigma-\int_{\partial B}\mathcal{S}_B\mathbf{g}_{a1}\cdot\frac{\partial\boldsymbol{\varphi}_b}{\partial\nu}\,d\sigma\,.$$

Subtracting (7.29) from (7.28) yields

$$\int_{\partial B}\boldsymbol{\varphi}_{a1}\cdot\mathbf{g}_b\,d\sigma$$

$$=\int_{\partial B}\frac{\partial(\boldsymbol{\varphi}_{a1}+\mathcal{S}_B\mathbf{g}_{a1})}{\partial\nu}\Big|_{-}\cdot(\mathcal{S}_B\mathbf{g}_a+\boldsymbol{\varphi}_a)\,d\sigma \tag{7.30}$$

$$-\int_{\partial B}\widetilde{\mathcal{S}}_B\mathbf{f}_{a1}\cdot\frac{\partial(\widetilde{\mathcal{S}}_B\mathbf{f}_b)}{\partial\widetilde{\nu}}\Big|_{-}\,d\sigma$$

$$-\int_{\partial B}\mathcal{S}_B\mathbf{g}_{a1}\cdot\left[\frac{\partial(\mathcal{S}_B\mathbf{g}_b)}{\partial\nu}\Big|_{-}-\frac{\partial(\mathcal{S}_B\mathbf{g}_b)}{\partial\nu}\Big|_{+}\right]\,d\sigma\,.$$

$$=\langle\boldsymbol{\varphi}_{a1}+\mathcal{S}_B\mathbf{g}_{a1},\boldsymbol{\varphi}_b+\mathcal{S}_B\mathbf{g}_b\rangle_B^{\lambda,\mu}-\langle\widetilde{\mathcal{S}}_B\mathbf{f}_{a1},\widetilde{\mathcal{S}}_B\mathbf{f}_b\rangle_B^{\widetilde{\lambda},\widetilde{\mu}}-\int_{\partial B}\mathcal{S}_B\mathbf{g}_{a1}\cdot\mathbf{g}_b\,d\sigma\,.$$

Note that $\beta-\widetilde{\beta}=1$. We compute in the same way using (7.24) to obtain that

$$\int_{\partial B}\boldsymbol{\varphi}_{a2}\cdot\mathbf{g}_b\,d\sigma=\langle\boldsymbol{\varphi}_{a2}+\mathcal{S}_B\mathbf{g}_{a2},\boldsymbol{\varphi}_b+\mathcal{S}_B\mathbf{g}_b\rangle_B^{\lambda,\mu} \tag{7.31}$$

$$-\langle\widetilde{\mathcal{S}}_B\mathbf{f}_{a2},\widetilde{\mathcal{S}}_B\mathbf{f}_b\rangle_B^{\widetilde{\lambda},\widetilde{\mu}}-\int_{\partial B}\mathcal{S}_B\mathbf{g}_{a2}\cdot\mathbf{g}_b\,d\sigma\,.$$

Adding (7.30) and (7.31), we get

$$\int_{\partial B} \boldsymbol{\varphi}_a \cdot \mathbf{g}_b \, d\sigma = \langle \boldsymbol{\varphi}_a + \mathcal{S}_B \mathbf{g}_a, \boldsymbol{\varphi}_b + \mathcal{S}_B \mathbf{g}_b \rangle_B^{\lambda,\mu} \tag{7.32}$$

$$- \langle \widetilde{\mathcal{S}}_B \mathbf{f}_a, \widetilde{\mathcal{S}}_B \mathbf{f}_b \rangle_B^{\widetilde{\lambda},\widetilde{\mu}} - \int_{\partial B} \mathcal{S}_B \mathbf{g}_a \cdot \mathbf{g}_b \, d\sigma \, .$$

Since

$$\int_{\partial B} \mathcal{S}_B \mathbf{g}_a \cdot \mathbf{g}_b \, d\sigma = \int_{\partial B} \mathbf{g}_a \cdot \mathcal{S}_B \mathbf{g}_b \, d\sigma \, ,$$

identity (7.32) obviously implies that

$$\langle a, Mb \rangle = \int_{\partial B} \boldsymbol{\varphi}_a \cdot \mathbf{g}_b \, d\sigma = \int_{\partial B} \boldsymbol{\varphi}_b \cdot \mathbf{g}_a \, d\sigma = \langle b, Ma \rangle \, ,$$

and the proof is complete. □

Theorem 7.6 (Positive-definiteness) *If $\widetilde{\mu} > \mu$ ($\widetilde{\mu} < \mu$, resp.), then M is positive (negative, resp.) definite on the space of symmetric matrices. Let κ be an eigenvalue of M. Then there are constants C_1 and C_2 depending only on $\lambda, \mu, \widetilde{\lambda}, \widetilde{\mu}$ and the Lipschitz character of B such that*

$$C_1 |B| \leq |\kappa| \leq C_2 |B| \, .$$

Proof. Let $\boldsymbol{\varphi} = ax$, as before. Since

$$\langle \boldsymbol{\varphi}, \boldsymbol{\varphi} \rangle_B^{\lambda,\mu} = [\lambda \mathrm{Tr}(a_{ij})^2 + 2\mu \sum_{ij} a_{ij}^2] |B| \, ,$$

we get from (7.16) that

$$\langle \bar{a}, Ma \rangle \geq 2\mu |B| \, \|a\|^2 \, ,$$

where $\|a\|^2 = \sum_{ij} a_{ij}^2$. On the other hand, we can obtain an upper bound for m_{pq}^{ij} from its definition. In fact, let $z \in B$. Since $\int_{\partial B} \mathbf{g}_i^j \, d\sigma = 0$, we have

$$m_{pq}^{ij} = \int_{\partial B} x_p \mathbf{e}_q \cdot \mathbf{g}_i^j(x) \, d\sigma = \int_{\partial B} (x_p - z_p) \mathbf{e}_q \cdot \mathbf{g}_i^j(x) \, d\sigma \, .$$

It then follows from (6.28) that

$$|m_{pq}^{ij}|^2 \leq \int_{\partial B} (x_p - z_p)^2 \, d\sigma \int_{\partial B} |\mathbf{g}_i^j|^2 \, d\sigma$$

$$\leq C \mathrm{diam}(B)^2 |\partial B| \left(\|x_j \mathbf{e}_i\|_{L^2(\partial B)}^2 + \|\nabla(x_j \mathbf{e}_i)\|_{L^2(\partial B)}^2 \right)$$

$$\leq C \mathrm{diam}(B)^2 |\partial B|^2 \, .$$

Thus, if B satisfies the geometric condition: $\mathrm{diam}(B)|\partial B| \leq C_0 |B|$, then we have

$$|m_{pq}^{ij}| \leq C|B|$$

where the constant C depends on $\lambda, \mu, \widetilde{\lambda}, \widetilde{\mu}$ and C_0. Observe that C_0 depends on the Lipschitz character of B. Hence

$$\langle \bar{a}, Ma \rangle \leq C|B| \|a\|^2 .$$

Therefore, there is a constant C depending on $\lambda, \mu, \widetilde{\lambda}, \widetilde{\mu}$ and the Lipschitz character of B such that

$$\mu|B| \|a\|^2 \leq \langle \bar{a}, Ma \rangle \leq C|B| \|a\|^2 . \tag{7.33}$$

Let κ be an eigenvalue of M and let the matrix a be its corresponding eigenvector. Then $\langle \bar{a}, Ma \rangle = \kappa \langle \bar{a}, a \rangle$ and

$$\langle \bar{a}, a \rangle = \frac{\widetilde{\mu}+\mu}{\widetilde{\mu}-\mu} \left| a - \frac{\mathrm{Tr}(a)}{d} I_d \right|^2 + \frac{d(\widetilde{\lambda}+\lambda)+2(\widetilde{\mu}+\mu)}{d(\widetilde{\lambda}-\lambda)+2(\widetilde{\mu}-\mu)} \left| \frac{\mathrm{Tr}(a)}{d} I_d \right|^2 . \tag{7.34}$$

Suppose that $\widetilde{\mu} > \mu$. Let

$$K_1 := \min \left(\frac{\widetilde{\mu}+\mu}{\widetilde{\mu}-\mu}, \frac{d(\widetilde{\lambda}+\lambda)+2(\widetilde{\mu}+\mu)}{d(\widetilde{\lambda}-\lambda)+2(\widetilde{\mu}-\mu)} \right) ,$$

$$K_2 := \max \left(\frac{\widetilde{\mu}+\mu}{\widetilde{\mu}-\mu}, \frac{d(\widetilde{\lambda}+\lambda)+2(\widetilde{\mu}+\mu)}{d(\widetilde{\lambda}-\lambda)+2(\widetilde{\mu}-\mu)} \right) .$$

Then

$$K_1|B| \|a\|^2 \leq \langle \bar{a}, a \rangle \leq K_2|B| \|a\|^2 ,$$

and therefore, estimates (7.33) imply that $\kappa > 0$ and

$$\frac{C_1}{K_2}|B| \leq \kappa \leq \frac{C_2}{K_1}|B| .$$

When $\widetilde{\mu} < \mu$, we obtain, by a word for word translation of the previous proof, that $\kappa < 0$ and similar upper and lower bounds for $|\kappa|$ hold. The proof is complete. $\square$

Theorem 7.6 shows that the eigenvalues of M carries information on the size of the corresponding domain. We now prove that some components of M also carry the same information.

If $(a_{ij}) = \frac{1}{2}(E_{ij} + E_{ji})$, $i \neq j$, then $\varphi = (x_j \mathbf{e}_i + x_i \mathbf{e}_j)/2$. Hence by (7.16), we obtain

$$m_{ij}^{ij} = \frac{\widetilde{\mu}-\mu}{\widetilde{\mu}+\mu} \left[Q_B^{\widetilde{\lambda},\widetilde{\mu}}(\widetilde{\mathcal{S}}_B \mathbf{f}_i^j) + Q_{\mathbb{R}^d \backslash B}^{\lambda,\mu}(\mathcal{S}_B \mathbf{g}_i^j) + \mu|B| \right] .$$

It then follows that

$$|m_{ij}^{ij}| \geq \mu \left| \frac{\mu - \widetilde{\mu}}{\mu + \widetilde{\mu}} \right| |B| .$$

Thus, we have the following corollary.

Corollary 7.7 *Suppose $i \neq j$. Then there exists a constant C depending only on $\lambda, \mu, \widetilde{\lambda}, \widetilde{\mu}$, and the Lipschitz character of B such that*

$$\mu \left| \frac{\mu - \widetilde{\mu}}{\mu + \widetilde{\mu}} \right| |B| \leq |m_{ij}^{ij}| \leq C|B| . \tag{7.35}$$

7.3 EMT's Under Linear Transforms

In this section we derive formulae for EMT's under linear transforms.

Lemma 7.8 *Let B be a bounded domain in $\mathbb{R}^d$ and let $(m_{pq}^{ij}(B))$ denote the EMT associated with B. Then*

$$m_{pq}^{ij}(\epsilon B) = \epsilon^d m_{pq}^{ij}(B) , \quad i,j,p,q = 1, \ldots, d .$$

Proof. Let $(\mathbf{f}_i^j, \mathbf{g}_i^j)$ and $(\boldsymbol{\varphi}_i^j, \boldsymbol{\psi}_i^j)$ be the solution of (7.10) on ∂B and $\partial(\epsilon B)$, respectively. We claim that

$$\boldsymbol{\psi}_i^j(\epsilon x) = \mathbf{g}_i^j(x), \quad x \in \partial B . \tag{7.36}$$

If $d = 3$, then (7.36) simply follows from the homogeneity. In fact, in three dimensions, the Kelvin matrix $\boldsymbol{\Gamma}(x)$ is homogeneous of degree -1. Thus for any $\mathbf{f}$,

$$\mathcal{S}_{\epsilon B}\mathbf{f}(\epsilon x) = \epsilon \mathcal{S}_B \mathbf{f}_\epsilon(x), \quad x \in \partial B ,$$

$$\frac{\partial(\mathcal{S}_{\epsilon B}\mathbf{f})}{\partial \nu}(\epsilon x) = \frac{\partial(\mathcal{S}_B \mathbf{f}_\epsilon)}{\partial \nu}(x) , \quad x \in \partial B ,$$

where $\mathbf{f}_\epsilon(x) = \mathbf{f}(\epsilon x)$. Then (7.36) follows from the uniqueness of the solution to (7.10).

In two dimensions, note first the easy to prove fact:

$$\boldsymbol{\Gamma}(\epsilon x) = \frac{A}{2\pi} \log \epsilon \, I_d + \boldsymbol{\Gamma}(x) .$$

Since the pair $((\boldsymbol{\varphi}_i^j)_\epsilon, (\boldsymbol{\psi}_i^j)_\epsilon)$ satisfies

$$\begin{cases} \widetilde{\mathcal{S}}_B(\boldsymbol{\varphi}_i^j)_\epsilon \Big|_- + \dfrac{A}{2\pi} \dfrac{\log \epsilon}{\epsilon} \displaystyle\int_{\partial B} (\boldsymbol{\varphi}_i^j)_\epsilon \, d\sigma - \mathcal{S}_B(\boldsymbol{\psi}_i^j)_\epsilon \Big|_+ = x_i \mathbf{e}_j |_{\partial B} , \\[2mm] \dfrac{\partial}{\partial \widetilde{\nu}} \widetilde{\mathcal{S}}_B(\boldsymbol{\varphi}_i^j)_\epsilon \Big|_- - \dfrac{\partial}{\partial \nu} \mathcal{S}_B(\boldsymbol{\psi}_i^j)_\epsilon \Big|_+ = \dfrac{\partial(x_i \mathbf{e}_j)}{\partial \nu} |_{\partial B} , \end{cases}$$

we have

$$\begin{cases} \widetilde{\mathcal{S}}_B \left[(\boldsymbol{\varphi}_i^j)_\epsilon - \mathbf{f}_i^j \right] \Big|_- - \mathcal{S}_B \left[(\boldsymbol{\psi}_i^j)_\epsilon - \mathbf{g}_i^j \right] \Big|_+ = \text{constant} , \\[2mm] \dfrac{\partial}{\partial \widetilde{\nu}} \widetilde{\mathcal{S}}_B \left[(\boldsymbol{\varphi}_i^j)_\epsilon - \mathbf{f}_i^j \right] \Big|_- - \dfrac{\partial}{\partial \nu} \mathcal{S}_B \left[(\boldsymbol{\psi}_i^j)_\epsilon - \mathbf{g}_i^j \right] \Big|_+ = 0 , \end{cases} \quad \text{on } \partial B .$$

We then obtain (7.36) from Lemma 6.14. Armed with this identity, we now write

$$m_{pq}^{ij}(\epsilon B) = \int_{\partial(\epsilon B)} x_p \mathbf{e}_q \cdot \mathbf{g}_i^j(\epsilon B) \, d\sigma$$

$$= \epsilon^d \int_{\partial B} x_p \mathbf{e}_q \cdot \mathbf{g}_i^j(B) \, d\sigma = \epsilon^d m_{pq}^{ij}(B) \, ,$$

to arrive at the desired conclusion.   □

Lemma 7.9 *Let $R = (r_{ij})$ be a unitary transform in $\mathbb{R}^d$ and let $\widehat{B}$ be a bounded Lipschitz domain in $\mathbb{R}^d$ and $B = R(\widehat{B})$. Let m_{pq}^{ij} and $\widehat{m}_{pq}^{ij}$, $i, j, p, q = 1, \ldots, d$, denote the EMT's associated with B and $\widehat{B}$, respectively. Then,*

$$m_{pq}^{ij} = \sum_{u,v=1}^{d} \sum_{k,l=1}^{d} r_{pu} r_{qv} r_{ik} r_{jl} \widehat{m}_{uv}^{kl} \, . \tag{7.37}$$

Proof. For $i, j = 1, \ldots, d$, let $(\mathbf{f}_i^j, \mathbf{g}_i^j)$ and $(\widehat{\mathbf{f}}_i^j, \widehat{\mathbf{g}}_i^j)$ be the solutions of (7.10) on ∂B and $\partial \widehat{B}$, respectively. It follows from Lemmas 6.9 (ii) and 6.12 that

$$\widetilde{\mathcal{S}}_{\widehat{B}}(R^{-1}(\mathbf{f}_i^j \circ R))\Big|_{-} - \mathcal{S}_{\widehat{B}}(R^{-1}(\mathbf{g}_i^j \circ R))\Big|_{+} = R^{-1}((x_i \mathbf{e}_j) \circ R)|_{\partial \widehat{B}} \, ,$$

$$\frac{\partial}{\partial \widetilde{\nu}} \widetilde{\mathcal{S}}_{\widehat{B}}(R^{-1}(\mathbf{f}_i^j \circ R))\Big|_{-} - \frac{\partial}{\partial \nu} \mathcal{S}_{\widehat{B}}(R^{-1}(\mathbf{g}_i^j \circ R))\Big|_{+} = \frac{\partial}{\partial \nu}(R^{-1}((x_i \mathbf{e}_j) \circ R))|_{\partial \widehat{B}} \, .$$

It is easy to see that

$$R^{-1}((x_i \mathbf{e}_j) \circ R) = R(x)_i R^{-1}(\mathbf{e}_j) = \sum_{k,l=1}^{d} r_{ik} r_{jl}(x_k \mathbf{e}_l) \, , \quad i, j = 1, \ldots, d \, .$$

It then follows from the uniqueness of the solution to the integral equation (7.4) that

$$R^{-1}(\mathbf{g}_i^j \circ R) = \sum_{k,l=1}^{d} r_{ik} r_{jl} \widehat{\mathbf{g}}_k^l \, , \quad i, j = 1, \ldots, d \, .$$

By (7.11) and a change of variables, we have

$$m_{pq}^{ij} = \int_{\partial B} x_p \mathbf{e}_q \cdot \mathbf{g}_i^j \, d\sigma$$

$$= \int_{\partial \widehat{B}} R^{-1}((x_p \mathbf{e}_q) \circ R) \cdot R^{-1}(\mathbf{g}_i^j \circ R) \, d\sigma$$

$$= \int_{\partial \widehat{B}} \sum_{u,v=1}^{d} r_{pu} r_{qv}(x_u \mathbf{e}_v) \cdot \sum_{k,l=1}^{d} r_{ik} r_{jl} \widehat{\mathbf{g}}_k^l \, d\sigma$$

$$= \sum_{u,v=1}^{d} \sum_{k,l=1}^{d} r_{pu} r_{qv} r_{ik} r_{jl} \widehat{m}_{uv}^{kl} \, .$$

The proof is complete. ☐

In two dimensions the unitary transform R is given by the rotation:

$$R = R_\theta = \begin{pmatrix} r_{11} & r_{12} \\ r_{21} & r_{22} \end{pmatrix} = \begin{pmatrix} \cos\theta & -\sin\theta \\ \sin\theta & \cos\theta \end{pmatrix} .$$

The following corollary follows from (7.37) after elementary but tedious computations.

Corollary 7.10 *Let* $B = R_\theta(\widehat{B})$, *and* (m_{pq}^{ij}) *and* $(\widehat{m}_{pq}^{ij})$ *denote the EMT's for* B *and* $\widehat{B}$, *respectively. Then,*

$$\begin{cases} m_{11}^{11} = \cos^4\theta\widehat{m}_{11}^{11} + \dfrac{1}{2}\sin^2(2\theta)\widehat{m}_{22}^{11} + \sin^2(2\theta)\widehat{m}_{12}^{12} + \sin^4\theta\widehat{m}_{22}^{22}, \\[2mm] m_{12}^{11} = \sin\theta\cos^3\theta\widehat{m}_{11}^{11} - \dfrac{1}{4}\sin(4\theta)\widehat{m}_{22}^{11} - \dfrac{1}{2}\sin(4\theta)\widehat{m}_{12}^{12} - \sin^3\theta\cos\theta\widehat{m}_{22}^{22}, \\[2mm] m_{22}^{11} = \dfrac{1}{2}\sin^2(2\theta)\widehat{m}_{11}^{11} + (1 - \dfrac{1}{2}\sin^2(2\theta))\widehat{m}_{22}^{11} - \sin^2(2\theta)\widehat{m}_{12}^{12} + \dfrac{1}{2}\sin^2(2\theta)\widehat{m}_{22}^{22}, \\[2mm] m_{12}^{12} = \dfrac{1}{2}\sin^2(2\theta)\widehat{m}_{11}^{11} - \dfrac{1}{2}\sin^2(2\theta)\widehat{m}_{22}^{11} + \cos^2(2\theta)\widehat{m}_{12}^{12} + \dfrac{1}{4}\sin^2(2\theta)\widehat{m}_{22}^{22}, \\[2mm] m_{22}^{12} = \sin^3\theta\cos\theta\widehat{m}_{11}^{11} + \dfrac{1}{4}\sin(4\theta)\widehat{m}_{22}^{11} + \dfrac{1}{2}\sin(4\theta)\widehat{m}_{12}^{12} - \sin\theta\cos^3\theta\widehat{m}_{22}^{22}, \\[2mm] m_{22}^{22} = \sin^4\theta\widehat{m}_{11}^{11} + \dfrac{1}{2}\sin^2(2\theta)\widehat{m}_{22}^{11} + \sin^2(2\theta)\widehat{m}_{12}^{12} + \cos^4\theta\widehat{m}_{22}^{22}. \end{cases}$$

$$(7.38)$$

Corollary 7.10 has an interesting consequence. If B is a disk, then $m_{pq}^{ij} = \widehat{m}_{pq}^{ij}$, $i, j, p, q = 1, 2$, for any θ. Thus we can observe from the first identity in (7.38) that

$$m_{11}^{11} = m_{22}^{22} = m_{22}^{11} + 2m_{12}^{12}. \tag{7.39}$$

It then follows from the second and the fifth identity in (7.38) that

$$m_{12}^{11} = m_{22}^{12} = 0.$$

Thus we have the following lemma.

Lemma 7.11 *If* B *is a disk, then the EMT* (m_{pq}^{ij}) *is isotropic and given by*

$$m_{pq}^{ij} = m_{22}^{11}\delta_{ij}\delta_{pq} + m_{12}^{12}(\delta_{ip}\delta_{jq} + \delta_{iq}\delta_{jp}), \quad i, j, p, q = 1, 2. \tag{7.40}$$

We also obtain the following lemma from Corollary 7.10.

Lemma 7.12 *Suppose that either* $m_{12}^{11} + m_{22}^{12}$ *or* $m_{11}^{11} - m_{22}^{22}$ *is not zero. Then*

$$\frac{m_{12}^{11} + m_{22}^{12}}{m_{11}^{11} - m_{22}^{22}} = \frac{1}{2}\tan 2\theta. \tag{7.41}$$

Proof. We can easily see from (7.38) that

$$m_{11}^{11} - m_{22}^{22} = \cos 2\theta(\widehat{m}_{11}^{11} - \widehat{m}_{22}^{22}), \quad m_{12}^{11} + m_{22}^{12} = \frac{1}{2}\sin 2\theta(\widehat{m}_{11}^{11} - \widehat{m}_{22}^{22}).$$

Thus we get (7.41). ☐

7.4 EMT's for Ellipses

In this section we compute the EMT associated with an ellipse. We suppose that the ellipse takes the form

$$B : \frac{x^2}{a^2} + \frac{y^2}{b^2} = 1 , \quad a, b > 0 . \tag{7.42}$$

The EMT's for general ellipses can be found using (7.38).

Suppose that B is an ellipse of the form (7.42). Let (λ, μ) and $(\widetilde{\lambda}, \widetilde{\mu})$ be the Lamé constants for $\mathbb{R}^2 \setminus \overline{B}$ and B, respectively. We will be looking for the solution to (7.13).

Let $\mathbf{u} = (u, v)$ be a solution to (7.13) and let $\mathbf{u}_e := \mathbf{u}|_{\mathbb{R}^2 \setminus B}$ and $\mathbf{u}_i := \mathbf{u}|_B$. By Theorem 6.20, there are functions φ_e and ψ_e holomorphic in $\mathbb{C} \setminus \overline{B}$ and φ_i and ψ_i holomorphic in B such that

$$2\mu(u_e + iv_e)(z) = \kappa\varphi_e(z) - z\overline{\varphi_e'(z)} - \overline{\psi_e(z)} , \quad z \in \mathbb{C} \setminus \overline{B} , \tag{7.43}$$

$$2\widetilde{\mu}(u_i + iv_i)(z) = \widetilde{\kappa}\varphi_i(z) - z\overline{\varphi_i'(z)} - \overline{\psi_i(z)} , \quad z \in B , \tag{7.44}$$

where

$$\kappa = \frac{\lambda + 3\mu}{\lambda + \mu} , \quad \widetilde{\kappa} = \frac{\widetilde{\lambda} + 3\widetilde{\mu}}{\widetilde{\lambda} + \widetilde{\mu}} ,$$

and

$$\begin{cases} \dfrac{1}{2\mu}\left(\kappa\varphi_e(z) - z\overline{\varphi_e'(z)} - \overline{\psi_e(z)} \right) = \dfrac{1}{2\widetilde{\mu}}\left(\widetilde{\kappa}\varphi_i(z) - z\overline{\varphi_i'(z)} - \overline{\psi_i(z)} \right) , \\ \varphi_e(z) + z\overline{\varphi_e'(z)} + \overline{\psi_e(z)} = \varphi_i(z) + z\overline{\varphi_i'(z)} + \overline{\psi_i(z)} + c \quad \text{on } \partial B , \end{cases} \tag{7.45}$$

where c is a constant. In order to find such $\varphi_e, \psi_e, \varphi_i, \psi_i$, we use elliptic coordinates as done in [216]. Let

$$r := \frac{1}{2}(a + b) , \quad m := \frac{a - b}{a + b} , \tag{7.46}$$

and define

$$z = x + iy = \omega(\zeta) := r\left(\zeta + \frac{m}{\zeta}\right) .$$

Then ω maps the exterior of the unit disk onto $\mathbb{C} \setminus \overline{B}$.

Lemma 7.13 *Suppose that $m > 0$. For given pair of complex numbers α and β, there are unique complex numbers A, B, C, E, F such that the functions $\varphi_e, \psi_e, \varphi_i,$ and ψ_i defined by*

$$\varphi_e \circ \omega(\zeta) = r\left[\alpha\zeta + \frac{A}{\zeta} \right] , \quad |\zeta| > 1 ,$$

$$\psi_e \circ \omega(\zeta) = r\left[\beta\zeta + \frac{B}{\zeta} + \frac{C\zeta}{\zeta^2 - m} \right] , \quad |\zeta| > 1 , \tag{7.47}$$

$$\varphi_i(z) = Ez , \quad z \in B ,$$

$$\psi_i(z) = Fz , \quad z \in B ,$$

satisfy the conditions (6.53) and (6.54). Here, the constant c in (6.54) can be taken to be zero. In fact, A, B, C, E, F are the unique solutions of the algebraic equations

$$\begin{cases} \dfrac{\kappa}{\mu}\alpha - \dfrac{1}{\mu}\left(\dfrac{\overline{A}}{m} + \overline{B}\right) = \dfrac{\widetilde{\kappa}E - \overline{E}}{\widetilde{\mu}} - \dfrac{m}{\widetilde{\mu}}\overline{F}\,, \\[2mm] \alpha + \left(\dfrac{\overline{A}}{m} + \overline{B}\right) = E + \overline{E} + m\overline{F}\,, \\[2mm] \dfrac{\kappa}{\mu}A - \dfrac{1}{\mu}(m\overline{\alpha} + \overline{\beta}) = m\dfrac{\widetilde{\kappa}E - \overline{E}}{\widetilde{\mu}} - \dfrac{1}{\widetilde{\mu}}\overline{F}\,, \\[2mm] A + (m\overline{\alpha} + \overline{\beta}) = m(E + \overline{E}) + \overline{F}\,, \\[2mm] (m^2 + 1)\alpha - \left(m + \dfrac{1}{m}\right)A + C = 0\,. \end{cases} \qquad (7.48)$$

Proof. Since

$$\frac{\omega(\zeta)}{\overline{\omega'(\zeta)}} = \frac{\zeta^2 + m}{\zeta(1 - m\zeta^2)}\,, \quad |\zeta| = 1\,,$$

we can check by elementary but tedious computations that the transmission conditions (7.45) are equivalent to the algebraic equations (7.48). Using Matlab, we can check that (7.48) has a unique solution A, B, C, E, F provided that $m > 0$. The proof is complete. $\square$

For a given pair of complex numbers α and β, let $\mathbf{u} = (u, v)$ be the solution defined by φ and ψ given by (7.47). Define

$$\begin{cases} m_{pq}(\alpha, \beta) := \displaystyle\int_{\partial B} \frac{\partial(x_p \mathbf{e}_q)}{\partial \nu} \cdot (u_e, v_e)\, d\sigma\,, \\[3mm] \widetilde{m}_{pq}(\alpha, \beta) := \displaystyle\int_{\partial B} \frac{\partial(x_p \mathbf{e}_q)}{\partial \widetilde{\nu}} \cdot (u_e, v_e)\, d\sigma\,. \end{cases} \qquad (7.49)$$

In order to compute the EMT m_{pq}^{ij} associated with B, we need to find the solution of (7.13) which behaves at infinity as $x_i \mathbf{e}_j$. Let $\alpha = \alpha_1 + i\alpha_2$, *etc*, and observe that the exterior solution $u_e + iv_e$ behaves at infinity as

$$u_e(z) + iv_e(z) = \frac{1}{2\mu}\Big[(\kappa\alpha_1 - \alpha_1 - \beta_1)x + (-\kappa\alpha_2 - \alpha_2 + \beta_2)y\Big]$$

$$+ \frac{i}{2\mu}\Big[(\kappa\alpha_2 + \alpha_2 + \beta_2)x + (\kappa\alpha_1 - \alpha_1 + \beta_1)y\Big]$$

$$+ O(|z|^{-1})\,. \qquad (7.50)$$

Therefore, to compute m_{pq}^{11} for example, we need to take $\alpha = \mu/(\kappa - 1)$ and $\beta = -\mu$. In view of (7.12) and (7.50), we get

$$\begin{cases} m_{pq}^{11} = -m_{pq}\left(\dfrac{\mu}{\kappa - 1}, -\mu\right) + \widetilde{m}_{pq}\left(\dfrac{\mu}{\kappa - 1}, -\mu\right)\,, \\[3mm] m_{pq}^{22} = -m_{pq}\left(\dfrac{\mu}{\kappa - 1}, \mu\right) + \widetilde{m}_{pq}\left(\dfrac{\mu}{\kappa - 1}, \mu\right)\,, \\[3mm] m_{pq}^{12} = -m_{pq}\left(\dfrac{i\mu}{\kappa + 1}, i\mu\right) + \widetilde{m}_{pq}\left(\dfrac{i\mu}{\kappa + 1}, i\mu\right)\,. \end{cases} \qquad (7.51)$$

We now compute $m_{pq}(\alpha, \beta)$. For $p, q = 1, 2$, let $a = a_{pq}$ and $b = b_{pq}$ be complex numbers such that $f(z) = az$ and $g(z) = bz$ satisfy

$$2\mu((x_p\mathbf{e}_q)_1 + i(x_p\mathbf{e}_q)_2) = \kappa f(z) - z\overline{f'(z)} - \overline{g(z)}, \quad z \in \mathbb{C}. \tag{7.52}$$

In fact, (a, b) is given by

$$(a, b) = \begin{cases} (\dfrac{\mu}{\kappa - 1}, -\mu) & \text{if } (p, q) = (1, 1), \\[2mm] (\dfrac{\mu}{\kappa - 1}, \mu) & \text{if } (p, q) = (2, 2), \\[2mm] (\dfrac{i\mu}{\kappa + 1}, i\mu) & \text{if } (p, q) = (1, 2). \end{cases} \tag{7.53}$$

Then, by (6.44), we get

$$\left(\left(\frac{\partial(x_p\mathbf{e}_q)}{\partial\nu} \right)_1 + i \left(\frac{\partial(x_p\mathbf{e}_q)}{\partial\nu} \right)_2 \right) d\sigma \tag{7.54}$$

$$= -i\partial\left[f(z) + z\overline{f'(z)} + \overline{g(z)} \right] = -i\partial\left[2\Re az + \bar{b}\bar{z} \right]. \tag{7.55}$$

Therefore

$$m_{pq}(\alpha, \beta) = \Re \int_{\partial B} \left(\left(\frac{\partial(x_p\mathbf{e}_q)}{\partial\nu} \right)_1 + i \left(\frac{\partial(x_p\mathbf{e}_q)}{\partial\nu} \right)_2 \right) (u_e - iv_e) \, d\sigma$$

$$- \Re\frac{-i}{2\mu} \int_{\partial B} \left[\kappa\overline{\varphi_e(z)} - \bar{z}\varphi_e'(z) - \psi_e(z) \right] \partial\left[2\Re az + \bar{b}\bar{z} \right]$$

$$= \Re\frac{-i}{2\tilde{\mu}} \int_{\partial B} \left[\tilde{\kappa}\overline{\varphi_i(z)} - \bar{z}\varphi_i'(z) - \psi_i(z) \right] \left[2\Re adz + \bar{b}d\bar{z} \right],$$

where the last equality comes from (7.45). It then follows from (7.47) that

$$m_{pq}(\alpha, \beta) = \Re\frac{-i}{2\tilde{\mu}} \int_{\partial B} \left[(\tilde{\kappa}\overline{E} - E)\bar{z} - Fz \right] \left[2\Re adz + \bar{b}d\bar{z} \right]$$

$$= \Re\frac{\pi}{\tilde{\mu}} \left[2\Re a(\tilde{\kappa}\overline{E} - E) + \bar{b}F \right]. \tag{7.56}$$

Following the same lines of proof, we get

$$\tilde{m}_{pq}(\alpha, \beta) = \Re\frac{\pi}{\mu} \left[2\Re\tilde{a}(\tilde{\kappa}\overline{E} - E) + \bar{\tilde{b}}F \right], \tag{7.57}$$

where $(\tilde{a}, \tilde{b})$ is defined by (7.53) with μ, κ replaced by $\tilde{\mu}, \tilde{\kappa}$.

Denote the solutions of (7.48), which depends on given α and β, by $A = A_1 + iA_2 = A(\alpha, \beta)$, etc. Then we obtain from (7.51), (7.53), (7.56), and (7.57) that

$$\frac{\widetilde{\mu}}{|B|}\Big[- m_{pq}(\alpha,\beta) + \widetilde{m}_{pq}(\alpha,\beta)\Big]$$

$$= \begin{cases} (\widetilde{\kappa} - 1)(\widetilde{\lambda} - \lambda + \widetilde{\mu} - \mu)E_1 - (\widetilde{\mu} - \mu)F_1 & \text{if } p = q = 1 , \\ (\widetilde{\mu} - \mu)F_2 & \text{if } p \neq q , \\ (\widetilde{\kappa} - 1)(\widetilde{\lambda} - \lambda + \widetilde{\mu} - \mu)E_1 + (\widetilde{\mu} - \mu)F_1 & \text{if } p = q = 2 . \end{cases} \qquad (7.58)$$

For given α, β, we solve the system of linear equations (7.48) to find $E(\alpha,\beta)$ and $F(\alpha,\beta)$, and using (7.51) and (7.58) we can find m_{pq}^{ij}, $i,j,p,q = 1,2$. In short, we have the following theorem.

Theorem 7.14 *Suppose that $0 < \widetilde{\lambda}, \widetilde{\mu} < \infty$. Let B be the ellipse of the form (7.42). Then,*

$$m_{11}^{12} = m_{22}^{12} = 0 , \qquad (7.59)$$

and $m_{11}^{11}, m_{22}^{22}, m_{22}^{11}, m_{12}^{12}$ can be computed using (7.48), (7.51), and (7.58). The remaining terms are determined by the symmetry properties (7.26). The EMT's for rotated ellipses can be found using (7.38).

Proof. It suffices to show (7.59). Since the coefficients of (7.48) are real, $E_1(\alpha,\beta) = F_1(\alpha,\beta) = 0$ if α and β are purely imaginary, and $E_2(\alpha,\beta) = F_2(\alpha,\beta) = 0$ if α and β are real. Thus (7.59) follows from (7.51) and (7.58). This completes the proof. $\square$

Since the meaning of (7.48) is not clear when $m = 0$, *i.e.*, when B is a disk, we now compute the EMT for a disk. If B is a disk of radius one, then we can easily check that (7.13) admits a unique solution $\mathbf{u} = (u,v)$ given by φ and ψ that are defined by

$$\varphi_e(z) = \alpha z + \frac{A}{z} , \quad |z| > 1 ,$$
$$\psi_e(z) = \beta z + \frac{B}{z} + \frac{C}{z^3} , \quad |z| > 1,$$
$$\varphi_i(z) = Ez , \quad |z| < 1 , \qquad (7.60)$$
$$\psi_i(z) = Fz , \quad |z| < 1 ,$$

where the coefficients A, B, C, E, F satisfy

$$\begin{cases} A = C = \dfrac{\widetilde{\mu} - \mu}{\kappa \widetilde{\mu} + \mu}\bar{\beta} , \\[2mm] B = \dfrac{\mu(\widetilde{\kappa} + 1)}{\mu - \widetilde{\mu}}E - \bar{\alpha} - \dfrac{\kappa\widetilde{\mu} + \mu}{\mu - \widetilde{\mu}}\alpha , \\[2mm] E = \dfrac{\widetilde{\mu}(\kappa + 1)}{(\widetilde{\kappa} - 1)\mu + 2\widetilde{\mu}}\Re\alpha + i\dfrac{\widetilde{\mu}(\kappa + 1)}{\mu(\widetilde{\kappa} + 1)}\Im\alpha , \\[2mm] F = \dfrac{\widetilde{\mu}(\kappa + 1)}{\kappa\widetilde{\mu} + \mu}\beta . \end{cases} \qquad (7.61)$$

We then obtain the following theorem from (7.51) and (7.58).

Theorem 7.15 *Let B be a disk. Then*

$$
\begin{cases}
m_{22}^{11} = |B|\mu \left[\dfrac{(\widetilde{\kappa}-1)(\kappa+1)(\widetilde{\lambda}-\lambda+\widetilde{\mu}-\mu)}{(\widetilde{\kappa}\mu+2\widetilde{\mu}-\mu)(\kappa-1)} - \dfrac{(\widetilde{\mu}-\mu)(\kappa+1)}{\kappa\widetilde{\mu}+\mu} \right] , \\[4mm]
m_{12}^{12} = |B|\mu \dfrac{(\kappa+1)(\widetilde{\mu}-\mu)}{\kappa\widetilde{\mu}+\mu} .
\end{cases}
$$

The remaining terms are determined by (7.40) and the symmetry properties (7.26).

7.5 EMT's for Elliptic Holes and Hard Ellipses

In this section we compute EMT's for elliptic holes and hard inclusions of elliptic shape. By a hole we mean that $\widetilde{\lambda} = \widetilde{\mu} = 0$ while by a hard inclusion we mean that $\widetilde{\mu} \to \infty$. In other words, Young's modulus tends to ∞, while Poisson's ratio tends to 0. Young's modulus, E, and Poisson's ratio, ν, are defined to be

$$
E = \frac{\mu(2\mu + d\lambda)}{\lambda + \mu}, \quad \nu = \frac{\lambda}{2(\lambda + \mu)} .
$$

We note that the EMT's associated with elliptic holes with $\widetilde{\lambda} = \widetilde{\mu} = 0$ were computed in [196] and [210].

Let us deal with the hard inclusions first. If $\widetilde{\mu} = \infty$, then we obtain from (7.48) that

$$
\begin{cases}
\kappa\alpha - \left(\dfrac{\overline{A}}{m} + \overline{B} \right) = 0 , \\[3mm]
\alpha + \left(\dfrac{\overline{A}}{m} + \overline{B} \right) = E + \overline{E} + m\overline{F} , \\[3mm]
\kappa A - (m\overline{\alpha} + \overline{\beta}) = 0 , \\[3mm]
A + (m\overline{\alpha} + \overline{\beta}) = m(E + \overline{E}) + \overline{F} ,
\end{cases}
$$

and hence

$$
\begin{cases}
E + \overline{E} + m\overline{F} = (\kappa + 1)\alpha , \\[3mm]
m(E + \overline{E}) + \overline{F} = \dfrac{\kappa + 1}{\kappa}(m\overline{\alpha} + \overline{\beta}) .
\end{cases}
$$

Thus we get

$$
\begin{cases}
E + \overline{E} = \dfrac{\kappa + 1}{1 - m^2} \left[\alpha - \dfrac{m^2}{\kappa}\overline{\alpha} - \dfrac{m}{\kappa}\overline{\beta} \right] , \\[3mm]
\overline{F} = \dfrac{\kappa + 1}{1 - m^2} \left[-m\alpha + \dfrac{m}{\kappa}\overline{\alpha} + \dfrac{1}{\kappa}\overline{\beta} \right] .
\end{cases}
$$

Observe that the first equation has a solution only when α and β are real. As $\widetilde{\mu} \to \infty$, $\widetilde{\kappa} \to 3$, and hence we obtain from (7.58) that

$$\frac{1}{|B|}\left[m_{pq}(\alpha,\beta) - \widetilde{m}_{pq}(\alpha,\beta)\right] = \begin{cases} -2E_1 + F_1 & \text{if } p = q = 1, \\ -F_2 & \text{if } p \neq q, \\ -2E_1 - F_1 & \text{if } p = q = 2. \end{cases} \tag{7.62}$$

If $\alpha = \mu/(\kappa - 1)$ and $\beta = -\mu$, then

$$\begin{cases} E + \overline{E} = \dfrac{\mu(\kappa + 1)}{1 - m^2}\left[\dfrac{1}{\kappa - 1} - \dfrac{m^2}{\kappa(\kappa - 1)} + \dfrac{m}{\kappa}\right], \\ \overline{F} = \dfrac{\mu(\kappa + 1)}{1 - m^2}\left[-\dfrac{m}{\kappa - 1} + \dfrac{m}{\kappa(\kappa - 1)} - \dfrac{1}{\kappa}\right]. \end{cases}$$

Thus we arrive at

$$m_{11}^{11} = -m_{pq}(\frac{\mu}{\kappa - 1}, -\mu) + \widetilde{m}_{pq}(\frac{\mu}{\kappa - 1}, -\mu) = |B|\frac{\mu(\kappa + 1)(m - 2\kappa + 1)}{(m - 1)\kappa(\kappa - 1)}.$$

Similarly we can compute m_{pq}^{ij} using (7.51) and (7.62). The result of computations is summarized in the following theorem.

Theorem 7.16 *Let B be the ellipse of the form (7.42) and suppose that $\widetilde{\mu} = \infty$. Then, in addition to (7.59),*

$$\begin{cases} m_{11}^{11} = |B|\dfrac{\mu(\kappa + 1)(m - 2\kappa + 1)}{(m - 1)\kappa(\kappa - 1)}, \\ m_{22}^{22} = |B|\dfrac{\mu(\kappa + 1)(1 - m - 2\kappa)}{(m - 1)\kappa(\kappa - 1)}, \\ m_{22}^{11} = |B|\dfrac{\kappa + 1}{\kappa(1 - \kappa)}, \\ m_{12}^{12} = |B|\dfrac{\mu(\kappa + 1)}{(1 + m)\kappa}. \end{cases} \tag{7.63}$$

The remaining terms are determined by the symmetry properties (7.26). The EMT's for rotated ellipses can be found using (7.38).

Let us now compute the EMT's for holes. To this end, we need to change the presentation of formula (7.58). By equating the first and third equations in (7.48), we obtain from (7.56) and (7.57) that

$$m_{pq}(\alpha,\beta) - \widetilde{m}_{pq}(\alpha,\beta)$$
$$= \frac{\pi}{(1 - m^2)\mu}\Re\left[2\Re(a - \widetilde{a})\left(\kappa(\alpha - mA) + (m^2\bar{\alpha} - \frac{\bar{A}}{m} + m\bar{\beta} - \bar{B})\right)\right.$$
$$\left. + (\bar{b} - \bar{\widetilde{b}})\left(\kappa(m\alpha - A) + (m\bar{\alpha} - \bar{A} + \bar{\beta} - m\bar{B})\right)\right]. \tag{7.64}$$

If $\widetilde{\lambda} = \widetilde{\mu} = 0$, then $E = F = 0$ in (7.48). Thus we get

$$\begin{cases} \alpha + \left(\dfrac{\overline{A}}{m} + \overline{B} \right) = 0 \,, \\[2mm] A + \left(m\overline{\alpha} + \overline{\beta} \right) = 0 \,, \\[2mm] (m^2 + 1)\alpha - \left(m + \dfrac{1}{m} \right) A + C = 0 \,. \end{cases} \tag{7.65}$$

Since $\widetilde{a} = \widetilde{b} = 0$, it follows from (7.64) and (7.65) that

$$m_{pq}(\alpha, \beta) = \frac{\pi(\kappa + 1)}{\mu} \Re \left[\bar{a}(2\Re a + \bar{b}m) + (m\alpha + \beta)(2\Re am + \bar{b}) \right] .$$

We now obtain the following theorem from (7.51) and (7.53) after elementary but tedious computations.

Theorem 7.17 *Let B be the ellipse of the form (7.42) and suppose that $\widetilde{\lambda} = \widetilde{\mu} = 0$. Then, in addition to (7.59),*

$$\begin{cases} m_{11}^{11} = -|B|\dfrac{\mu(\kappa + 1)}{(\kappa - 1)^2}[2(1 + m^2) - 4m(\kappa - 1) + (\kappa - 1)^2] \,, \\[2mm] m_{22}^{22} = -|B|\dfrac{\mu(\kappa + 1)}{(\kappa - 1)^2}[2(1 + m^2) + 4m(\kappa - 1) + (\kappa - 1)^2] \,, \\[2mm] m_{22}^{11} = |B|\dfrac{\mu(\kappa + 1)}{(\kappa - 1)^2}[-2(1 + m^2) + (\kappa - 1)^2] \,, \\[2mm] m_{12}^{12} = -|B|\mu(\kappa + 1) \,. \end{cases} \tag{7.66}$$

The remaining terms are determined by the symmetry properties (7.26). The EMT's for rotated ellipses can be found using (7.38).

As an immediate consequence of Theorem 7.16 and Theorem 7.17, we get the following result.

Corollary 7.18 *Let B be a disk. If $\widetilde{\lambda} = \widetilde{\mu} = 0$, then*

$$\begin{cases} m_{22}^{11} = |B|\dfrac{\mu(\kappa + 1)}{(\kappa - 1)^2}[-2 + (\kappa - 1)^2] \,, \\[2mm] m_{12}^{12} = -|B|\mu(\kappa + 1) \,. \end{cases}$$

If $\widetilde{\mu} = \infty$, then

$$\begin{cases} m_{22}^{11} = |B|\dfrac{\kappa + 1}{\kappa(1 - \kappa)} \,, \\[2mm] m_{12}^{12} = |B|\dfrac{\mu(\kappa + 1)}{\kappa} \,. \end{cases}$$

8

Derivation of Full Asymptotic Expansions

We suppose that the elastic medium occupies a bounded domain Ω in $\mathbb{R}^d$, with a connected Lipschitz boundary $\partial\Omega$. Let the constants (λ, μ) denote the background Lamé coefficients, that are the elastic parameters in the absence of any inclusions. Suppose that the elastic inclusion D in Ω is given by

$$D = \epsilon B + z \,, \tag{8.1}$$

where B is a bounded Lipschitz domain in $\mathbb{R}^d$. We assume that there exists $c_0 > 0$ such that

$$\inf_{x \in D} \operatorname{dist}(x, \partial\Omega) > c_0 \,.$$

Suppose that D has the pair of Lamé constants $(\widetilde{\lambda}, \widetilde{\mu})$ satisfying (6.18) and (6.19).

The purpose of this chapter is to establish a complete asymptotic formula for the displacement vector in terms of the reference Lamé constants, the location of the inclusion and its geometry. Our derivation is rigorous, and based on layer potential techniques. The asymptotic expansions in this chapter are valid for an elastic inclusion with Lipschitz boundaries.

8.1 Full Asymptotic Expansions

We first observe that if D is of the form (8.1), then the Lipschitz character of D, and hence the constant C in (6.28), depends on ϵ. However, for such a domain, we can obtain the following lemma by scaling the integral equation (6.22) and the estimate (6.28).

Lemma 8.1 *For any given* $(\mathbf{F}, \mathbf{G}) \in W_1^2(\partial D) \times L^2(\partial D)$*, let* $(\mathbf{f}, \mathbf{g}) \in L^2(\partial D) \times L^2(\partial D)$ *be the solution of (6.22). Then there exists a constant C depending only on λ, μ, $\widetilde{\lambda}$, $\widetilde{\mu}$, and the Lipschitz character of B, but not on ϵ, such that*

$$\|\mathbf{g}\|_{L^2(\partial D)} \le C\big(\epsilon^{-1}\|\mathbf{F}\|_{L^2(\partial D)} + \|\frac{\partial \mathbf{F}}{\partial T}\|_{L^2(\partial D)} + \|\mathbf{G}\|_{L^2(\partial D)}\big) \,. \tag{8.2}$$

Here $\partial/\partial T$ denotes the tangential derivative.

Proof.   Assuming without loss of generality that $z = 0$, we scale $x = \epsilon y$, $y \in B$. Let $\mathbf{f}_\epsilon(y) = \mathbf{f}(\epsilon y)$, $y \in \partial B$, etc. Let (φ, ψ) be the solution to the integral equation

$$\begin{cases} \widetilde{\mathcal{S}}_B \varphi|_- - \mathcal{S}_B \psi|_+ = \epsilon^{-1} \mathbf{F}_\epsilon & \text{on } \partial B , \\ \dfrac{\partial}{\partial \widetilde{\nu}} \widetilde{\mathcal{S}}_B \varphi \Big|_- - \dfrac{\partial}{\partial \nu} \mathcal{S}_B \psi \Big|_+ = \mathbf{G}_\epsilon & \text{on } \partial B . \end{cases}$$

Following the lines of the proof of (7.36), we can show that

$$\mathbf{g}_\epsilon = \psi .$$

It then follows from (6.28) that

$$\|\mathbf{g}_\epsilon\|_{L^2(\partial B)} = \|\psi_\epsilon\|_{L^2(\partial B)} \le C\left(\|\epsilon^{-1} \mathbf{F}_\epsilon\|_{W_1^2(\partial B)} + \|\mathbf{G}_\epsilon\|_{L^2(\partial B)} \right) ,$$

where C does not depend on ϵ. By scaling back using $x = \epsilon y$, we obtain (8.2). This completes the proof. $\square$

Let $\mathbf{u}$ be the solution of (6.21). In this chapter, we derive an asymptotic formula for $\mathbf{u}$ as ϵ goes to 0 in terms of the background solution $\mathbf{U}$. The background solution is the solution of (6.13).

Recall that $\mathbf{u}$ is represented as

$$\mathbf{u}(x) = \mathbf{U}(x) - N_D \psi(x) , \quad x \in \partial\Omega , \tag{8.3}$$

where ψ is defined by (6.28). See (6.38). Let $\mathbf{H}$ be the function defined in (6.27). For a given integer n, define $\mathbf{H}^{(n)}$ by

$$\mathbf{H}^{(n)}(x) := \sum_{|\alpha|=0}^{n} \frac{1}{\alpha!} \partial^\alpha \mathbf{H}(z)(x-z)^\alpha$$

$$= \left(\sum_{|\alpha|=0}^{n} \frac{1}{\alpha!} \partial^\alpha H_1(z)(x-z)^\alpha , \dots , \sum_{|\alpha|=0}^{n} \frac{1}{\alpha!} \partial^\alpha H_d(z)(x-z)^\alpha \right)$$

$$= \sum_{j=1}^{d} \sum_{|\alpha|=0}^{n} \frac{1}{\alpha!} \partial^\alpha H_j(z)(x-z)^\alpha \mathbf{e}_j .$$

Define φ_n and ψ_n in $L^2(\partial D)$ by

$$\begin{cases} \widetilde{\mathcal{S}}_D \varphi_n|_- - \mathcal{S}_D \psi_n|_+ = \mathbf{H}^{(n)}|_{\partial D} , \\ \dfrac{\partial}{\partial \widetilde{\nu}} \widetilde{\mathcal{S}}_D \varphi_n \Big|_- - \dfrac{\partial}{\partial \nu} \mathcal{S}_D \psi_n \Big|_+ = \dfrac{\partial \mathbf{H}^{(n)}}{\partial \nu} \Big|_{\partial D} , \end{cases}$$

and set
$$\varphi := \varphi_n + \varphi_R, \quad \text{and} \quad \psi := \psi_n + \psi_R \,.$$

Since (φ_R, ψ_R) is the solution of the integral equation (6.22) with $\mathbf{F} = \mathbf{H} - \mathbf{H}^{(n)}$ and $\mathbf{G} = \partial(\mathbf{H} - \mathbf{H}^{(n)})/\partial\nu$, it follows from (8.2) that

$$\|\psi_R\|_{L^2(\partial D)} \le C\big(\epsilon^{-1}\|\mathbf{H} - \mathbf{H}^{(n)}\|_{L^2(\partial D)} + \|\nabla(\mathbf{H} - \mathbf{H}^{(n)})\|_{L^2(\partial D)}\big) \,. \quad (8.4)$$

By (6.30) we get

$$\epsilon^{-1}\|\mathbf{H} - \mathbf{H}^{(n)}\|_{L^2(\partial D)} + \|\nabla(\mathbf{H} - \mathbf{H}^{(n)})\|_{L^2(\partial D)}$$
$$\le |\partial D|^{1/2}\Big[\epsilon^{-1}\|\mathbf{H} - \mathbf{H}^{(n)}\|_{L^\infty(\partial D)} + \|\nabla(\mathbf{H} - \mathbf{H}^{(n)})\|_{L^\infty(\partial D)}\Big]$$
$$\le \|\mathbf{H}\|_{\mathcal{C}^{n+1}(\overline{D})}\epsilon^n|\partial D|^{1/2}$$
$$\le C\|\mathbf{g}\|_{L^2(\partial\Omega)}\epsilon^n|\partial D|^{1/2} \,.$$

It then follows from (8.4) that

$$\|\psi_R\|_{L^2(\partial D)} \le C\|\mathbf{g}\|_{L^2(\partial\Omega)}\epsilon^n|\partial D|^{1/2} \,, \quad (8.5)$$

where C is independent of ϵ.

By (8.3) we get

$$\mathbf{u}(x) = \mathbf{U}(x) - N_D\psi_n(x) - N_D\psi_R(x) \,, \quad x \in \partial\Omega \,. \quad (8.6)$$

The first two terms in (8.6) are the main terms in our asymptotic expansion and the last term is the error term. We claim that the error term is $O(\epsilon^{n+d})$. In fact, since $\psi, \psi_n \in L^2_\Psi(\partial D)$, in particular, $\int_{\partial D} \psi \, d\sigma = \int_{\partial D} \psi_n \, d\sigma = 0$, we get $\int_{\partial D} \psi_R \, d\sigma = 0$. It then follows from (8.5) that, for $x \in \partial\Omega$,

$$|N_D\psi_R(x)| = \left|\int_{\partial D} \Big(\mathbf{N}(x - y) - \mathbf{N}(x - z)\Big)\psi_R(y) \, d\sigma(y)\right|$$
$$\le C\epsilon|\partial D|^{1/2}\|\psi_R\|_{L^2(\partial D)}$$
$$\le C\|\mathbf{g}\|_{L^2(\partial\Omega)}\epsilon^{n+d} \,.$$

In order to expand out the second term in (8.6), we first define some auxiliary functions. Let $D_0 := D - z$, the translate of D by $-z$. For multi-index $\alpha \in \mathbb{N}^d$ and $j = 1, \dots, d$, define φ_α^j and ψ_α^j by

$$\begin{cases} \widetilde{\mathcal{S}}_{D_0}\varphi_\alpha^j\big|_- - \mathcal{S}_{D_0}\psi_\alpha^j\big|_+ = x^\alpha\mathbf{e}_j\big|_{\partial D_0} \,, \\ \dfrac{\partial}{\partial\nu}\widetilde{\mathcal{S}}_{D_0}\varphi_\alpha^j\bigg|_- - \dfrac{\partial}{\partial\nu}\mathcal{S}_{D_0}\psi_\alpha^j\bigg|_+ = \dfrac{\partial(x^\alpha\mathbf{e}_j)}{\partial\nu}\big|_{\partial D_0} \,. \end{cases} \quad (8.7)$$

Then the linearity and the uniqueness of the solution to (8.7) yield

$$\psi_n(x) = \sum_{j=1}^{d} \sum_{|\alpha|=0}^{n} \frac{1}{\alpha!} \partial^\alpha H_j(z) \psi_\alpha^j(x-z) \,, \quad x \in \partial D \,.$$

Recall that $D_0 = \epsilon B$ and $(\mathbf{f}_\alpha^j, \mathbf{g}_\alpha^j)$ is the solution of (7.4). Then, following the same lines of the proof of (7.36), we can see that

$$\psi_\alpha^j(x) = \epsilon^{|\alpha|-1} \mathbf{g}_\alpha^j(\epsilon^{-1}x) \,,$$

and hence

$$\psi_n(x) = \sum_{j=1}^{d} \sum_{|\alpha|=0}^{n} \frac{1}{\alpha!} \partial^\alpha H_j(z) \epsilon^{|\alpha|-1} \mathbf{g}_\alpha^j(\epsilon^{-1}(x-z)) \,, \quad x \in \partial D \,.$$

We thus get

$$N_D\psi_n(x) = \sum_{j=1}^{d} \sum_{|\alpha|=0}^{n} \frac{1}{\alpha!} \partial^\alpha H_j(z) \epsilon^{|\alpha|+d-2} \int_{\partial B} \mathbf{N}(x, z+\epsilon y) \mathbf{g}_\alpha^j(y) \, d\sigma(y). \quad (8.8)$$

We now consider the asymptotic expansion of $\mathbf{N}(x, z+\epsilon y)$ as $\epsilon \to 0$. We remind the reader that $x \in \partial\Omega$ and $z + \epsilon y \in \partial D$. By (6.37) we have the following relation:

$$\left(-\frac{1}{2}I + \mathcal{K}_\Omega\right)\left[\mathbf{N}(\cdot, \epsilon y + z)\right](x) = \mathbf{\Gamma}(x - z - \epsilon y) \,, \quad x \in \partial\Omega, \quad \text{modulo } \Psi \,.$$

Since

$$\mathbf{\Gamma}(x - \epsilon y) = \sum_{|\beta|=0}^{+\infty} \frac{1}{\beta!} \epsilon^{|\beta|} \partial^\beta(\mathbf{\Gamma}(x)) y^\beta \,,$$

we get, modulo Ψ,

$$\left(-\frac{1}{2}I + \mathcal{K}_\Omega\right)\left[\mathbf{N}(\cdot, \epsilon y + z)\right](x) = \sum_{|\beta|=0}^{+\infty} \frac{1}{\beta!} \epsilon^{|\beta|} \partial^\beta(\mathbf{\Gamma}(x-z)) y^\beta$$

$$= \left(-\frac{1}{2}I + \mathcal{K}_\Omega\right)\left[\sum_{|\beta|=0}^{+\infty} \frac{1}{\beta!} \epsilon^{|\beta|} \partial_z^\beta \mathbf{N}(\cdot, z) y^j\right](x) \,.$$

Since $\mathbf{N}(\cdot, w) \in L_\Psi^2(\partial\Omega)$ for all $w \in \Omega$, we have the following asymptotic expansion of the Neumann function.

Lemma 8.2 *For $x \in \partial\Omega$, $z \in \Omega$, $y \in \partial B$, and $\epsilon \to 0$,*

$$\mathbf{N}(x, \epsilon y + z) = \sum_{|\beta|=0}^{+\infty} \frac{1}{\beta!} \epsilon^{|\beta|} \partial_z^\beta \mathbf{N}(x, z) y^\beta \,.$$

It then follows from (8.8) that

$$N_D \psi_n(x)$$
$$= \sum_{j=1}^{d} \sum_{|\alpha|=0}^{n} \frac{1}{\alpha!} \partial^\alpha H_j(z) \epsilon^{|\alpha|+d-2} \sum_{|\beta|=0}^{+\infty} \frac{1}{\beta!} \epsilon^{|\beta|} \partial_z^\beta \mathbf{N}(x,z) \int_{\partial B} y^\beta \mathbf{g}_\alpha^j(y) \, d\sigma(y) \ .$$

Note that $\sum_{j=1}^{d} \sum_{|\alpha|=l} \frac{1}{\alpha!} \partial^\alpha H_j(z) \mathbf{g}_\alpha^j$ is the solution of (6.28) when the right-hand side is given by the function

$$\mathbf{u} = \sum_{j=1}^{d} \sum_{|\alpha|=l} \frac{1}{\alpha!} \partial^\alpha H_j(z) x^\alpha \mathbf{e}_j \ .$$

Moreover, this function is a solution of $\mathcal{L}_{\lambda,\mu} \mathbf{u} = 0$ in B and therefore, $\partial \mathbf{u} / \partial \nu|_{\partial B} \in L_\Psi^2(\partial B)$. Hence, by Theorem 6.13, we obtain that

$$\sum_{j=1}^{d} \sum_{|\alpha|=l} \frac{1}{\alpha!} \partial^\alpha H_j(z) \mathbf{g}_\alpha^j \in L_\Psi^2(\partial B) \ .$$

In particular, we have

$$\sum_{j=1}^{d} \sum_{|\alpha|=l} \frac{1}{\alpha!} \partial^\alpha H_j(z) \int_{\partial B} \mathbf{g}_\alpha^j(y) \, d\sigma(y) = 0 \quad \forall \, l \ .$$

On the other hand, $\mathbf{g}_0^j = 0$ by Lemma 6.14. We finally obtain by combining these facts with the above identity that

$$N_D \psi_n(x) = \sum_{j=1}^{d} \sum_{|\alpha|=1}^{n} \sum_{|\beta|=1}^{+\infty} \frac{\epsilon^{|\alpha|+|\beta|+d-2}}{\alpha!\beta!} \partial^\alpha H_j(z) \partial_z^\beta \mathbf{N}(x,z) \int_{\partial B} y^\beta \mathbf{g}_\alpha^j(y) \, d\sigma(y) \ .$$
$$(8.9)$$

We then obtain from the definition of the elastic moment tensors, (8.6), and (8.9) that

$$\mathbf{u}(x) = \mathbf{U}(x) - \sum_{j=1}^{d} \sum_{|\alpha|=1}^{n} \sum_{|\beta|=1}^{n-|\alpha|+1} \frac{\epsilon^{|\alpha|+|\beta|+d-2}}{\alpha!\beta!} \partial^\alpha H_j(z) \partial_z^\beta \mathbf{N}(x,z) M_{\alpha\beta}^j \quad (8.10)$$
$$+ O(\epsilon^{n+d}) \ , \quad x \in \partial\Omega \ .$$

Observe that formula (8.10) still uses the function $\mathbf{H}$ which depends on ϵ. Therefore the remaining task is to transform this formula into a formula which is expressed using only the background solution $\mathbf{U}$.

By (6.7), $\mathbf{U} = -\mathcal{S}_\Omega(\mathbf{g}) + \mathcal{D}_\Omega(\mathbf{U}|_{\partial\Omega})$ in Ω. Thus substitution of (8.10) into (6.27) yields that, for any $x \in \Omega$,

$$\mathbf{H}(x) = -\mathcal{S}_\Omega(\mathbf{g})(x) + \mathcal{D}_\Omega(\mathbf{u}|_{\partial\Omega})(x)$$

$$= \mathbf{U}(x) - \sum_{j=1}^{d} \sum_{|\alpha|=1}^{n} \sum_{|\beta|=1}^{n-|\alpha|+1} \frac{\epsilon^{|\alpha|+|\beta|+d-2}}{\alpha!\beta!} \partial^\alpha H_j(z) \mathcal{D}_\Omega(\partial_z^\beta \mathbf{N}(.,z))(x) M_{\alpha\beta}^j$$

$$+ O(\epsilon^{n+d}) . \tag{8.11}$$

In (8.11) the remainder $O(\epsilon^{n+d})$ is uniform in the $\mathcal{C}^k$-norm on any compact subset of Ω for any k and therefore

$$\partial^\gamma \mathbf{H}(z) + \sum_{j=1}^{d} \sum_{|\alpha|=1}^{n} \sum_{|\beta|=1}^{n-|\alpha|+1} \epsilon^{|\alpha|+|\beta|+d-2} \partial^\alpha H_j(z) P_{\alpha\beta\gamma}^j = \partial^\gamma \mathbf{U}(z) + O(\epsilon^{n+d}) ,$$

$$\tag{8.12}$$

for all $\gamma \in \mathbb{N}^d$ with $|\gamma| \leq n$, where

$$P_{\alpha\beta\gamma}^j = \frac{1}{\alpha!\beta!} \partial^\gamma \mathcal{D}_\Omega(\partial_z^\beta \mathbf{N}(.,z))(x)|_{x=z} M_{\alpha\beta}^j .$$

We now introduce a linear transform which transforms $\partial^\alpha \mathbf{H}(z)$ to $\partial^\alpha \mathbf{U}(z)$. Let

$$N := d \sum_{k=1}^{n} \frac{(k+1)(k+2)}{2} ,$$

and define the linear transform $\mathcal{P}_\epsilon$ on $\mathbb{R}^N$ by

$$\mathcal{P}_\epsilon : (\mathbf{v}_\gamma)_{\gamma\in\mathbb{N}^d,|\gamma|\leq n} \mapsto \left(\mathbf{v}_\gamma + \sum_{j=1}^{d} \sum_{|\alpha|=1}^{n} \sum_{|\beta|=1}^{n-|\alpha|+1} \epsilon^{|\alpha|+|\beta|+d-2} v_\alpha^j P_{\alpha\beta\gamma}^j \right)_{\gamma\in\mathbb{N}^d,|\gamma|\leq n} .$$

Observe that

$$\mathcal{P}_\epsilon = I - \epsilon^d \mathcal{P}_1 - \ldots - \epsilon^{n+d-1} \mathcal{P}_n ,$$

where the definitions of $\mathcal{P}_j$ are obvious. Since ϵ is small, $\mathcal{P}_\epsilon$ is invertible. We now define $\mathcal{Q}_i$, $i = 1, \ldots, n-1$, by

$$\mathcal{P}_\epsilon^{-1} = I + \epsilon^d \mathcal{Q}_1 + \ldots + \epsilon^{n+d-1} \mathcal{Q}_n + O(\epsilon^{n+d}) .$$

It then follows from (8.12) that

$$((\partial^\gamma \mathbf{H})(z))_{|\gamma|\leq n} = (I + \sum_{i=1}^{n-d} \epsilon^{i+2} \mathcal{Q}_i)(((\partial^\gamma \mathbf{U})(z))_{|\gamma|\leq n}) + O(\epsilon^n) ,$$

which yields the main result of this chapter.

Theorem 8.3 *Let* $\mathbf{u}$ *be the solution of (6.19) and* $\mathbf{U}$ *is the background solution. The following pointwise asymptotic expansion on* $\partial\Omega$ *holds:*

$$\mathbf{u}(x) = \mathbf{U}(x)$$

$$-\sum_{j=1}^{d} \sum_{|\alpha|=1}^{n} \sum_{|\beta|=1}^{n-|\alpha|+1} \frac{\epsilon^{|\alpha|+|\beta|+d-2}}{\alpha!\beta!} \left((I + \sum_{i=1}^{n-d} \epsilon^{i+2} \mathcal{Q}_i)((\partial^\gamma \mathbf{U})(z)) \right)_\alpha^j \partial_z^\beta \mathbf{N}(x,z) M_{\alpha\beta}^j$$

$$+ O(\epsilon^{n+d}), \quad x \in \partial\Omega.$$

$$(8.13)$$

The operator $\mathcal{Q}_j$ describes the interaction between the inclusion and $\partial\Omega$. It is interesting to compare (8.13) with formula (7.8) in the free space. In (7.8) no $\mathcal{Q}_j$ involves. It is because the free space does not have any boundary.

If $n = d$, (8.13) takes simpler form: for $x \in \partial\Omega$,

$$\mathbf{u}(x) = \mathbf{U}(x) - \sum_{j=1}^{d} \sum_{|\alpha|=1}^{d} \sum_{|\beta|=1}^{d+1-|\alpha|} \frac{\epsilon^{|\alpha|+|\beta|+d-2}}{\alpha!\beta!} (\partial^\alpha U_j)(z) \partial_z^\beta \mathbf{N}(x,z) M_{\alpha\beta}^j + O(\epsilon^{2d}).$$

$$(8.14)$$

Observe that no $\mathcal{Q}_j$ appears in (8.14). This is because D is well-separated from $\partial\Omega$.

The coefficient of the leading-order term, namely, the ϵ^d-term of the expansion is

$$\sum_{j,p,q=1}^{d} (\partial_p U_j)(z) \partial_{z_q} N_{ij}(x,z) m_{pq}^{ij}, \quad i = 1,\dots,d.$$

By Theorem 7.6, this term is bounded below and above by constant multiples of

$$\|\nabla U(z)\| \left[\sum_{q,j=1}^{d} |\partial_{z_q} N_{ij}(x,z)|^2 \right]^{1/2},$$

for $i = 1,\dots,d$, and those constants are independent of ϵ.

When there are multiple well-separated inclusions

$$D_s = \epsilon B_s + z_s, \quad s = 1, \cdots, m,$$

where $|z_s - z_{s'}| > 2c_0$ for some $c_0 > 0$, $s \neq s'$, then by iterating formula (8.14), we obtain the following theorem.

Theorem 8.4 *The following asymptotic expansion holds uniformly for* $x \in \partial\Omega$:

$$\mathbf{u}(x) = \mathbf{U}(x)$$

$$-\sum_{s=1}^{m} \sum_{j=1}^{d} \sum_{|\alpha|=1}^{d} \sum_{|\beta|=1}^{d+1-|\alpha|} \frac{\epsilon^{|\alpha|+|\beta|+d-2}}{\alpha!\beta!} (\partial^\alpha U_j)(z_s) \partial_z^\beta \mathbf{N}(x,z_s)(M^s)_{\alpha\beta}^j$$

$$+ O(\epsilon^{2d}),$$

where $(M^s)_{\alpha\beta}^j$ *are EMT's corresponding to* B_s, $s = 1, \cdots, m$.

Detection of Inclusions

As in the previous chapter, assume that the elastic inclusion D in Ω is given by $D = \epsilon B + z$, where B is a bounded Lipschitz domain in $\mathbb{R}^d$. In this chapter, we propose a method to detect the elastic moment tensors and the center z of D via a finite number of pairs of tractions and displacements caused by tractions measured on $\partial\Omega$. The reconstructed EMT will provide information on the size and some geometric features of the inclusion. Our methods use the asymptotic formula derived in the previous chapter.

9.1 Detection of EMT's

Given a traction $\mathbf{g} \in L^2_\Psi(\partial\Omega)$, let $\mathbf{H}[\mathbf{g}]$ be defined by

$$\mathbf{H}[\mathbf{g}](x) = -\mathcal{S}_\Omega(\mathbf{g})(x) + \mathcal{D}_\Omega(\mathbf{f})(x), \quad x \in \mathbb{R}^d \setminus \partial\Omega, \quad \mathbf{f} := \mathbf{u}|_{\partial\Omega}, \quad (9.1)$$

where $\mathbf{u}$ is the solution to (6.21), $\mathcal{S}_\Omega$ and $\mathcal{D}_\Omega$ are the single and double layer potentials for the Lamé system on $\partial\Omega$.

Theorem 9.1 *For $x \in \mathbb{R}^d \setminus \overline{\Omega}$,*

$$\mathbf{H}[\mathbf{g}](x) = -\sum_{j=1}^{d} \sum_{|\alpha|=1}^{d} \sum_{|\beta|=1}^{d+1-|\alpha|} \frac{\epsilon^{|\alpha|+|\beta|+d-2}}{\alpha!\beta!} (\partial^\alpha U_j)(z)\partial^\beta \mathbf{\Gamma}(x-z)M^j_{\alpha\beta}$$
$$+ O\left(\frac{\epsilon^{2d}}{|x|^{d-1}}\right), \quad (9.2)$$

where $\mathbf{U} = (U_1, \ldots, U_d)$ is the background solution, i.e., the solution to (6.13), $M^j_{\alpha\beta}$ are the elastic moment tensors associated with B, and $\mathbf{\Gamma}$ is the Kelvin matrix of fundamental solutions corresponding to the Lamé parameters (λ, μ).

Proof. Since $|\nabla\mathbf{\Gamma}(x-y)| = O(|x|^{1-d})$ as $|x| \to \infty$ for each y in a fixed bounded set, substituting (8.14) into (9.1) yields

$$\mathbf{H}[\mathbf{g}](x) = -\mathcal{S}_\Omega(\mathbf{g})(x) + \mathcal{D}_\Omega(\mathbf{U}|_{\partial\Omega})(x)$$

$$- \sum_{j=1}^{d} \sum_{|\alpha|=1}^{d} \sum_{|\beta|=1}^{d+1-|\alpha|} \frac{\epsilon^{|\alpha|+|\beta|+d-2}}{\alpha!\beta!} (\partial^\alpha U_j)(z) \mathcal{D}_\Omega(\partial_z^\beta \mathbf{N}(\cdot,z))(x) M_{\alpha\beta}^j$$

$$+ O(\frac{\epsilon^{2d}}{|x|^{d-1}}) \,.$$

But $\partial\mathbf{U}/\partial\nu = \mathbf{g}$ on $\partial\Omega$. Therefore, it follows from (6.8) that

$$-\mathcal{S}_\Omega(\mathbf{g})(x) + \mathcal{D}_\Omega(\mathbf{U}|_{\partial\Omega})(x) = 0 \quad \text{for } x \in \mathbb{R}^d \setminus \overline{\Omega} \,.$$

From (6.9) and (6.37), we now obtain

$$\mathcal{D}_\Omega(\mathbf{N}(\cdot,z))|_+(x) = (-\frac{1}{2}I + \mathcal{K}_\Omega)(\mathbf{N}(\cdot,z))(x) = \mathbf{\Gamma}(x-z) \,, \ x \in \partial\Omega \,, \text{ modulo } \Psi \,.$$

By $\mathcal{D}_\Omega(\mathbf{N}(\cdot,z))(x) = O(|x|^{1-d})$ and $\mathbf{\Gamma}(x-z) - \mathbf{\Gamma}(z) = O(|x|^{1-d})$ as $|x| \to \infty$, we have the identity

$$\mathcal{D}_\Omega(\mathbf{N}(\cdot,z))(x) = \mathbf{\Gamma}(x-z) - \mathbf{\Gamma}(z) \,, \quad x \in \mathbb{R}^d \setminus \overline{\Omega} \,,$$

from which we conclude that

$$\mathcal{D}_\Omega(\partial_z^\beta \mathbf{N}(\cdot,z))(x) = \partial_z^\beta \mathcal{D}_\Omega(\mathbf{N}(\cdot,z))(x) = \partial_z^\beta \mathbf{\Gamma}(x-z) \,, \quad |\beta| \geq 1 \,,$$

and hence (9.2) is immediate. This completes the proof. $\square$

If $\mathbf{g} = \partial\mathbf{U}/\partial\nu|_{\partial\Omega}$ where $\mathbf{U}$ is linear, then $\partial^\alpha\mathbf{U} = 0$ if $|\alpha| > 1$ and therefore,

$$\mathbf{H}[\mathbf{g}](x) = -\sum_{j=1}^{d} \sum_{|\alpha|=1}^{d} \sum_{|\beta|=1}^{d} \frac{\epsilon^{|\beta|+d-1}}{\beta!} (\partial^\alpha U_j)(z) \partial^\beta \mathbf{\Gamma}(x-z) M_{\alpha\beta}^j + O(\frac{\epsilon^{2d}}{|x|^{d-1}}) \,.$$

Since $\partial^\beta \mathbf{\Gamma}(x-z) = O(|x|^{-d+2-|\beta|})$ as $|x| \to \infty$ if $|\beta| \geq 1$, we get

$$\mathbf{H}[\mathbf{g}](x) = -\epsilon^d \sum_{j=1}^{d} \sum_{|\alpha|=1}^{d} \sum_{|\beta|=1}^{d} (\partial^\alpha U_j)(z) \partial^\beta \mathbf{\Gamma}(x-z) M_{\alpha\beta}^j + O(\frac{\epsilon^{d+1}}{|x|^d})$$

$$+ O(\frac{\epsilon^{2d}}{|x|^{d-1}}) \,,$$

or equivalently, for $k = 1, \ldots, d$,

$$H_k[\mathbf{g}](x) = -\epsilon^d \sum_{i,j,p,q=1}^{d} (\partial_i U_j)(z) \partial_p \Gamma_{kq}(x-z) m_{pq}^{ij} + O(\frac{\epsilon^{d+1}}{|x|^d})$$

$$+ O(\frac{\epsilon^{2d}}{|x|^{d-1}}) \,. \tag{9.3}$$

Since $\partial_p\Gamma_{kq}(x-z) = \partial_p\Gamma_{kq}(x) + O(|x|^d)$, we obtain from (9.3) that

$$H_k[\mathbf{g}](x) = -\epsilon^d \sum_{i,j,p,q=1}^{d} (\partial_i U_j)(z)\partial_p\Gamma_{kq}(x)m_{pq}^{ij} + O(\frac{\epsilon^d}{|x|^d}) + O(\frac{\epsilon^{2d}}{|x|^{d-1}}) \,. \quad (9.4)$$

For a general $\mathbf{g}$, we have the following formula:

$$H_k[\mathbf{g}](x) = -\epsilon^d \sum_{i,j,p,q=1}^{d} (\partial_i U_j)(z)\partial_p\Gamma_{kq}(x-z)m_{pq}^{ij} + O(\frac{\epsilon^{d+1}}{|x|^{d-1}}) \,,$$

from which the following expansion is obvious:

$$H_k[\mathbf{g}](x) = -\epsilon^d \sum_{i,j,p,q=1}^{d} (\partial_i U_j)(z)\partial_p\Gamma_{kq}(x)m_{pq}^{ij} + O(\frac{\epsilon^d}{|x|^d}) + O(\frac{\epsilon^{d+1}}{|x|^{d-1}}) \,. \quad (9.5)$$

Finally, (9.4) and (9.5) yield the following far-field relations.

Theorem 9.2 *If* $|x| = O(\epsilon^{-1})$, *then, for* $k = 1,\ldots,d$,

$$|x|^{d-1}H_k[\mathbf{g}](x) = -\epsilon^d|x|^{d-1} \sum_{i,j,p,q=1}^{d} (\partial_i U_j)(z)\partial_p\Gamma_{kq}(x)m_{pq}^{ij} \,. \quad (9.6)$$

Moreover, if $\mathbf{U}$ *is linear, then for all* x *with* $|x| = O(\epsilon^{-d})$,

$$|x|^{d-1}H_k[\mathbf{g}](x) = -\epsilon^d|x|^{d-1} \sum_{i,j,p,q=1}^{d} (\partial_i U_j)(z)\partial_p\Gamma_{kq}(x)m_{pq}^{ij} \,. \quad (9.7)$$

Both identities hold modulo $O(\epsilon^{2d})$.

Theorem 9.3 (Reconstruction of EMT) *For* $u,v = 1,\ldots,d$, *let*

$$\mathbf{g}_{uv} := \frac{\partial}{\partial\nu}\left(\frac{x_u\mathbf{e}_v + x_v\mathbf{e}_u}{2}\right)\Big|_{\partial\Omega},$$

and define

$$h_{kl}^{uv} := \lim_{t\to\infty} t^{d-1}H_k[\mathbf{g}_{uv}](t\mathbf{e}_l), \quad k,l,u,v = 1,\ldots,d \,. \quad (9.8)$$

Then the entries m_{kl}^{uv}, $u,v,k,l = 1,\ldots,d$, *of the EMT can be recovered, modulo* $O(\epsilon^d)$, *as follows: for* $u,v,k,l = 1,\ldots,d$,

$$\epsilon^d m_{kl}^{uv} = \begin{cases} -\dfrac{2\omega_d\mu(\lambda + 2\mu)}{\lambda + (d-2)\mu}\left[\dfrac{\lambda+\mu}{\mu}\displaystyle\sum_{j=1}^{d} h_{jj}^{uv} + h_{kk}^{uv}\right] \\ \qquad\qquad\qquad\qquad\qquad \text{if } k = l \,, \\ -\omega_d(\lambda + 2\mu)h_{kl}^{uv} \quad \text{if } k \neq l \,, \end{cases} \quad (9.9)$$

modulo $O(\epsilon^{2d})$, *where* $\omega_d = 2\pi$ *if* $d = 2$ *and* $\omega_d = 4\pi$ *if* $d = 3$.

Proof. Easy computations show that

$$\partial_p \Gamma_{kq}(x) = \frac{A}{\omega_d}\frac{\delta_{kq}x_p}{|x|^d} - \frac{B}{\omega_d}\frac{\delta_{kp}x_q + \delta_{pq}x_k}{|x|^d} + \frac{dB}{\omega_d}\frac{x_k x_q x_p}{|x|^{d+2}} . \tag{9.10}$$

If $x = te_l$, $t \in \mathbb{R}$, $l = 1, \ldots, d$, then

$$\partial_p \Gamma_{kq}(te_l) = \frac{1}{\omega_d t^{d-1}}\Big[A\delta_{kq}\delta_{pl} - B(\delta_{kp}\delta_{ql} + \delta_{kl}\delta_{pq}) + dB\delta_{kl}\delta_{ql}\delta_{pl} \Big]$$

$$:= \frac{1}{\omega_d t^{d-1}} e_{klpq} . \tag{9.11}$$

The background solution $\mathbf{U}$ corresponding to $\mathbf{g}_{uv}$ is given by $\mathbf{U}(x) = \frac{1}{2}(x_u \mathbf{e}_v + x_v \mathbf{e}_u)$ and hence

$$\partial_i U_j(z) = \frac{1}{2}(\delta_{iu}\delta_{jv} + \delta_{iv}\delta_{ju}) . \tag{9.12}$$

Therefore the right-hand side of (9.7) equals

$$-\frac{\epsilon^d}{2\omega_d}\sum_{i,j,p,q=1}^{d}(\delta_{iu}\delta_{jv} + \delta_{iv}\delta_{ju})e_{klpq}m_{pq}^{ij} = -\frac{\epsilon^d}{\omega_d}\sum_{p,q=1}^{d}e_{klpq}m_{pq}^{uv} .$$

The last equality is valid because of the symmetry of the EMT, in particular, $m_{pq}^{uv} = m_{pq}^{vu}$. It then follows from (9.7) that if $t = O(\epsilon^{-1})$, then, modulo $O(\epsilon^{2d})$,

$$t^{d-1}H_k[\mathbf{g}_{uv}](te_l) = \begin{cases} -\dfrac{\epsilon^d}{\omega_d}\Big[(A + (d-2)B)m_{kk}^{uv} - B\displaystyle\sum_{i\neq k}m_{ii}^{vu}\Big] & \text{if } k = l , \\[2ex] -\dfrac{\epsilon^d}{\omega_d}(-B + A)m_{kl}^{uv} & \text{if } k \neq l . \end{cases} \tag{9.13}$$

By solving (9.13) for m_{pq}^{ij}, we obtain (9.9). This completes the proof. □

Once we determine the EMT $\epsilon^d m_{pq}^{ij}$ associated with D, then we can estimate the size of D by Corollary 7.7.

Theorem 9.4 (Size estimation) *For $i \neq j$,*

$$|D| \approx \left| \frac{\mu + \widetilde{\mu}}{\mu(\mu - \widetilde{\mu})} \right| |\epsilon^d m_{ij}^{ij}|$$

if $\widetilde{\mu}$ is known. If $\widetilde{\mu}$ is unknown, then $|(\mu + \widetilde{\mu})/(\mu(\mu - \widetilde{\mu}))|$ is assumed to be $1/\mu$.

9.2 Representation of the EMT's by Ellipses

Suppose $d = 2$. As in the electrostatic case, the reconstructed EMT carries an information about the inclusion other than the size. In order to visualize this information, we now describe a method to find an ellipse which represents the reconstructed EMT in the two-dimensional case.

For an ellipse D centered at the origin, let $m_{pq}^{ij}(D)$ be the EMT associated with D. Let $\widehat{D}$ be the ellipse of the form

$$\widehat{D} : \frac{x^2}{a^2} + \frac{y^2}{b^2} = 1$$

such that

$$D = R_\theta(\widehat{D}) \,,$$

for some θ. Then by Theorem 7.14, m_{pq}^{ij} are determined by θ, $|D|$, and m defined by (7.46).

Let M_{pq}^{ij}, $i, j, p, q = 1, 2$, be the EMT determined by the method of the previous section. Our goal is to find an ellipse D so that

$$m_{pq}^{ij}(D) = M_{pq}^{ij}, \quad i, j, p, q = 1, 2 \,. \tag{9.14}$$

Observe that the collection of two-dimensional EMT's has six degrees of freedom while the collection of ellipses has only three. So the equation (9.14) may not have a solution. Thus instead we seek to find an ellipse D so that $m_{pq}^{ij}(D)$ best fits M_{pq}^{ij} for $i, j, p, q = 1, 2$.[1] We can achieve this goal in the following steps.

Representation by ellipses with knowledge of $(\widetilde{\lambda}, \widetilde{\mu})$. Suppose that the Lamé constants $(\widetilde{\lambda}, \widetilde{\mu})$ of the inclusion D are known.

Step 1: First we set a tolerance τ. If both $|M_{12}^{11} + M_{22}^{12}|$ and $|M_{11}^{11} - M_{22}^{22}|$ are smaller than τ, then represent the EMT by the disk of the size found in the previous section. If either $|M_{12}^{11} + M_{22}^{12}|$ or $|M_{11}^{11} - M_{22}^{22}|$ is larger than τ, then first find the angle of rotation θ by solving (7.41), namely,

$$\frac{M_{12}^{11} + M_{22}^{12}}{M_{11}^{11} - M_{22}^{22}} = \frac{1}{2}\tan 2\theta \,, \quad 0 \le \theta < \frac{\pi}{2} \,. \tag{9.15}$$

Step 2: We then compute $\widehat{M}_{pq}^{ij}$ by reversing the rotation by θ found in (9.15) using formula (7.37). Since it suffices to replace r_{ij} with $(-1)^{i+j}r_{ij}$ in (7.37), we get

$$\widehat{M}_{pq}^{ij} = \sum_{u,v=1}^{2} \sum_{k,l=1}^{2} (-1)^{i+j+u+v+k+l+p+q} r_{pu} r_{qv} r_{ik} r_{jl} M_{uv}^{kl} \,, \tag{9.16}$$

---

[1] It would be interesting and useful to find a class of domains that can represent the reconstructed EMT in a unique and canonical way.

where
$$\begin{pmatrix} r_{11} & r_{12} \\ r_{21} & r_{22} \end{pmatrix} = \begin{pmatrix} \cos\theta & -\sin\theta \\ \sin\theta & \cos\theta \end{pmatrix}.$$

Step 3: The ideal next step would be to use (7.51) and (7.58) for finding $|D|$ and m that produce the entries $\widehat{m}_{pq}^{ij}$ that minimize

$$|\widehat{m}_{11}^{11} - \widehat{M}_{11}^{11}| + |\widehat{m}_{22}^{22} - \widehat{M}_{22}^{22}| + |\widehat{m}_{22}^{11} - \widehat{M}_{22}^{11}| + |\widehat{m}_{12}^{12} - \widehat{M}_{12}^{12}|. \qquad (9.17)$$

But it is not so clear how to minimize (9.17) since $\widehat{m}_{pq}^{ij}$ is a nonlinear function of m, defined by (7.46), and $|D|$. So we propose a different method to find $|D|$ and m.

The relation (7.39) suggests that $2(\widehat{m}_{22}^{11} + 2\widehat{m}_{12}^{12}) - (\widehat{m}_{11}^{11} + \widehat{m}_{22}^{22})$ carries an information on the size of m, the ratio of long and short axes. On the other hand, (7.35) shows that m_{12}^{12} carries an information on $|D|$, the size of D. So, we solve

$$2(\widehat{m}_{22}^{11} + 2\widehat{m}_{12}^{12}) - (\widehat{m}_{11}^{11} + \widehat{m}_{22}^{22}) = 2(\widehat{M}_{22}^{11} + 2M_{12}^{12}) - (M_{11}^{11} + \widehat{M}_{22}^{22})$$
$$\widehat{m}_{12}^{12} = \widehat{M}_{12}^{12},$$

$$(9.18)$$

using (7.51) and (7.58). Numerical tests show that (9.18) may have multiple solutions. Among the solutions found by solving (9.18), we choose a one which minimizes (9.17).

Representation by ellipses without knowledge of $(\widetilde{\lambda}, \widetilde{\mu})$. Suppose that the Lamé constants $(\widetilde{\lambda}, \widetilde{\mu})$ of the inclusion D are unknown. Then Step 1 and Step 2 are the same as before. Instead of Step 3, we use Step 3'.

Step 3': If the reconstructed M_{pq}^{ij} is negative-definite on symmetric matrices, then $\widetilde{\mu} < \mu$ by Theorem 7.6. So, set $\widetilde{\lambda} = \widetilde{\mu} = 0$ and solve (9.18) for m and $|D|$ using (7.66). If the reconstructed M_{pq}^{ij} is positive-definite on symmetric matrices, then set $\widetilde{\mu} = \infty$ and solve (9.18) for m and $|D|$ using (7.63). Among the solutions found by solving (9.18), we choose a one which minimizes (9.17).

9.3 Detection of the Location

Having found $\epsilon^d m_{kp}^{uv}$, we now proceed to find the location z of D. We propose two methods, one using only linear solutions and the other using quadratic solutions.

Detection of the location – Linear method. In view of (9.3) and (9.12), we have

$$-\epsilon^d \sum_{p,q=1}^{d} \partial_p \Gamma_{kq}(x - z) m_{pq}^{uv} = H_k[\mathbf{g}_{uv}](x) + O(\frac{\epsilon^{d+1}}{|x|^d}) + O(\frac{\epsilon^{2d}}{|x|^{d-1}}), \qquad (9.19)$$

for $k, u, v = 1, \ldots, d$. Since $m_{pq}^{uv} = m_{qp}^{vu}$, $p, q, u, v = 1, \ldots, d$, we can symmetrize (9.19) to obtain

$$
-\frac{\epsilon^d}{2} \sum_{p,q=1}^{d} \left[\partial_p \Gamma_{kq}(x-z) + \partial_q \Gamma_{kp}(x-z) \right] m_{pq}^{uv}
$$

$$
= H_k[\mathbf{g}_{uv}](x) + O(\frac{\epsilon^{d+1}}{|x|^d}) + O(\frac{\epsilon^{2d}}{|x|^{d-1}}) . \tag{9.20}
$$

Let V be the space of $d \times d$ symmetric matrices and define a linear transform P on V by

$$
P((a_{pq})) = (\sum_{p,q=1}^{d} a_{pq} \epsilon^d m_{pq}^{uv}) .
$$

Then by Theorem 7.6, P is invertible on V. Let (n_{pq}^{ij}) be the matrix for P^{-1} on V, namely,

$$
P^{-1}((a_{pq})) = (\sum_{p,q=1}^{d} a_{pq} n_{pq}^{ij}) , \quad (a_{pq}) \in V . \tag{9.21}
$$

It then follows from (9.20) that

$$
-\frac{1}{2} \left[\partial_p \Gamma_{kq}(x-z) + \partial_q \Gamma_{kp}(x-z) \right]
$$

$$
= \sum_{i,j=1}^{d} H_k[\mathbf{g}_{ij}](x) n_{ij}^{pq} + O(\frac{\epsilon}{|x|^d}) + O(\frac{\epsilon^d}{|x|^{d-1}}) , \quad k = 1, \ldots, d . \tag{9.22}
$$

Observe from (9.10) that

$$
\sum_{p=1}^{d} \partial_p \Gamma_{kp}(x-z) = \frac{(-B+A)}{\omega_d} \frac{x_k - z_k}{|x-z|^d} = \frac{1}{\omega_d(2\mu+\lambda)} \frac{x_k - z_k}{|x-z|^d} ,
$$

for $k = 1, \ldots, d$. Hence we obtain from (9.22) that

$$
\frac{x_k - z_k}{|x-z|^d} = -\omega_d(2\mu+\lambda) \sum_{i,j=1}^{d} H_k[\mathbf{g}_{ij}](x) \sum_{p=1}^{d} n_{ij}^{pp} + O(\frac{\epsilon}{|x|^d})
$$

$$
+ O(\frac{\epsilon^d}{|x|^{d-1}}) . \tag{9.23}
$$

Multiplying both sides of (9.23) by $|x|^{d-1}$, we arrive at the following formula. If $|x| = O(\epsilon^{-d+1})$, then

$$
\frac{x_k - z_k}{|x-z|} = -\omega_d(2\mu+\lambda)|x|^{d-1} \sum_{i,j=1}^{d} H_k[\mathbf{g}_{ij}](x) \sum_{p=1}^{d} n_{ij}^{pp} + O(\epsilon^d) , \tag{9.24}
$$

for $k = 1, \ldots, d$. Formula (9.24) says that we can recover $(x_k - z_k)/|x - z|$, $k = 1, \ldots, d$, from $H_k[\mathbf{g}_{ij}]$.

We now use an idea from Theorem 5.2 to recover the center z from $(x - z)/|x - z|$. Fix k and freeze x_l, $l \neq k$, so that $\sum_{l \neq k} |x_l| = O(\epsilon^{-d+1})$. Then consider

$$\omega_d(2\mu + \lambda)|x|^{d-1} \sum_{i,j=1}^{d} H_k[\mathbf{g}_{ij}](x) \sum_{p=1}^{d} n_{ij}^{pp}$$

as a function of x_k. In fact, for

$$x = x_k \mathbf{e}_k + \sum_{l \neq k} x_l \mathbf{e}_l ,$$

define

$$\Phi_k(x_k) = \omega_d(2\mu + \lambda)|x|^{d-1} \sum_{i,j=1}^{d} H_k[\mathbf{g}_{ji}](x) \sum_{p=1}^{d} n_{ij}^{pp} . \qquad (9.25)$$

We then find the unique zero of Φ_k, say z_k^* and therefore the point $(z_1^*, \ldots, z_d^*)$ is the center z within a precision of $O(\epsilon^d)$.

Detection of the location – Quadratic method. This method uses the relation (9.6). In view of (9.6) and (9.11), we get

$$t^{d-1} H_k[\mathbf{g}](t\mathbf{e}_l) = -\frac{1}{\omega_d} \sum_{i,j,p,q=1}^{d} (\partial_i U_j)(z) e_{klpq} \epsilon^d m_{pq}^{ij} , \quad \text{modulo } O(\epsilon^{d+1}) .$$

$$(9.26)$$

Since $m_{pq}^{ij} = m_{qp}^{ij}$, we get

$$\sum_{i,j,p,q=1}^{d} (\partial_i U_j)(z) e_{klpq} \epsilon^d m_{pq}^{ij} = \frac{1}{2} \sum_{p,q=1}^{d} (e_{klpq} + e_{klqp}) \sum_{i,j=1}^{d} (\partial_i U_j)(z) \epsilon^d m_{pq}^{ij} .$$

Since $e_{klpq} + e_{klqp} = e_{lkpq} + e_{lkqp}$, we can define a linear transform T on V by

$$T((a_{pq})) := \frac{1}{2} \left(\sum_{p,q=1}^{d} (e_{klpq} + e_{klqp}) a_{pq} \right) .$$

We claim that T is invertible. To prove this suppose that $T((a_{pq}))_{kl} = 0$, $k, l = 1, \ldots, d$. If $k = l$, then

$$(A + (d-1)B)a_{kk} + \sum_{p \neq k} a_{pp} = 0 , \quad k = 1, \ldots, d .$$

Since $A + (d-1)B \neq -1$, we get $a_{kk} = 0$, $k = 1, \ldots, d$. On the other hand, if $k \neq l$, then

$$(-A + B)(a_{kl} + a_{lk}) = 0 ,$$

and hence $a_{kl} = 0$ since (a_{pq}) is symmetric. Therefore $(a_{pq})_{p,q=1}^d = 0$ and T is invertible on V.

It then follows from (9.26) that

$$-\sum_{i,j=1}^d (\partial_i U_j)(z)\epsilon^d m_{pq}^{ij} = \omega_d T^{-1}(t^{d-1}H_k[\mathbf{g}](t\mathbf{e}_l))_{pq} , \quad \text{modulo } O(\epsilon^{d+1}) .$$

(9.27)

We then apply second-order homogeneous solutions for $\mathbf{U}$. In fact, in the two-dimensional case, take

$$\mathbf{U}(x) = (2x_1 x_2, x_1^2 - x_2^2) ,$$

and $\mathbf{g} = \partial\mathbf{U}/\partial\nu$. Then using (9.27), we can determine $(\partial_i U_j)(z)$, thus z, from the elastic moment tensor m_{pq}^{ij} and the limit value of $H_k[\mathbf{g}]$ at $t \to \infty$. In the three-dimensional case, we apply two homogeneous polynomials:

$$\mathbf{U}(x) = (2x_1 x_2, x_1^2 - x_2^2, 0), \quad (2x_1 x_3, 0, x_1^2 - x_3^2) .$$

9.4 Numerical Results

In this section we summarize the algorithms described in detail in previous sections and show some results of numerical experiments. The first algorithm is to identify a disk which represents the reconstructed EMT by using (7.7). We call this algorithm the disk identification algorithm. The second one is to find an ellipse which can represent the reconstructed EMT by using the method described in Sect. 9.2. We call this algorithm the ellipse identification algorithm. It is worth emphasizing that both of these recovery methods are non-iterative direct algorithms. We only present them in two dimensions even though they work in the three-dimensional case. Details of the implementation of the proposed algorithms can be found in [166]. When comparing these two algorithms, it turns out that the ellipse reconstruction algorithm performs far better in estimating the size and orientation of the inclusion. But unlike the disk reconstruction algorithm, the ellipse reconstruction method requires Lamé constants not only for the background but also of the inclusion.

The proposed identification algorithms do not rely on a forward solver while iterative algorithms require a sequence of forward solutions. Solutions of the elastostatic problem obtained by a second-order finite-difference forward solver are used only for generation of numerical simulations. In Example 1, effectiveness and stability of the algorithms for a disk inclusion are numerically demonstrated. Validity of the asymptotic expansions for the radius and the centers has been checked under various physical configurations in Example 2. Example 3 shows that the disk reconstruction algorithm provides pretty good disk approximations even for domains with non-circular inclusions. Example 4 shows that the ellipse recovery method gives perfect reconstruction results for elliptic inclusions and fairly good approximations for general domains in the sense that it provides correct estimations on the major and minor axes, and the orientation.

Disk reconstruction procedure.

Step R: Compute $\epsilon^2 m_{kl}^{uv}$ using formulae (9.8),

$$h_{kl}^{uv} := \lim_{t \to \infty} t H_k[\mathbf{g}_{uv}](t\mathbf{e}_l) ,$$

and m_{kl}^{uv} in (9.9) for $u \le v$, $k \le l$, $u \le k$, and $v \le l$,

$$\epsilon^2 m_{kl}^{uv} = \begin{cases} -\pi\mu \left[\dfrac{\lambda}{2\mu} \sum\limits_{j=1}^{2} h_{jj}^{uv} + h_{kk}^{uv} \right] , & k = l , \\[4mm] -\pi(\lambda + \mu) h_{kl}^{uv} , & k \neq l . \end{cases}$$

Then the computed radius is given by

$$r_c = \sqrt{\frac{|m_{12}^{12}|}{\pi}} .$$

Step C1: Compute the matrix $(n_{ij}^{pq})_{i,j,p,q=1,2}$, defined in (9.21),

$$\left(\sum_{p,q=1}^{2} a_{pq} n_{pq}^{ij} \right) := P^{-1}((a_{pq})) \quad \text{where} \quad P((a_{pq})) = \left(\sum_{p,q=1}^{2} a_{pq} \epsilon^2 m_{pq}^{uv} \right) .$$

Then find the unique zero z_k^*, $k = 1, 2$, defined in (9.25),

$$\Phi_k(x_k) = 2\pi(\mu + \lambda)|x| \sum_{i,j=1}^{2} H_k[\mathbf{g}_{ji}](x) \sum_{p=1}^{2} n_{ij}^{pp} ,$$

by Newton's method with $H_k[\mathbf{g}_{ji}](x)$ and $(\partial/\partial x_k) H_k[\mathbf{g}_{ji}](x)$. In the iteration, the other coordinate x_{2-k} is frozen to a constant larger than $O(\epsilon^{-2})$ and we just choose x_{2-k} to be 10^3.

Step C2: Compute the center point z using (9.27):

$$- \sum_{i,j=1}^{2} (\partial_i U_j)(z) \epsilon^2 m_{pq}^{ij} = 2\pi T^{-1}(t H_k[\mathbf{g}](t\mathbf{e}_l))_{pq} ,$$

where

$$T(a_{pq}) := \frac{1}{2} \sum_{p,q=1}^{2} (e_{klpq} + e_{klqp}) a_{pq} , \quad \mathbf{U}(x) = (2x_1 x_2, x_1^2 - x_2^2) .$$

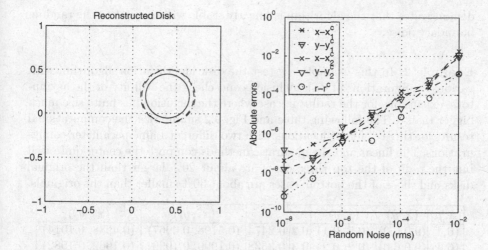

Fig. 9.1. The dashed-dotted circle represents the solution by the linear method and the dashed circle by the quadratic method. The right-hand plot shows the perturbation error due to the random boundary noise.

Example 1: In [166], the following reconstruction procedure is implemented for two-dimensional domains using Matlab and its performance is tested using a circular inclusion. The disk is centered at $(0.4, 0.2)$ and is of radius $r = 0.2$. The Lamé constants of the disk are $(\widetilde{\lambda}, \widetilde{\mu}) = (9, 6)$ while the background Lamé constants are $(\lambda, \mu) = (6, 4)$. The functions $\mathbf{u}^{1,1}$, $\mathbf{u}^{1,2}$, $\mathbf{u}^{2,2}$, and $\mathbf{u}^q$ denote the inhomogeneous solutions with the same boundary values of the corresponding homogeneous solutions, $\mathbf{U}^{1,1} = (2x, 0)$, $\mathbf{U}^{1,2} = (y, x)$, $\mathbf{U}^{2,2} = (0, y)$, $\mathbf{U}^q = (2xy, x^2 - y^2)$, respectively.

The following table summarizes a computational result of the algorithm using the forward solutions on a 128×128 mesh. The radius r^c is the computed radius in Step R, (x_1^c, y_1^c) is the center obtained by the linear method in Step C1, and (x_2^c, y_2^c) is the one obtained by the quadratic method in Step C2.

(λ, μ)	$(\lambda, \widetilde{\mu})$	(x, y)	r	r^c	(x_1^c, y_1^c)	(x_2^c, y_2^c)
(6,4)	(9,6)	(0.4, 0.2)	0.25	0.3036	(0.4110, 0.1961)	(0.3983, 0.1985)

The left-hand diagram in Fig. 9.1 shows the original disk as a solid curve; the dashed-dotted circle is the reconstructed disk by the linear disk reconstruction method and the dashed circle is by the quadratic reconstruction method. In order to check the stability of the algorithm, we add random white noise to the Neumann and Dirichlet boundary data. Since computational results for radius and centers have some errors even without noise, we compare the difference between those with and without noise. We plot the absolute perturbation error of the reconstructed values with respect to white random noise level measured in the root mean square sense. The right-hand plot in Fig. 9.1

demonstrates that the algorithm is linearly stable with respect to the random boundary noise.

Example 2: In this example, we test the disk identification algorithm with various configurations of disk inclusions and check the validity of the asymptotic expansions for the radius in case where the inclusion has finite size much bigger than 0. The following table and Fig. 9.2 summarize the computational results for three different locations with two different Lamé parameter configurations. The linear and the quadratic methods compute the center quite well but the radii of the top three cases are about 20% larger than the original disks and those of the bottom cases are about 30% smaller than the originals.

(λ, μ)	$(\tilde{\lambda}, \tilde{\mu})$	(x, y)	r	r^c	(x_1^c, y_1^c)	(x_2^c, y_2^c)
(6,4)	(9.0,6.0)	(0.5, 0.1)	0.2	0.2474	(0.5198, 0.0967)	(0.4988, 0.1014)
(6,4)	(9.0,6.0)	(0.2, 0.2)	0.3	0.3638	(0.1999, 0.1999)	(0.1962, 0.1982)
(6,4)	(9.0,6.0)	(-0.3, -0.3)	0.5	0.5931	(-0.2974, -0.2972)	(-0.2947, -0.2981)
(6,4)	(7.0,4.5)	(0.5, 0.1)	0.2	0.1371	(0.5203, 0.0967)	(0.4995, 0.1009)
(6,4)	(7.0,4.5)	(0.2, 0.2)	0.3	0.2029	(0.2003, 0.2003)	(0.1969, 0.1977)
(6,4)	(7.0,4.5)	(-0.3, -0.3)	0.5	0.3366	(-0.3006, -0.3005)	(-0.2990, -0.2995)

In order to check the validity of the asymptotic expansion, we compute the radii by the disk reconstruction method for various combinations of radii and Lamé parameters while fixing the center of inclusion at $(0.4, 0.2)$. We use three different computational grids to check the computational accuracy of our forward and inverse solvers. In Fig. 9.3, the dotted line is used for the results on 48×48, the dashed line on 64×64, and the solid line on 128×128 grid; the computational results on the three different grids seem to be almost identical. The figure also shows that the computed radius is not identical but proportional to the original value. The ratio between the computed and the original radius is independent of the radius, which is strong evidence of a missing second-order asymptotic expansion term for the radius. It is worth noting that the asymptotic expansion of EMT in (9.9) is correct up to $O(\epsilon^{2d})$, which gives a valid expression for the radius up to second-order accuracy in the two-dimensional case.

Example 3 (General domain cases): We now test the disk reconstruction algorithm with non-circular shape inclusions even though the algorithm has been derived for circular inclusions. The computational results on the 64×64 grid show fairly good agreement with their circular approximations. It is also worth mentioning that $(\lambda_0, \mu_0) = (6, 4)$, $(\lambda_1, \mu_1) = (9, 6)$ gives about 20% bigger results and $(\lambda_0, \mu_0) = (4, 6)$, $(\lambda_1, \mu_1) = (6, 9)$ about 50% bigger than originally, in disk cases shown in Fig. 9.4, therefore the computed results are bigger than the inclusions, especially for the three lower examples.

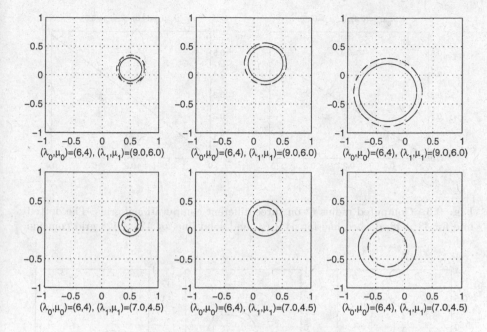

Fig. 9.2. Reconstruction results. Dashed-dotted circles by the linear method and dashed circles by the quadratic method. The three upper cases have stiff inclusions with $(\widetilde{\lambda}, \widetilde{\mu}) = (9, 6)$, $(\lambda, \mu) = (6, 4)$ and the three lower cases with $(\widetilde{\lambda}, \widetilde{\mu}) = (6, 4)$, $(\lambda, \mu) = (9, 6)$. We use the notation λ_0, μ_0 for λ, μ and λ_1, μ_1 for $\widetilde{\lambda}, \widetilde{\mu}$.

We now summarize the ellipse identification algorithm.

Ellipse reconstruction procedure. Let M_{pq}^{ij} be the reconstructed EMT. Given a tolerance τ, if both $|M_{12}^{11} + M_{22}^{12}|$ and $|M_{11}^{11} - M_{22}^{22}|$ are smaller than τ, then find the disk of the size found in the previous subsection. If either $|M_{12}^{11} + M_{22}^{12}|$ or $|M_{11}^{11} - M_{22}^{22}|$ is larger than τ, then

(E1): Determine the angle of rotation θ by solving (9.15), namely

$$\frac{M_{12}^{11} + M_{22}^{12}}{M_{11}^{11} - M_{22}^{22}} = \frac{1}{2}\tan 2\theta, \quad 0 \le \theta < \frac{\pi}{2}.$$

(E2): Using the angle θ found in (E1), solve (9.16) to find $\widehat{M}_{pq}^{ij}$:

$$\widehat{M}_{pq}^{ij} = \sum_{u,v=1}^{2} \sum_{k,l=1}^{2} (-1)^{u+k+p+i} r_{pu} r_{vq} r_{ik} r_{lj} M_{uv}^{kl},$$

where

$$\begin{pmatrix} r_{11} & r_{12} \\ r_{21} & r_{22} \end{pmatrix} = \begin{pmatrix} \cos\theta & -\sin\theta \\ \sin\theta & \cos\theta \end{pmatrix}.$$

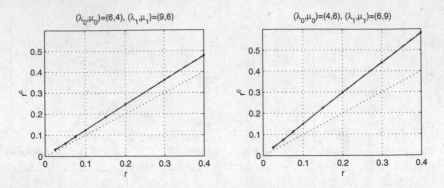

Fig. 9.3. Computed radius r^c on three different computational grids. The dotted line for 48×48, dashed line for 64×64, and solid line for 128×128 grid coincide well.

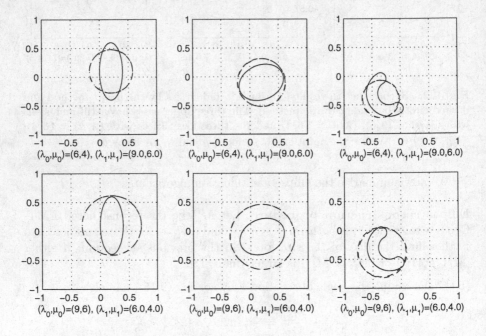

Fig. 9.4. Reconstruction of general shape inclusion.

(E3): Find $|D|$ and m by solving (9.18):

$$2(\widehat{m}_{22}^{11} + 2\widehat{m}_{12}^{12}) - (\widehat{m}_{11}^{11} + \widehat{m}_{22}^{22}) = 2(\widehat{M}_{22}^{11} + 2M_{12}^{12}) - (M_{11}^{11} + \widehat{M}_{22}^{22}),$$
$$\widehat{m}_{12}^{12} = \widehat{M}_{12}^{12}.$$

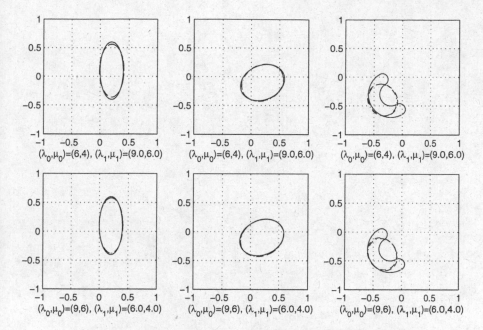

Fig. 9.5. Computed ellipses for various inclusions marked with solid curves. The centers of dotted ellipses are computed by the linear method and dashed-dotted ones by the quadratic method.

The relation between $|D|$, m and $\widehat{m}_{pq}^{ij}$ is given by (7.48), (7.51), and (7.58) if the Lamé constants $(\widetilde{\lambda}, \widetilde{\mu})$ are known. Otherwise, it is given by (7.63) (resp. (7.66)) if the reconstructed EMT is negative-definite (resp. positive-definite).

(E4): Among the solutions in (E3), choose a one which minimizes the quantity given in (9.17):

$$\left|\widehat{m}_{11}^{11} - \widehat{M}_{11}^{11}\right| + \left|\widehat{m}_{22}^{22} - \widehat{M}_{22}^{22}\right| + \left|\widehat{m}_{22}^{11} - \widehat{M}_{22}^{11}\right| + \left|\widehat{m}_{12}^{12} - \widehat{M}_{12}^{12}\right|.$$

Example 4: In this example, we test the algorithm using the same domains as in Example 3. Fig. 9.5 shows the reconstructed ellipses when their Lamé constants $(\widetilde{\lambda}, \widetilde{\mu})$ are known. It is not surprising that the ellipse recovery method gives perfect size information for the disks and ellipses, as shown in the first diagram in Fig. 9.5, since the information on the Lamé constants $(\widetilde{\lambda}, \widetilde{\mu})$ is used.

Part III

Detection of Small Electromagnetic Inclusions

Suppose that an electromagnetic medium occupies a bounded domain Ω in $\mathbb{R}^d$, with a connected Lipschitz boundary $\partial\Omega$. Suppose that Ω contains a finite number m of small inclusions $\{D_s\}_{s=1}^m$, each of the form $D_s = \delta B_s + z_s$, where B_s, $s = 1, \ldots, m$, is a Lipschitz bounded domain in $\mathbb{R}^d$ containing the origin[2]. We assume that the domains $\{D_s\}_{s=1}^m$ are separated from each other and from the boundary. More precisely, we assume that there exists a constant $c_0 > 0$ such that

$$|z_s - z_{s'}| \geq 2c_0 > 0 \,, \forall\, s \neq s' \quad \text{and} \quad \text{dist}(z_s, \partial\Omega) \geq 2c_0 > 0, \forall\, s \,,$$

that δ, the common order of magnitude of the diameters of the inclusions, is sufficiently small, that these inclusions are disjoint, and that their distance to $\mathbb{R}^d \setminus \overline{\Omega}$ is larger than c_0. Let μ_0 and ε_0 denote the permeability and the permittivity of the background medium Ω, and assume that $\mu_0 > 0$ and $\varepsilon_0 > 0$ are positive constants. Let $\mu_s > 0$ and $\varepsilon_s > 0$ denote the permeability and the permittivity of the s-th inclusion, D_s, which are also assumed to be positive constants. Introduce the piecewise-constant magnetic permeability

$$\mu_\delta(x) = \begin{cases} \mu_0 \,, & x \in \Omega \setminus \overline{\cup_{s=1}^m D_s} \,, \\ \mu_s \,, & x \in D_s, \ s = 1 \ldots m \,. \end{cases}$$

If we allow the degenerate case $\delta = 0$, then the function $\mu_0(x)$ equals the constant μ_0. The piecewise constant electric permittivity, $\varepsilon_\delta(x)$, is defined analogously.

Let the electric field u denote the solution to the Helmholtz equation

$$\nabla \cdot (\frac{1}{\mu_\delta}\nabla u) + \omega^2 \varepsilon_\delta u = 0 \ \text{in} \ \Omega \,,$$

---

[2] We use δ instead of ϵ for the small parameter to avoid possible confusion with the notation for the permittivity.

with the boundary condition $u = f \in W^2_{\frac{1}{2}}(\partial\Omega)$, where $\omega > 0$ is a given frequency. This problem can be written as

$$
\begin{cases}
(\Delta + \omega^2 \varepsilon_0 \mu_0)u = 0 & \text{in } \Omega \setminus \overline{\cup_{s=1}^m D_s} \,, \\
(\Delta + \omega^2 \varepsilon_s \mu_s)u = 0 & \text{in } D_s \,, s = 1,\ldots,m \,, \\
\dfrac{1}{\mu_s}\dfrac{\partial u}{\partial \nu}\bigg|_{-} - \dfrac{1}{\mu_0}\dfrac{\partial u}{\partial \nu}\bigg|_{+} = 0 & \text{on } \partial D_s \,, s = 1,\ldots,m \,, \\
u\big|_{-} - u\big|_{+} = 0 & \text{on } \partial D_s, s = 1,\ldots,m \,, \\
u = f & \text{on } \partial\Omega \,.
\end{cases}
$$

In this part we are concerned with the detection of unknown electromagnetic inclusions D_s, $s = 1,\ldots,m$, by means of a finite number of voltage-to-current pairs $(f, \partial u/\partial \nu|_{\partial\Omega})$ measured on $\partial\Omega$.

Possible applications of this problem include acoustical sounding of biological media, underwater acoustics [250], and electromagnetic induction tomography for imaging electrical property variations in the earth [262].

The use of the formal equivalence between electromagnetics and linear acoustics, by term-to-term replacing permittivity and permeability by compressibility and volume density of mass, and the scalar electric field by the scalar acoustic pressure characteristic of compressional waves inside fluid media opens up the investigation below to many other applications, being said that the type of materials and of geometrical configurations investigated and the range of values that are allowed to be taken by the two sets of parameters in either discipline may differ considerably in practice.

A number of algorithms have been proposed for the numerical treatment of inverse problems for the above Helmholtz equation. In the design of such schemes, two fundamental difficulties have to be overcome: nonlinearity and ill-posedness. The existing attempts to solve inverse problems for the Helmholtz equation include linearized inversion schemes [161, 220, 134, 11], methods based on nonlinear regularization techniques [179, 54, 103, 130, 107, 135], and a continuation method in frequency [75, 88, 76, 42, 41]. The third technique is a promising new approach based on recursive linearization on the frequency that makes use of multi-frequency data.

Following our approach throughout this book, we design efficient and robust algorithms to reconstruct the location and certain geometric features of the electromagnetic inclusions D_s.

Results similar to those presented in this part have been obtained in the context of the full (time-harmonic) Maxwell equations in [31].

Most of the results presented in Part III are from [19] and [20].

10

Well-Posedness

In this chapter, by using the theory of collectively compact operators, we prove existence and uniqueness of a solution to the Helmholtz equation

$$
\begin{cases}
(\Delta + \omega^2 \varepsilon_0 \mu_0)u = 0 & \text{in } \Omega \setminus \overline{\cup_{s=1}^{m} D_s}\,, \\[2mm]
(\Delta + \omega^2 \varepsilon_s \mu_s)u = 0 & \text{in } D_s\,, s = 1, \ldots, m\,, \\[2mm]
\dfrac{1}{\mu_s}\dfrac{\partial u}{\partial \nu}\Big|_{-} - \dfrac{1}{\mu_0}\dfrac{\partial u}{\partial \nu}\Big|_{+} = 0 & \text{on } \partial D_s\,, s = 1, \ldots, m\,, \\[2mm]
u\big|_{-} - u\big|_{+} = 0 & \text{on } \partial D_s, s = 1, \ldots, m\,, \\[2mm]
u = f & \text{on } \partial \Omega\,,
\end{cases}
\tag{10.1}
$$

assuming that

$$
\omega^2 \varepsilon_0 \mu_0 \text{ is not an eigenvalue for the operator } -\Delta \text{ in } L^2(\Omega)
$$

with homogeneous Dirichlet boundary conditions. $\qquad$ (10.2)

10.1 Existence and Uniqueness of a Solution

In order to define the natural weak formulation of the problem (10.1), let a_δ denote the sesquilinear form

$$
a_\delta(u, v) = \int_\Omega \frac{1}{\mu_\delta} \nabla u \cdot \nabla v - \omega^2 \int_\Omega \varepsilon_\delta u v \,,
\tag{10.3}
$$

defined on $W_0^{1,2}(\Omega) \times W_0^{1,2}(\Omega)$. Let b be a given conjugate-linear functional on $W_0^{1,2}(\Omega)$. Our assumption (10.2) is that the variational problem

$$
a_0(u, v) = b(v) \text{ for all } v \in W_0^{1,2}(\Omega)
$$

has a unique solution. The following lemma from [259] shows that the assumption (10.2) also leads to the unique solvability of (10.1).

Lemma 10.1 *Suppose (10.2) is satisfied, and let $a_\delta, 0 \leq \delta$, be the sesquilinear forms introduced by (10.3). There exists a constant $0 < \delta_0$, such that given any $0 \leq \delta < \delta_0$, and any bounded, conjugate-linear functional, b, on $W_0^{1,2}(\Omega)$, there is a unique $u \in W_0^{1,2}(\Omega)$ satisfying $a_\delta(u,v) = b(v)$ for all $v \in W_0^{1,2}(\Omega)$. Furthermore, there exists a constant C, independent of δ and b, such that*

$$\|u\|_{W^{1,2}(\Omega)} \leq C \sup_{v \in W_0^{1,2}(\Omega), \|v\|_{W^{1,2}(\Omega)}=1} |b(v)| .$$

Proof.   In order to prove this lemma it is convenient to introduce a decomposition of a_δ. Pick a fixed positive constant, λ, with $\lambda > \omega^2$, and write a_δ as $a_\delta = A_\delta + B_\delta$, where

$$A_\delta(u,v) = \int_\Omega \frac{1}{\mu_\delta} \nabla u \cdot \nabla v + (\lambda - \omega^2) \int_\Omega \varepsilon_\delta uv$$

and

$$B_\delta(u,v) = -\lambda \int_\Omega \varepsilon_\delta uv .$$

Suppose (10.2) is satisfied. Then the sesquilinear form A_δ is uniformly continuous and uniformly coercive on $W_0^{1,2}(\Omega) \times W_0^{1,2}(\Omega)$. It is also convenient to introduce a family of bounded linear operators $K_\delta : W_0^{1,2}(\Omega) \to W_0^{1,2}(\Omega)$ by

$$A_\delta(K_\delta u, v) = B_\delta(u,v) = -\lambda \int_\Omega \varepsilon_\delta uv ,$$

for all u and v in $W_0^{1,2}(\Omega)$.

Let δ_n be a sequence converging to zero. We first show that the linear operators $\{K_{\delta_n}\}$ are compact and K_{δ_n} converges pointwise to K_0 as δ_n approaches 0. We remind the reader that the operators $\{K_{\delta_n}\}$ are collectively compact iff the set $\{K_{\delta_n}(u) : n \geq 1, u \in W_0^{1,2}(\Omega), \|u\|_{W^{1,2}(\Omega)} \leq 1\}$ is relatively compact (its closure is compact) in $W_0^{1,2}(\Omega)$.

Fix $u \in W_0^{1,2}(\Omega)$, then

$$A_{\delta_n}((K_{\delta_n} - K_0)u, v) = B_{\delta_n}(u,v) - B_0(u,v) + A_0(K_0 u, v) - A_{\delta_n}(K_0 u, v) ,$$

for all $v \in W_0^{1,2}(\Omega)$. We easily see that

$$\sup_{v \in W_0^{1,2}(\Omega), \|v\|_{W^{1,2}(\Omega)}=1} |B_0(u,v) - B_{\delta_n}(u,v)| \to 0 , \qquad (10.4)$$

as $\delta_n \to 0$. It is also clear that

$$\sup_{v \in W_0^{1,2}(\Omega), \|v\|_{W^{1,2}(\Omega)}=1} |A_0(K_0 u, v) - A_{\delta_n}(K_0 u, v)| \to 0 , \qquad (10.5)$$

as $\delta_n \to 0$. A combination of (10.4) and (10.5) yields

$$\sup_{v \in W_0^{1,2}(\Omega), ||v||_{W^{1,2}(\Omega)}=1} |A_{\delta_n}((K_{\delta_n} - K_0)u, v)| \to 0 \,,$$

as $\delta_n \to 0$. Since A_{δ_n} is uniformly coercive on $W_0^{1,2}(\Omega) \times W_0^{1,2}(\Omega)$, it follows now that

$$||(K_{\delta_n} - K_0)u||_{W^{1,2}(\Omega)} \to 0 \,,$$

as $\delta_n \to 0$. This verifies the pointwise convergence of the operators $\{K_{\delta_n}\}$.

Let $K_{\tau_m}(u_m)$ be any sequence from the set

$$\left\{ K_{\delta_n}(u) : n \geq 1, u \in W_0^{1,2}(\Omega), ||u||_{W^{1,2}(\Omega)} \leq 1 \right\} .$$

In order to verify the collective compactness of the operators $\{K_{\delta_n}\}$ we need to show that the sequence $K_{\tau_m}(u_m)$ contains a convergent subsequence. By extraction of a subsequence (still referred to as $K_{\tau_m}(u_m)$) we may assume that either: (1) $\tau_m = \tau$ is constant (*i.e.*, independent of m) or: (2) $\tau_m \to 0$ as $m \to +\infty$. We may also assume that u_m converges weakly to some $u_\infty \in W_0^{1,2}(\Omega)$. We introduce the sequence $u'_m = u_m - u_\infty$. Clearly $||u'_m||_{W^{1,2}(\Omega)} \leq 2$ and u'_m converges weakly to zero. Since the imbedding $W_0^{1,2}(\Omega) \hookrightarrow L^2(\Omega)$ is compact, this gives that u'_m has a subsequence (still referred to as u'_m) converging strongly to zero in $L^2(\Omega)$. From the definition of K_{τ_m} it follows immediately that

$$\sup_{v \in W_0^{1,2}(\Omega), ||v||_{W^{1,2}(\Omega)}=1} |A_{\tau_m}(K_{\tau_m} u'_m, v)| = \sup_{v \in W_0^{1,2}(\Omega), ||v||_{W^{1,2}(\Omega)}=1} |B_{\tau_m}(u'_m, v)|$$

$$\leq C||u'_m||_{L^2(\Omega)} \,.$$

Since A_{τ_m} is uniformly coercive and since $||u'_m||_{L^2(\Omega)} \to 0$, we conclude from the above estimate that $||K_{\tau_m} u'_m||_{W^{1,2}(\Omega)} \to 0$, which is exactly what we are aiming at.

We want now to solve the variational problem:

Find $u \in W_0^{1,2}(\Omega)$ such that $a_\delta(u, v) = b(v)$ for all $v \in W_0^{1,2}(\Omega)$. (10.6)

This problem can be rewritten as

$$A_\delta(u, v) + B_\delta(u, v) = b(v) \text{ for all } v \in W_0^{1,2}(\Omega) \,,$$

or as

$$A_\delta((I + K_\delta)u_\delta, v) = b(v) \text{ for all } v \in W_0^{1,2}(\Omega) \,. \quad (10.7)$$

Since A_δ is uniformly continuous and coercive on $W_0^{1,2}(\Omega)$, it now follows that the variational problem (10.7) is equivalent to the problem of finding $u \in W_0^{1,2}(\Omega)$ such that

$$(I + K_\delta)u = F_\delta \,.$$

Here the function $F_\delta \in W_0^{1,2}(\Omega)$ is defined by $A_\delta(F_\delta, v) = b(v)$ for all $v \in W_0^{1,2}(\Omega)$, and therefore satisfies

$$||F_\delta||_{W^{1,2}(\Omega)} \leq C \sup_{v \in W_0^{1,2}(\Omega), ||v||_{W^{1,2}(\Omega)}=1} |b(v)| . \qquad (10.8)$$

By the same arguments as we just went through earlier in this proof, the original variational problem to find $U \in W_0^{1,2}(\Omega)$ such that $a_0(U, v) = b(v)$ for all $v \in W_0^{1,2}(\Omega)$ is thus equivalent to the problem to find $U \in W_0^{1,2}(\Omega)$ such that $(I + K_0)U = F_0$ with $F_0 \in W_0^{1,2}(\Omega)$ defined by $A_0(F_0, v) = b(v)$ for all $v \in W_0^{1,2}(\Omega)$. The fact that this problem has a unique solution (assumption (10.2)) implies that $I + K_0$ is an invertible operator. For any sequence δ_n converging to zero we have already verified that the operators $\{K_{\delta_n}\}$ are collectively compact and converge pointwise to K_0. From the theory of collectively compact operators [35] (see Theorem A.4 in Appendix A.3) it follows that there exists a constant $0 < \delta_0$, such that given any $0 \leq \delta < \delta_0$, the operator $I + K_\delta$ is invertible with

$$||(I + K_\delta)^{-1}F_\delta||_{W^{1,2}(\Omega)} \leq C||F_\delta||_{W^{1,2}(\Omega)} \qquad (10.9)$$

for some constant C, independent of δ. It then follows from (10.8) and (10.9) that the variational problem (10.6) has a unique solution $u \in W_0^{1,2}(\Omega)$ satisfying

$$||u||_{W^{1,2}(\Omega)} \leq C \sup_{v \in W_0^{1,2}(\Omega), ||v||_{W^{1,2}(\Omega)}=1} |b(v)| .$$

Thus the proof of Lemma 10.1 is complete. $\square$

Suppose that there exists a constant $c_0 > 0$ such that

$$|z_s - z_{s'}| \geq 2c_0 > 0 , \forall s \neq s' \quad \text{and} \quad \text{dist}(z_s, \partial\Omega) \geq 2c_0 > 0, \forall s . \qquad (10.10)$$

Using the above lemma we can show as in [259] that the following holds.

Proposition 10.2 *Suppose (10.10) and (10.2) are satisfied. There exists $0 < \delta_0$ such that, given an arbitrary $f \in W_{\frac{1}{2}}^2(\partial\Omega)$, and any $0 < \delta < \delta_0$, the boundary value problem (10.1) has a unique weak solution u in $W^{1,2}(\Omega)$. The constant δ_0 depends on the domains $\{B_s\}_{s=1}^m$, Ω, the constants $\{\mu_s, \varepsilon_s\}_{s=0}^m$, and c_0, but is otherwise independent of the points $\{z_s\}_{s=1}^m$. Moreover, let U denote the unique weak solution to the boundary value problem:*

$$\begin{cases} (\Delta + \omega^2 \varepsilon_0 \mu_0)U = 0 & \text{in } \Omega , \\ U = f & \text{on } \partial\Omega . \end{cases}$$

There exists a constant C, independent of δ and f, such that

$$||u - U||_{W^{1,2}(\Omega)} \leq C\delta^{\frac{d}{2}}||f||_{W_{\frac{1}{2}}^2(\partial\Omega)} .$$

The constant C depends on the domains $\{B_s\}_{s=1}^m$, Ω, the constants $\{\mu_s, \varepsilon_s\}_{s=0}^m$, and c_0, but is otherwise independent of the points $\{z_s\}_{s=1}^m$.

Proof.   The function $u - U$ is in $W_0^{1,2}(\Omega)$, and for any $v \in W_0^{1,2}(\Omega)$

$$a_\delta(u - U, v) = \int_\Omega \frac{1}{\mu_\delta} \nabla(u - U) \cdot \nabla v - \omega^2 \int_\Omega \varepsilon_\delta(u - U)v$$

$$= \sum_{s=1}^m \int_{\delta B_s + z_s} \left[\left(\frac{1}{\mu_0} - \frac{1}{\mu_s} \right) \nabla U \cdot \nabla v + \omega^2(\varepsilon_s - \varepsilon_0)Uv \right] .$$

Next

$$\left| \int_{\delta B_s + z_s} \left[\left(\frac{1}{\mu_0} - \frac{1}{\mu_s} \right) \nabla U \cdot \nabla v + \omega^2(\varepsilon_s - \varepsilon_0)Uv \right] \right|$$

is bounded by

$$C\|U\|_{W^{1,2}(\delta B_s + z_s)}\|v\|_{W^{1,2}(\Omega)} .$$

Since the inclusions are away from the boundary $\partial\Omega$, standard elliptic regularity results give that

$$\|U\|_{W^{1,\infty}(\delta B_s + z_s)} \leq C\|U\|_{W^{1,2}(\Omega)} \leq C\|f\|_{W^{\frac{2}{3}}_{\frac{1}{2}}(\partial\Omega)} ,$$

and so

$$\|U\|_{W^{1,2}(\delta B_s + z_s)} \leq \|U\|_{W^{1,\infty}(\delta B_s + z_s)} \delta^{\frac{d}{2}} |B_s|^{\frac{1}{2}} \leq C\delta^{\frac{d}{2}} \|f\|_{W^{\frac{2}{3}}_{\frac{1}{2}}(\partial\Omega)} .$$

From Lemma 10.1 it then follows immediately that

$$\|u - U\|_{W^{1,2}(\Omega)} \leq C\delta^{\frac{d}{2}} \|f\|_{W^{\frac{2}{3}}_{\frac{1}{2}}(\partial\Omega)} ,$$

exactly as desired. $\square$

11

Representation of Solutions

In this chapter we derive representation formulae for the solution to the transmission problem

$$\nabla \cdot (\frac{1}{\mu_\delta} \nabla u) + \omega^2 \varepsilon_\delta u = 0 \quad \text{in } \Omega, \qquad (11.1)$$

similar to (2.17) and (2.19).

11.1 Preliminary Results

We begin this chapter by deriving the outgoing fundamental solution to the Helmholtz equation. We refer to the books of Colton and Kress [90], [91] and the one of Nédélec [224] for detailed treatments of the Helmholtz equation, emphasizing existence and uniqueness results for the exterior problem.

Let Ω be a bounded domain in $\mathbb{R}^d, d = 2$ or 3, with a connected Lipschitz boundary, a permeability equal to $\mu_0 > 0$, and a permittivity equal to $\varepsilon_0 > 0$. Consider a bounded domain $D \subset\subset \Omega$ with a connected Lipschitz boundary, a permeability $0 < \mu \neq \mu_0 < +\infty$, and a permittivity $0 < \varepsilon \neq \varepsilon_0 < +\infty$. Let $k_0 := \omega\sqrt{\varepsilon_0\mu_0}$ and $k := \omega\sqrt{\varepsilon\mu}$, where $\omega > 0$ is a given frequency.

A fundamental solution $\Gamma_k(x)$ to the Helmholtz operator $\Delta + k^2$ in $\mathbb{R}^d$ is a solution (in the sense of distributions) of

$$(\Delta + k^2)\Gamma_k = \delta_0, \qquad (11.2)$$

where δ_0 is the Dirac mass at 0. Solutions are not unique, since we can add to a solution any plane wave (of the form $e^{ik\theta \cdot x}, \theta \in \mathbb{R}^d : |\theta| = 1$) or any combination of such plane waves. We need to specify the behavior of the solutions at infinity. It is natural to look for radial solutions of the form $\Gamma_k(x) = w_k(r)$ that is subject to the extra Sommerfeld radiation condition or outgoing wave condition

$$\left| \frac{dw_k}{dr} - ikw_k \right| \leq Cr^{-(d+1)/2} \quad \text{at infinity.} \qquad (11.3)$$

If $d = 3$, equation (11.2) becomes

$$\frac{1}{r^2}\frac{d}{dr}r^2\frac{dw_k}{dr} + k^2 w_k = 0, \quad r > 0,$$

whose solution is

$$w_k(r) = c_1\frac{e^{ikr}}{r} + c_2\frac{e^{-ikr}}{r}.$$

It is easy to check that the Sommerfeld radiation condition (11.3) leads to $c_2 = 0$ and then (11.2) leads to $c_1 = -1/(4\pi)$.

If $d = 2$, equation (11.2) becomes

$$\frac{1}{r}\frac{d}{dr}r\frac{dw_k}{dr} + k^2 w_k = 0, \quad r > 0.$$

This is a Bessel equation whose solutions are not elementary functions. It is known that the Hankel functions of the first and second kind of order 0, $H_0^{(1)}(kr)$ and $H_0^{(2)}(kr)$, form a basis for the solution space. At infinity $(r \to +\infty)$, only $H_0^{(1)}(kr)$ satisfies the outgoing radiation condition (11.3). At the origin $(r \to 0)$, $H_0^{(1)}(kr)$ behaves like $H_0^{(1)}(kr) \sim (2i/\pi)\log(r)$. The following lemma holds.

Lemma 11.1 *The outgoing fundamental solution $\Gamma_k(x)$ to the operator $\Delta + k^2$ is given by*

$$\Gamma_k(x) = \begin{cases} -\dfrac{i}{4}H_0^1(k|x|), & d = 2, \\[2mm] -\dfrac{e^{ik|x|}}{4\pi|x|}, & d = 3, \end{cases}$$

for $x \neq 0$, where H_0^1 is the Hankel function of the first kind of order 0.

Let for $x \neq 0$

$$\Gamma_0(x) := \Gamma(x) = \begin{cases} \dfrac{1}{2\pi}\log|x|, & d = 2, \\[2mm] -\dfrac{1}{4\pi|x|}, & d = 3. \end{cases}$$

For a bounded domain D in $\mathbb{R}^d$ and $k > 0$ let $\mathcal{S}_D^k$ and $\mathcal{D}_D^k$ be the single and double layer potentials defined by Γ_k, that is,

$$\mathcal{S}_D^k\varphi(x) = \int_{\partial D} \Gamma_k(x-y)\varphi(y)\,d\sigma(y), \quad x \in \mathbb{R}^d,$$

$$\mathcal{D}_D^k\varphi(x) = \int_{\partial D} \frac{\partial\Gamma_k(x-y)}{\partial\nu_y}\varphi(y)\,d\sigma(y), \quad x \in \mathbb{R}^d \setminus \partial D,$$

for $\varphi \in L^2(\partial D)$. Because $\Gamma_k - \Gamma_0$ is a smooth function, we can easily prove from (2.12) and (2.13) that

$$\frac{\partial(\mathcal{S}_D^k\varphi)}{\partial\nu}\bigg|_\pm (x) = \left(\pm\frac{1}{2}I + (\mathcal{K}_D^k)^*\right)\varphi(x) \quad \text{a.e. } x \in \partial D , \tag{11.4}$$

$$(\mathcal{D}_D^k\varphi)\bigg|_\pm (x) = \left(\mp\frac{1}{2}I + \mathcal{K}_D^k\right)\varphi(x) \quad \text{a.e. } x \in \partial D , \tag{11.5}$$

for $\varphi \in L^2(\partial D)$, where $\mathcal{K}_D^k$ is the operator defined by

$$\mathcal{K}_D^k\varphi(x) = \text{p.v.} \int_{\partial D} \frac{\partial\Gamma_k(x-y)}{\partial\nu_y}\varphi(y)d\sigma(y) , \tag{11.6}$$

and $(\mathcal{K}_D^k)^*$ is the L^2-adjoint of $\mathcal{K}_D^k$, that is,

$$(\mathcal{K}_D^k)^*\varphi(x) = \text{p.v.} \int_{\partial D} \frac{\partial\Gamma_k(x-y)}{\partial\nu_x}\varphi(y)d\sigma(y) . \tag{11.7}$$

The singular integral operators $\mathcal{K}_D^k$ and $(\mathcal{K}_D^k)^*$ are bounded on $L^2(\partial D)$.

We will need the following important result from the theory of the Helmholtz equation. For its proof we refer to [91] (Lemma 2.11).

Lemma 11.2 (Rellich's lemma) *Let $R_0 > 0$ and $B_R(0) = \{|x| < R\}$. Let u satisfy the Helmholtz equation $\Delta u + k_0^2 u = 0$ for $|x| > R_0$. Assume, furthermore, that*

$$\lim_{R\to+\infty} \int_{\partial B_R(0)} |u(x)|^2 \, d\sigma(x) = 0 .$$

Then, $u \equiv 0$ for $|x| > R_0$.

Note that the assertion of this lemma does not hold if k_0 is imaginary or $k_0 = 0$.

Now we can prove the following uniqueness result.

Lemma 11.3 *Suppose $d = 2$ or 3. Let D be a bounded Lipschitz domain in $\mathbb{R}^d$. Let $u \in W_{\text{loc}}^{1,2}(\mathbb{R}^d \setminus \overline{D})$ satisfy*

$$\begin{cases} \Delta u + k_0^2 u = 0 & \text{in } \mathbb{R}^d \setminus \overline{D} , \\[2mm] \left|\dfrac{\partial u}{\partial|x|} - ik_0 u\right| = O\left(|x|^{-(d+1)/2}\right) & \text{as } |x| \to +\infty \quad \text{uniformly in } \dfrac{x}{|x|} , \\[2mm] \Im \displaystyle\int_{\partial D} \overline{u}\dfrac{\partial u}{\partial\nu} \, d\sigma = 0 . \end{cases}$$

Then, $u \equiv 0$ in $\mathbb{R}^d \setminus \overline{D}$.

Proof. Let $B_R(0) = \{|x| < R\}$. For R large enough, $D \subset B_R(0)$. Notice first that by multiplying $\Delta u + k_0^2 u = 0$ by $\overline{u}$ and integrating by parts over $B_R(0) \setminus \overline{D}$ we arrive at

$$\Im \int_{\partial B_R(0)} \overline{u}\frac{\partial u}{\partial\nu} \, d\sigma = 0 ,$$

since

$$\Im \int_{\partial D} \overline{u} \frac{\partial u}{\partial \nu} \, d\sigma = 0 \, .$$

But

$$\Im \int_{\partial B_R(0)} \overline{u} \left(\frac{\partial u}{\partial \nu} - i k_0 u \right) d\sigma = -k_0 \int_{\partial B_R(0)} |u|^2 \, .$$

Applying the Cauchy–Schwarz inequality,

$$\left| \Im \int_{\partial B_R(0)} \overline{u} \left(\frac{\partial u}{\partial \nu} - i k_0 u \right) d\sigma \right| \le \left(\int_{\partial B_R(0)} |u|^2 \right)^{1/2} \left(\int_{\partial B_R(0)} \left| \frac{\partial u}{\partial \nu} - i k_0 u \right|^2 d\sigma \right)^{1/2}$$

and using the Sommerfeld radiation condition

$$\left| \frac{\partial u}{\partial |x|} - i k_0 u \right| = O \left(|x|^{-(d+1)/2} \right) \quad \text{as } |x| \to +\infty \, ,$$

we get

$$\left| \Im \int_{\partial B_R(0)} \overline{u} \left(\frac{\partial u}{\partial \nu} - i k_0 u \right) d\sigma \right| \le \frac{C}{R} \left(\int_{\partial B_R(0)} |u|^2 \right)^{1/2} \, ,$$

for some positive constant C independent of R. Consequently, we obtain that

$$\left(\int_{\partial B_R(0)} |u|^2 \right)^{1/2} \le \frac{C}{R} \, ,$$

which indicates by the Rellich's Lemma that $u \equiv 0$ in $\mathbb{R}^d \setminus \overline{B_R(0)}$. Hence, by the unique continuation property for $\Delta + k_0^2$, we can conclude that $u \equiv 0$ up to the boundary ∂D. This finishes the proof. $\square$

11.2 Representation Formulae

We now present two representations of the solution of (11.8) similar to the representation formula (2.39) for the transmission problem for the harmonic equation. Let $f \in W_{\frac{1}{2}}^2(\partial \Omega)$, and let u and U denote the solutions to the Helmholtz equations

$$\begin{cases} \nabla \cdot (\frac{1}{\mu_\delta} \nabla u) + \omega^2 \varepsilon_\delta u = 0 & \text{in } \Omega \, , \\ u = f & \text{on } \partial \Omega \, , \end{cases} \tag{11.8}$$

and

$$\begin{cases} \Delta U + \omega^2 \varepsilon_0 \mu_0 U = 0 & \text{in } \Omega \, , \\ U = f & \text{on } \partial \Omega \, . \end{cases} \tag{11.9}$$

The following theorem is of importance to us for establishing our representation formulae.

Theorem 11.4 *Suppose that k_0^2 is not a Dirichlet eigenvalue for $-\Delta$ on D. For each $(F, G) \in W_1^2(\partial D) \times L^2(\partial D)$, there exists a unique solution $(f, g) \in L^2(\partial D) \times L^2(\partial D)$ to the system of integral equations*

$$
\begin{cases}
\mathcal{S}_D^k f - \mathcal{S}_D^{k_0} g = F \\
\dfrac{1}{\mu} \dfrac{\partial (\mathcal{S}_D^k f)}{\partial \nu}\bigg|_- - \dfrac{1}{\mu_0} \dfrac{\partial (\mathcal{S}_D^{k_0} g)}{\partial \nu}\bigg|_+ = G
\end{cases}
\quad on \ \partial D . \qquad (11.10)
$$

Furthermore, there exists a constant C independent of F and G such that

$$
\|f\|_{L^2(\partial D)} + \|g\|_{L^2(\partial D)} \leq C \left(\|F\|_{W_1^2(\partial D)} + \|G\|_{L^2(\partial D)} \right), \qquad (11.11)
$$

where in the three-dimensional case the constant C can be chosen independently of k_0 and k if k_0 and k go to zero.

Proof.   We only give the proof for $d = 3$ and $\mu_0 \neq \mu$ leaving the general case to the reader. Let $X := L^2(\partial D) \times L^2(\partial D)$ and $Y := W_1^2(\partial D) \times L^2(\partial D)$, and define the operator $T : X \to Y$ by

$$
T(f, g) := \left(\mathcal{S}_D^k f - \mathcal{S}_D^{k_0} g, \frac{1}{\mu} \frac{\partial (\mathcal{S}_D^k f)}{\partial \nu}\bigg|_- - \frac{1}{\mu_0} \frac{\partial (\mathcal{S}_D^{k_0} g)}{\partial \nu}\bigg|_+ \right).
$$

We also define T_0 by

$$
T_0(f, g) := \left(\mathcal{S}_D^0 f - \mathcal{S}_D^0 g, \frac{1}{\mu} \frac{\partial (\mathcal{S}_D^0 f)}{\partial \nu}\bigg|_- - \frac{1}{\mu_0} \frac{\partial (\mathcal{S}_D^0 g)}{\partial \nu}\bigg|_+ \right).
$$

We can easily see that $\mathcal{S}_D^{k_0} - \mathcal{S}_D^0 : L^2(\partial D) \to W_1^2(\partial D)$ is a compact operator, and so is $\frac{\partial}{\partial \nu} \mathcal{S}_D^{k_0}|_\pm - \frac{\partial}{\partial \nu} \mathcal{S}_D^0|_\pm : L^2(\partial D) \to L^2(\partial D)$. Therefore, $T - T_0$ is a compact operator from X into Y. It can be proved that $T_0 : X \to Y$ is invertible. In fact, a solution (f, g) of the equation $T_0(f, g) = (F, G)$ is given by

$$
f = g + (\mathcal{S}_D^0)^{-1}(F)
$$
$$
g = \frac{\mu_0 \mu}{\mu_0 - \mu} (\lambda I + (\mathcal{K}_D^0)^*)^{-1} \left(G + \frac{1}{\mu} (\frac{1}{2} I - (\mathcal{K}_D^0)^*)((\mathcal{S}_D^0)^{-1}(F)) \right),
$$

where $\lambda = (\mu + \mu_0)/(2(\mu - \mu_0))$. Recall now that the invertibility of $\mathcal{S}_D^0$ and $\lambda I + (\mathcal{K}_D^0)^*$ was proved in Theorems 2.8 and 2.13. Thus we see, by the Fredholm alternative, that it is enough to prove that T is injective.

Suppose that $T(f, g) = 0$. Then the function u defined by

$$
u(x) := \begin{cases}
\mathcal{S}_D^{k_0} g(x) & \text{if } x \in \mathbb{R}^d \setminus D , \\
\mathcal{S}_D^k f(x) & \text{if } x \in D ,
\end{cases}
$$

is the unique solution of the transmission problem (11.8) with Ω replaced by $\mathbb{R}^d$ and the Dirichlet boundary condition replaced by the Sommerfeld radiation condition

$$\left| \frac{\partial u}{\partial |x|}(x) - ik_0 u(x) \right| = O(|x|^{-(d+1)/2}) , \quad |x| \to \infty .$$

Using the fact that

$$\int_{\partial D} \frac{\partial u}{\partial \nu}\Big|_+ \bar{u}\, d\sigma = \frac{\mu_0}{\mu} \int_{\partial D} \frac{\partial u}{\partial \nu}\Big|_- \bar{u}\, d\sigma = \frac{\mu_0}{\mu} \int_D (|\nabla u|^2 - k^2 |u|^2)\, dx ,$$

we find that

$$\Im \int_{\partial D} \frac{\partial u}{\partial \nu}\Big|_+ \bar{u}\, d\sigma = 0 ,$$

which gives, by applying Lemma 11.3, that $u \equiv 0$ in $\mathbb{R}^d \setminus D$. Now u satisfies $(\Delta + k^2)u = 0$ in D and $u = \partial u/\partial \nu = 0$ on ∂D. By the unique continuation property of $\Delta + k^2$, we readily get $u \equiv 0$ in D, and hence in $\mathbb{R}^d$. In particular, $\mathcal{S}_D^{k_0} g = 0$ on ∂D. Since $(\Delta + k_0^2)\mathcal{S}_D^{k_0} g = 0$ in D and k_0^2 is not a Dirichlet eigenvalue for $-\Delta$ on D, we have $\mathcal{S}_D^{k_0} g = 0$ in D, and hence in $\mathbb{R}^d$. It then follows from the jump relation (11.4) that

$$g = \frac{\partial (\mathcal{S}_D^{k_0} g)}{\partial \nu}\Big|_+ - \frac{\partial (\mathcal{S}_D^{k_0} g)}{\partial \nu}\Big|_- = 0 \quad \text{on } \partial D .$$

On the other hand, $\mathcal{S}_D^k f$ satisfies $(\Delta + k^2)\mathcal{S}_D^k f = 0$ in $\mathbb{R}^d \setminus \overline{D}$ and $\mathcal{S}_D^k f = 0$ on ∂D. It then follows from Lemma 11.3 (see also Theorem 3.7 of [91]), that $\mathcal{S}_D^k f = 0$. Then, in the same way as above, we can conclude that $f = 0$. This finishes the proof of solvability of (11.10). The estimate (11.11) is an easy consequence of solvability and the closed graph theorem. Finally, it can be easily proved in the three-dimensional case that if k_0 and k go to zero, then the constant C in (11.11) can be chosen independently of k_0 and k. We leave the details to the reader. $\square$

The following representation formula holds.

Theorem 11.5 *Suppose that k_0^2 is not a Dirichlet eigenvalue for $-\Delta$ on D. Let u be the solution of (11.8) and $g := \frac{\partial u}{\partial \nu}|_{\partial \Omega}$. Define*

$$H(x) := -\mathcal{S}_\Omega^{k_0}(g)(x) + \mathcal{D}_\Omega^{k_0}(f)(x) , \quad x \in \mathbb{R}^d \setminus \partial \Omega , \tag{11.12}$$

and let $(\varphi, \psi) \in L^2(\partial D) \times L^2(\partial D)$ be the unique solution of

$$\begin{cases} \mathcal{S}_D^k \varphi - \mathcal{S}_D^{k_0} \psi = H \\ \dfrac{1}{\mu} \dfrac{\partial (\mathcal{S}_D^k \varphi)}{\partial \nu}\Big|_- - \dfrac{1}{\mu_0} \dfrac{\partial (\mathcal{S}_D^{k_0} \psi)}{\partial \nu}\Big|_+ = \dfrac{1}{\mu_0} \dfrac{\partial H}{\partial \nu} \end{cases} \quad \text{on } \partial D . \tag{11.13}$$

Then u can be represented as

$$u(x) = \begin{cases} H(x) + \mathcal{S}_D^{k_0}\psi(x) \,, & x \in \Omega \setminus \overline{D} \,, \\ \mathcal{S}_D^k \varphi(x) \,, & x \in D \,. \end{cases} \tag{11.14}$$

Moreover, there exists $C > 0$ independent of H such that

$$\|\varphi\|_{L^2(\partial D)} + \|\psi\|_{L^2(\partial D)} \le C \left(\|H\|_{L^2(\partial D)} + \|\nabla H\|_{L^2(\partial D)} \right) . \tag{11.15}$$

Proof. Note that u defined by (11.14) satisfies the differential equations and the transmission condition on ∂D in (11.8). Thus in order to prove (11.14), it suffices to prove that $\partial u / \partial \nu = g$ on $\partial \Omega$. Let $f := u|_{\partial \Omega}$ and consider the following transmission problem:

$$\begin{cases} (\Delta + k_0^2)v = 0 & \text{in } (\Omega \setminus \overline{D}) \cup (\mathbb{R}^d \setminus \overline{\Omega}) \,, \\ (\Delta + k^2)v = 0 & \text{in } D \,, \\ v|_- - v|_+ = 0 \,, & \dfrac{1}{\mu}\dfrac{\partial v}{\partial \nu}\bigg|_- - \dfrac{1}{\mu_0}\dfrac{\partial v}{\partial \nu}\bigg|_+ = 0 \quad \text{on } \partial D \,, \\ v|_- - v|_+ = f, & \dfrac{\partial v}{\partial \nu}\bigg|_- - \dfrac{\partial v}{\partial \nu}\bigg|_+ = g \quad \text{on } \partial \Omega \,, \\ \left| \dfrac{\partial v}{\partial |x|}(x) - ik_0 v(x) \right| = O(|x|^{-(d+1)/2}) \,, & |x| \to \infty \,. \end{cases} \tag{11.16}$$

We claim that (11.16) has a unique solution. In fact, if $f = g = 0$, then we can show as before that $v = 0$ in $\mathbb{R}^d \setminus \overline{D}$. Thus

$$v = \frac{\partial v}{\partial \nu}\bigg|_- = 0 \quad \text{on } \partial D \,.$$

By the unique continuation for the operator $\Delta + k^2$, we have $v = 0$ in D, and hence $v \equiv 0$ in $\mathbb{R}^d$. Note that v_p, $p = 1, 2$, defined by

$$v_1(x) = \begin{cases} u(x) \,, & x \in \Omega \,, \\ 0 \,, & x \in \mathbb{R}^d \setminus \overline{\Omega} \,, \end{cases} \qquad v_2(x) = \begin{cases} H(x) + \mathcal{S}_D^{k_0}\psi(x) \,, & x \in \Omega \setminus \overline{D} \,, \\ \mathcal{S}_D^k\varphi(x) \,, & x \in D \,, \end{cases}$$

are two solutions of (11.16), and hence $v_1 \equiv v_2$. This finishes the proof. $\square$

Proposition 11.6 *For each $n \in \mathbb{N}$ there exists C_n independent of D such that*

$$\|H\|_{\mathcal{C}^n(\overline{D})} \le C_n \|f\|_{W_{\frac{1}{2}}^2(\partial \Omega)} \,.$$

Proof. Let $g := \partial u / \partial \nu |_{\partial \Omega}$. By the definition (11.12), it is easy to see that

$$\|H\|_{C^n(\overline{D})} \le C\left(\|g\|_{W^2_{-\frac{1}{2}}(\partial\Omega)} + \|f\|_{W^2_{\frac{1}{2}}(\partial\Omega)} \right),$$

where the constant C depends only on n and $\text{dist}(D, \partial\Omega)$. Therefore it is enough to show that

$$\|g\|_{W^2_{-\frac{1}{2}}(\partial\Omega)} \le C\|f\|_{W^2_{\frac{1}{2}}(\partial\Omega)}$$

for some C independent of D.

Let φ be a C^∞-cutoff function which is 0 in a neighborhood of D and 1 in a neighborhood of $\partial\Omega$. Let $v \in W^2_{\frac{1}{2}}(\partial\Omega)$ and define $\tilde{v} \in W^{1,2}(\Omega)$ to be the unique solution to $\Delta\tilde{v} = 0$ in Ω and $\tilde{v} = v$ on $\partial\Omega$. Let $\langle\ ,\ \rangle_{\frac{1}{2}, -\frac{1}{2}}$ denote the $W^2_{\frac{1}{2}} - W^2_{-\frac{1}{2}}$ pairing on $\partial\Omega$. Then

$$\langle v, g \rangle_{\frac{1}{2}, -\frac{1}{2}} = \int_\Omega \Delta(\varphi u)\tilde{v}\,dx + \int_\Omega \nabla(\varphi u) \cdot \nabla\tilde{v}\,dx$$

$$= \int_\Omega \Delta\varphi u\tilde{v}\,dx + 2\int_\Omega \nabla\varphi \cdot \nabla u\tilde{v}\,dx - k_0^2 \int_\Omega \varphi u\tilde{v}\,dx$$

$$+ \int_\Omega \nabla(\varphi u) \cdot \nabla\tilde{v}\,dx \,.$$

Therefore, it follows from the Cauchy–Schwartz inequality that

$$|\langle v, g \rangle_{\frac{1}{2}, -\frac{1}{2}}| \le C\|u\|_{W^{1,2}(\Omega\backslash\overline{D})}\|\tilde{v}\|_{W^{1,2}(\Omega)} \le C\|u\|_{W^{1,2}(\Omega\backslash\overline{D})}\|v\|_{W^2_{\frac{1}{2}}(\partial\Omega)} \,.$$

Since $v \in W^2_{\frac{1}{2}}(\partial\Omega)$ is arbitrary, we get

$$\|g\|_{W^2_{-\frac{1}{2}}(\partial\Omega)} \le C\|u\|_{W^{1,2}(\Omega\backslash\overline{D})} \,. \tag{11.17}$$

Note that the constant C depends only on $\text{dist}(D, \partial\Omega)$. On the other hand, since k_0^2 is not a Dirichlet eigenvalue for the Helmholtz equation (11.8) in Ω we can prove that

$$\|u\|_{W^{1,2}(\Omega)} \le C\|f\|_{W^2_{\frac{1}{2}}(\partial\Omega)} \,,$$

where C depends only on $\omega^2, \mu_0, \mu, \varepsilon_0$, and ε. It then follows from (11.17) that

$$\|g\|_{W^2_{-\frac{1}{2}}(\partial\Omega)} \le C\|f\|_{W^2_{\frac{1}{2}}(\partial\Omega)} \,.$$

This completes the proof. $\square$

We now transform the representation formula (11.14) into the one using the Green's function and the background solution U, that is, the solution of (11.9).

Let $G_{k_0}(x, y)$ be the Dirichlet Green's function for $\Delta + k_0^2$ in Ω, *i.e.*, for each $y \in \Omega$, G is the solution of

$$\begin{cases} (\Delta + k_0^2) G_{k_0}(x, y) = \delta_y(x) , & x \in \Omega , \\ G_{k_0}(x, y) = 0 , & x \in \partial\Omega . \end{cases}$$

Then,

$$U(x) = \int_{\partial\Omega} \frac{\partial G_{k_0}(x, y)}{\partial \nu_y} f(y) d\sigma(y) , \quad x \in \Omega .$$

Introduce one more notation. For a Lipschitz domain $D \subset\subset \Omega$ and $\varphi \in L^2(\partial D)$, let

$$G_D^{k_0} \varphi(x) := \int_{\partial D} G_{k_0}(x, y) \varphi(y) \, d\sigma(y) , \quad x \in \overline{\Omega} .$$

Our second representation formula is the following.

Theorem 11.7 *Let* ψ *be the function defined in (11.13). Then*

$$\frac{\partial u}{\partial \nu}(x) = \frac{\partial U}{\partial \nu}(x) + \frac{\partial (G_D^{k_0} \psi)}{\partial \nu}(x) , \quad x \in \partial\Omega . \tag{11.18}$$

To prove Theorem 11.7 we first observe an easy identity. If $x \in \mathbb{R}^d \setminus \Omega$ and $z \in \Omega$ then

$$\int_{\partial\Omega} \Gamma_{k_0}(x - y) \left. \frac{\partial G_{k_0}(z, y)}{\partial \nu_y} \right|_{\partial\Omega} d\sigma(y) - \Gamma_{k_0}(x - z) . \tag{11.19}$$

As a consequence of (11.19), we have

$$\left(\frac{1}{2} I + (\mathcal{K}_\Omega^{k_0})^* \right) \left(\left. \frac{\partial G_{k_0}(z, \cdot)}{\partial \nu_y} \right|_{\partial\Omega} \right)(x) = \frac{\partial \Gamma_{k_0}(x - z)}{\partial \nu_x} , \tag{11.20}$$

for all $x \in \partial\Omega$ and $z \in \Omega$.

Our second observation is the following.

Lemma 11.8 *If* k_0^2 *is not a Dirichlet eigenvalue for* $-\Delta$ *on* Ω, *then* $(1/2) I + (\mathcal{K}_\Omega^{k_0})^* : L^2(\partial\Omega) \to L^2(\partial\Omega)$ *is injective.*

Proof. Suppose that $\varphi \in L^2(\partial\Omega)$ and $\left((1/2) I + (\mathcal{K}_\Omega^{k_0})^* \right) \varphi = 0$. Define

$$u(x) := \mathcal{S}_\Omega^{k_0} \varphi(x) , x \in \mathbb{R}^d \setminus \overline{\Omega} .$$

Then u is a solution of $(\Delta + k_0^2)u = 0$ in $\mathbb{R}^d \setminus \overline{\Omega}$, and satisfies the Sommerfeld radiation condition

$$\left| \frac{\partial u}{\partial |x|} - i k_0 u \right| = O\left(|x|^{-(d+1)/2} \right) \quad \text{as } |x| \to +\infty ,$$

and the Neumann boundary condition

$$\left.\frac{\partial u}{\partial \nu}\right|_{\partial \Omega} = \left(\frac{1}{2}I + (\mathcal{K}_\Omega^{k_0})^*\right)\varphi = 0 \, .$$

Therefore, by Lemma 11.3, we obtain $\mathcal{S}_\Omega^{k_0}\varphi(x) = 0$, $x \in \mathbb{R}^d \setminus \overline{\Omega}$. Since k_0^2 is not a Dirichlet eigenvalue for $-\Delta$ on Ω, we can prove that $\varphi \equiv 0$ in the same way as before. This completes the proof. $\square$

With these two observations available we are now ready to prove Theorem 11.7.

Proof of Theorem 11.7. Let $g := \partial u/\partial \nu$ and $g_0 := \partial U/\partial \nu$ on $\partial \Omega$ for convenience. By the divergence theorem, we get

$$U(x) = -\mathcal{S}_\Omega^{k_0}(g_0)(x) + \mathcal{D}_\Omega^{k_0}(f)(x) \, , \quad x \in \Omega \, .$$

It then follows from (11.12) that

$$H(x) = -\mathcal{S}_\Omega^{k_0}(g)(x) + \mathcal{S}_\Omega^{k_0}(g_0)(x) + U(x) \, , \quad x \in \Omega \, .$$

Consequently, substituting (11.14) into the above equation, we see that for $x \in \Omega$

$$H(x) = -\mathcal{S}_\Omega^{k_0}\left(\left.\frac{\partial H}{\partial \nu}\right|_{\partial \Omega} + \left.\frac{\partial (\mathcal{S}_D^{k_0}\psi)}{\partial \nu}\right|_{\partial \Omega}\right)(x) + \mathcal{S}_\Omega^{k_0}(g_0)(x) + U(x) \, .$$

Therefore the jump formula (11.4) yields

$$\begin{aligned}
\frac{\partial H}{\partial \nu} &= -\left(-\frac{1}{2}I + (\mathcal{K}_\Omega^{k_0})^*\right)\left(\left.\frac{\partial H}{\partial \nu}\right|_{\partial \Omega} + \left.\frac{\partial (\mathcal{S}_D^{k_0}\psi)}{\partial \nu}\right|_{\partial \Omega}\right) \\
&\quad + \left(\frac{1}{2}I + (\mathcal{K}_\Omega^{k_0})^*\right)(g_0) \quad \text{on } \partial \Omega \, .
\end{aligned} \tag{11.21}$$

By (11.20), we have for $x \in \partial \Omega$

$$\begin{aligned}
\frac{\partial (\mathcal{S}_D^{k_0}\psi)}{\partial \nu}(x) &= \int_{\partial D} \frac{\partial \Gamma_{k_0}(x-y)}{\partial \nu_x}\psi(y)\, d\sigma(y) \\
&= \left(\frac{1}{2}I + (\mathcal{K}_\Omega^{k_0})^*\right)\left(\left.\frac{\partial (G_D^{k_0}\psi)}{\partial \nu}\right|_{\partial \Omega}\right)(x) \, .
\end{aligned} \tag{11.22}$$

Thus we obtain

$$\begin{aligned}
&\left(-\frac{1}{2}I + (\mathcal{K}_\Omega^{k_0})^*\right)\left(\left.\frac{\partial (\mathcal{S}_D^{k_0}\psi)}{\partial \nu}\right|_{\partial \Omega}\right) \\
&= \left(\frac{1}{2}I + (\mathcal{K}_\Omega^{k_0})^*\right)\left(\left(-\frac{1}{2}I + (\mathcal{K}_\Omega^{k_0})^*\right)\left(\left.\frac{\partial (G_D^{k_0}\psi)}{\partial \nu}\right|_{\partial \Omega}\right)\right) \quad \text{on } \partial \Omega \, .
\end{aligned}$$

It then follows from (11.21) that

$$\left(\frac{1}{2}I + (\mathcal{K}_\Omega^{k_0})^*\right)\left(\frac{\partial H}{\partial \nu}\bigg|_{\partial\Omega} + \left(-\frac{1}{2}I + (\mathcal{K}_\Omega^{k_0})^*\right)\left(\frac{\partial(G_D^{k_0}\psi)}{\partial \nu}\bigg|_{\partial\Omega}\right) - g_0\right) = 0$$

on $\partial\Omega$ and hence, by Lemma 11.8, we arrive at

$$\frac{\partial H}{\partial \nu}\bigg|_{\partial\Omega} + \left(-\frac{1}{2}I + (\mathcal{K}_\Omega^{k_0})^*\right)\left(\frac{\partial(G_D^{k_0}\psi)}{\partial \nu}\bigg|_{\partial\Omega}\right) - g_0 = 0 \quad \text{on } \partial\Omega. \quad (11.23)$$

By substituting this equation into (11.14), we get

$$\frac{\partial u}{\partial \nu} = \frac{\partial U}{\partial \nu} - \left(-\frac{1}{2}I + (\mathcal{K}_\Omega^{k_0})^*\right)\left(\frac{\partial(G_D^{k_0}\psi)}{\partial \nu}\bigg|_{\partial\Omega}\right) + \frac{\partial(S_D^{k_0}\psi)}{\partial \nu} \quad \text{on } \partial\Omega.$$

Finally, using (11.22) we conclude that (11.18) holds and the proof is then complete. □

Observe that, by (11.4), (11.23) is equivalent to

$$\frac{\partial}{\partial \nu}\left(H + \mathcal{S}_\Omega^{k_0}\left(\frac{\partial(G_D^{k_0}\psi)}{\partial \nu}\bigg|_{\partial\Omega}\right) - U\right)\bigg|_{-} = 0 \quad \text{on } \partial\Omega.$$

On the other hand, by (11.19),

$$\mathcal{S}_\Omega^{k_0}\left(\frac{\partial(G_D^{k_0}\psi)}{\partial \nu}\bigg|_{\partial\Omega}\right)(x) = \mathcal{S}_D^{k_0}\psi(x), \quad x \in \partial\Omega.$$

Thus, by (11.14), we obtain

$$H(x) + \mathcal{S}_\Omega^{k_0}\left(\frac{\partial(G_D^{k_0}\psi)}{\partial \nu}\bigg|_{\partial\Omega}\right)(x) - U(x) = 0, \quad x \in \partial\Omega.$$

Then, by the unique continuation for $\Delta + k_0^2$, we obtain the following Lemma.

Lemma 11.9 *We have*

$$H(x) = U(x) - \mathcal{S}_\Omega^{k_0}\left(\frac{\partial(G_D^{k_0}\psi)}{\partial \nu}\bigg|_{\partial\Omega}\right)(x), \quad x \in \Omega. \quad (11.24)$$

Derivation of Asymptotic Formulae

Suppose that the domain D is of the form $D = \delta B + z$, and let u be the solution of (11.8). The function U is the background solution as before. In this chapter we derive an asymptotic expansion of $\partial u/\partial \nu$ on $\partial \Omega$ as $\delta \to 0$ in terms of the background solution U. The leading-order term in this asymptotic formula has been derived in by Vogelius and Volkov in [259] (see also [25] where the second-order term in the asymptotic expansions of solutions to the Helmholtz equation is obtained). The proof of our asymptotic expansion is radically different from the variational ones in [259, 25]. It is based on layer potential techniques and the decomposition formula (11.14) of the solution to the Helmholtz equation. For simplicity, although the asymptotic expansions are valid in the two-dimensional case we only consider $d = 3$ in what follows.

12.1 Asymptotic Expansion

We first derive an estimate of the form (11.15) with a constant C independent of δ.

Proposition 12.1 Let $D = \delta B + z$ and $(\varphi, \psi) \in L^2(\partial D) \times L^2(\partial D)$ be the unique solution of (11.13). There exists $\delta_0 > 0$ such that for all $\delta \leq \delta_0$, there exists a constant C independent of δ such that

$$\|\varphi\|_{L^2(\partial D)} + \|\psi\|_{L^2(\partial D)} \leq C\left(\delta^{-1}\|H\|_{L^2(\partial D)} + \|\nabla H\|_{L^2(\partial D)}\right). \qquad (12.1)$$

Proof. After the scaling $x = z + \delta y$, (11.13) takes the form

$$\begin{cases} \mathcal{S}_B^{k\delta}\varphi_\delta - \mathcal{S}_B^{k_0\delta}\psi_\delta = \dfrac{1}{\delta}H_\delta \\[2mm] \dfrac{1}{\mu}\dfrac{\partial(\mathcal{S}_B^{k\delta}\varphi_\delta)}{\partial \nu}\bigg|_- - \dfrac{1}{\mu_0}\dfrac{\partial(\mathcal{S}_B^{k_0\delta}\psi_\delta)}{\partial \nu}\bigg|_+ = \dfrac{1}{\delta\mu_0}\dfrac{\partial H_\delta}{\partial \nu} \end{cases} \quad \text{on } \partial B\,,$$

where $\varphi_\delta(y) = \varphi(z + \delta y)$, $y \in \partial B$, *etc*, and the single layer potentials $\mathcal{S}_B^{k\delta}$ and $\mathcal{S}_B^{k_0\delta}$ are defined by the fundamental solutions $\Gamma_{k\delta}$ and $\Gamma_{k_0\delta}$, respectively. It then follows from Theorem 11.4 that for δ small enough the following estimate holds:

$$\|\varphi_\delta\|_{L^2(\partial B)} + \|\psi_\delta\|_{L^2(\partial B)} \le C\delta^{-1}\|H_\delta\|_{W_1^2(\partial B)},$$

for some constant C independent of δ. By scaling back, we obtain (12.1). $\square$

Let H be the function defined in (11.12). Fix $n \in \mathbb{N}$, define

$$H_n(x) = \sum_{|i|=0}^{n} \frac{\partial^i H(z)}{i!}(x - z)^i,$$

and let (φ_n, ψ_n) be the unique solution of

$$\begin{cases} \mathcal{S}_D^k\varphi_n - \mathcal{S}_D^{k_0}\psi_n = H_{n+1} \\ \dfrac{1}{\mu}\dfrac{\partial(\mathcal{S}_D^k\varphi_n)}{\partial\nu}\bigg|_{-} - \dfrac{1}{\mu_0}\dfrac{\partial(\mathcal{S}_D^{k_0}\psi_n)}{\partial\nu}\bigg|_{+} = \dfrac{1}{\mu_0}\dfrac{\partial H_{n+1}}{\partial\nu} \end{cases} \quad \text{on } \partial D. \qquad (12.2)$$

Then $(\varphi - \varphi_n, \psi - \psi_n)$ is the unique solution of (12.2) with the right-hand sides defined by $H - H_{n+1}$. Therefore, by (12.1), we get

$$\begin{aligned} &\|\varphi - \varphi_n\|_{L^2(\partial D)} + \|\psi - \psi_n\|_{L^2(\partial D)} \\ &\le C\left(\delta^{-1}\|H - H_{n+1}\|_{L^2(\partial D)} + \|\nabla(H - H_{n+1})\|_{L^2(\partial D)}\right). \end{aligned} \qquad (12.3)$$

By the definition of H_{n+1}, we have

$$\begin{aligned} \|H - H_{n+1}\|_{L^2(\partial D)} &\le C|\partial D|^{1/2}\|H - H_{n+1}\|_{L^\infty(\partial D)} \\ &\le C|\partial D|^{1/2}\delta^{n+2}\|H\|_{C^{n+2}(\overline{D})}, \end{aligned}$$

and

$$\|\nabla(H - H_{n+1})\|_{L^2(\partial D)} \le C|\partial D|^{1/2}\delta^{n+1}\|H\|_{C^{n+1}(\overline{D})}.$$

It then follows from (12.3) and Proposition 11.6 that

$$\|\varphi - \varphi_n\|_{L^2(\partial D)} + \|\psi - \psi_n\|_{L^2(\partial D)} \le C|\partial D|^{1/2}\delta^{n+1}. \qquad (12.4)$$

By (11.18), we obtain

$$\frac{\partial u}{\partial\nu}(x) = \frac{\partial U}{\partial\nu}(x) + \frac{\partial(G_D^{k_0}\psi_n)}{\partial\nu}(x) + \frac{\partial(G_D^{k_0}(\psi - \psi_n))}{\partial\nu}(x), \quad x \in \partial\Omega.$$

Since $\text{dist}(D, \partial\Omega) \ge c_0$, we get

$$\sup_{x\in\partial\Omega,\, y\in\partial D}\left|\frac{\partial G_{k_0}}{\partial\nu}(x, y)\right| \le C$$

for some C. Hence, for each $x \in \partial\Omega$, we have from (12.4)

$$\left| \frac{\partial(G_D^{k_0}(\psi - \psi_n))}{\partial\nu}(x) \right| \leq \left[\int_{\partial D} \left| \frac{\partial G_{k_0}(x, y)}{\partial\nu_x} \right|^2 d\sigma(y) \right]^{1/2} \|\psi - \psi_n\|_{L^2(\partial D)}$$

$$\leq C|\partial D|^{1/2}|\partial D|^{1/2}\delta^{n+1} \leq C'\delta^{n+d} ,$$

where C and C' are independent of $x \in \partial\Omega$ and δ. Thus we conclude that

$$\frac{\partial u}{\partial\nu}(x) = \frac{\partial U}{\partial\nu}(x) + \frac{\partial(G_D^{k_0}\psi_n)}{\partial\nu}(x) + O(\delta^{n+d}) , \quad \text{uniformly in } x \in \partial\Omega . \quad (12.5)$$

For each multi-index i, define (φ_i, ψ_i) to be the unique solution to

$$\begin{cases} S_B^{k\delta}\varphi_i - S_B^{k_0\delta}\psi_i = x^i \\ \dfrac{1}{\mu}\dfrac{\partial(S_B^{k\delta}\varphi_i)}{\partial\nu}\bigg|_{-} - \dfrac{1}{\mu_0}\dfrac{\partial(S_B^{k_0\delta}\psi_i)}{\partial\nu}\bigg|_{+} = \dfrac{1}{\mu_0}\dfrac{\partial x^i}{\partial\nu} \end{cases} \quad \text{on } \partial B . \quad (12.6)$$

Then, we claim that

$$\varphi_n(x) = \sum_{|i|=0}^{n+1} \delta^{|i|-1}\frac{\partial^i H(z)}{i!}\varphi_i(\delta^{-1}(x - z)) ,$$

$$\psi_n(x) = \sum_{|i|=0}^{n+1} \delta^{|i|-1}\frac{\partial^i H(z)}{i!}\psi_i(\delta^{-1}(x - z)) .$$

In fact, the expansions follow from the uniqueness of the solution to the integral equation (11.10) and the relation

$$S_D^{k_0}\left(\sum_{|i|=0}^{n+1} \delta^{|i|-1}\frac{\partial^i H(z)}{i!}\varphi_i(\delta^{-1}(\cdot - z)) \right)(x)$$

$$= \sum_{|i|=0}^{n+1} \delta^{|i|}\frac{\partial^i H(z)}{i!}(S_B^{k_0\delta}\varphi_i)(\delta^{-1}(x - z)) ,$$

for $x \in \partial D$. It then follows from (12.5) that

$$\frac{\partial u}{\partial\nu}(x) = \frac{\partial U}{\partial\nu}(x) + \sum_{|i|=0}^{n+1} \delta^{|i|-1}\frac{\partial^i H(z)}{i!}\frac{\partial}{\partial\nu}G_D^{k_0}(\psi_i(\delta^{-1}(\cdot - z)))(x)$$

$$+ O(\delta^{n+d}) ,$$

$$\quad (12.7)$$

uniformly in $x \in \partial\Omega$. Note that

$$G_D^{k_0}(\psi_i(\delta^{-1}(\cdot - z)))(x) = \int_{\partial D} G_{k_0}(x,y)\psi_i(\delta^{-1}(y-z))\,d\sigma(y)$$

$$= \delta^{d-1}\int_{\partial B} G_{k_0}(x,\delta w + z)\psi_i(w)\,d\sigma(w)\,.$$

Moreover, for x near $\partial\Omega$, $z \in \Omega$, $w \in \partial B$, and sufficiently small δ, we have

$$G_{k_0}(x,\delta w + z) = \sum_{|j|=0}^{\infty} \frac{\delta^{|j|}}{j!}\partial_z^j G_{k_0}(x,z)w^j\,.$$

Therefore, we get

$$G_D^{k_0}(\psi_i(\delta^{-1}(\cdot - z)))(x) = \sum_{|j|=0}^{\infty} \frac{\delta^{|j|+d-1}}{j!}\partial_z^j G_{k_0}(x,z)\int_{\partial B} w^j \psi_i(w)\,d\sigma(w)\,.$$

Define, for multi-indices i and j in $\mathbb{N}^d$,

$$W_{ij} := \int_{\partial B} w^j \psi_i(w)\,d\sigma(w)\,. \tag{12.8}$$

Then we obtain the following theorem from (12.7).

Theorem 12.2 *The following pointwise asymptotic expansion on $\partial\Omega$ holds:*

$$\frac{\partial u}{\partial \nu}(x) = \frac{\partial U}{\partial \nu}(x) + \delta^{d-2}\sum_{|j|=0}^{n+1}\sum_{|i|=0}^{n-|j|+1} \frac{\delta^{|i|+|j|}}{i!j!}\partial^i H(z)\frac{\partial\partial_z^j G_{k_0}(x,z)}{\partial\nu_x}W_{ij} \tag{12.9}$$

$$+ O(\delta^{n+d})\,,$$

where the remainder $O(\delta^{d+n})$ is dominated by $C\delta^{d+n}\|f\|_{W_{\frac{1}{2}}^2(\partial\Omega)}$ for some C independent of $x \in \partial\Omega$.

In view of (11.18), we obtain the following expansion:

$$\frac{\partial(G_D^{k_0}\psi)}{\partial \nu}(x) = \delta^{d-2}\sum_{|j|=0}^{n+1}\sum_{|i|=0}^{n-|j|+1} \frac{\delta^{|i|+|j|}}{i!j!}\partial^i H(z)\frac{\partial\partial_z^j G_{k_0}(x,z)}{\partial\nu_x}W_{ij} \tag{12.10}$$

$$+ O(\delta^{n+d})\,.$$

Observe that ψ_i, and hence, W_{ij} depends on δ, and so does H. Thus the formula (12.9) is not a genuine asymptotic formula. However, since it is simple and has some potential applicability in solving the inverse problem for the Helmholtz equation, we made a record of it as a theorem.

Observe that by the definition (12.6) of ψ_i, $\|\psi_i\|_{L^2(\partial B)}$ is bounded, and hence

$$|W_{ij}| \leq C_{ij}\,, \quad \forall\, i,j\,,$$

where the constant C_{ij} is independent of δ. Since δ is small, we can derive an asymptotic expansion of (φ_i, ψ_i) using their definition (12.6). Let us briefly explain this. Let

$$T_\delta \begin{bmatrix} f \\ g \end{bmatrix} := \begin{bmatrix} \mathcal{S}_B^{k\delta} f - \mathcal{S}_B^{k_0\delta} g \\ \dfrac{1}{\mu} \left. \dfrac{\partial(\mathcal{S}_B^{k\delta} f)}{\partial \nu} \right|_- - \dfrac{1}{\mu_0} \left. \dfrac{\partial(\mathcal{S}_B^{k_0\delta} g)}{\partial \nu} \right|_+ \end{bmatrix} \quad \text{on } \partial B \, ,$$

and let T_0 be the operator when $\delta = 0$. Then the solution (φ_i, ψ_i) of the integral equation (12.6) is given by

$$\begin{bmatrix} \varphi_i \\ \psi_i \end{bmatrix} = \left[I + T_0^{-1}(T_\delta - T_0) \right]^{-1} T_0^{-1} \begin{bmatrix} x^i \\ \dfrac{1}{\mu_0} \dfrac{\partial x^i}{\partial \nu} \end{bmatrix} . \tag{12.11}$$

By expanding $T_\delta - T_0$ in a power series of δ, we can derive the expansions of ψ_i and W_{ij}. Let, for $i, j \in \mathbb{N}^d$, $(\widehat{\varphi}_i, \widehat{\psi}_i)$ be the leading-order term in the expansion of (φ_i, ψ_i). Then $(\widehat{\varphi}_i, \widehat{\psi}_i)$ is the solution of the system of the integral equations

$$\begin{cases} \mathcal{S}_B^0 \widehat{\varphi}_i - \mathcal{S}_B^0 \widehat{\psi}_i = x^i \\ \dfrac{1}{\mu} \left. \dfrac{\partial(\mathcal{S}_B^0 \widehat{\varphi}_i)}{\partial \nu} \right|_- - \dfrac{1}{\mu_0} \left. \dfrac{\partial(\mathcal{S}_B^0 \widehat{\psi}_i)}{\partial \nu} \right|_+ = \dfrac{1}{\mu_0} \dfrac{\partial x^i}{\partial \nu} \end{cases} \quad \text{on } \partial B \, . \tag{12.12}$$

As a simplest case, let us now take $n = 1$ in (12.9) to find the leading-order term in the asymptotic expansion of $\partial u/\partial \nu|_{\partial \Omega}$ as $\delta \to 0$. We first investigate the dependence of W_{ij} on δ for $|i| \le 1$ and $|j| \le 1$. If $|i| \le 1$, then both sides of the first equation in (12.12) are harmonic in B, and hence

$$\mathcal{S}_B^0 \widehat{\varphi}_i - \mathcal{S}_B^0 \widehat{\psi}_i = x^i \quad \text{in } B \, .$$

Therefore we get

$$\left. \dfrac{\partial(\mathcal{S}_B^0 \widehat{\varphi}_i)}{\partial \nu} \right|_- - \left. \dfrac{\partial(\mathcal{S}_B^0 \widehat{\psi}_i)}{\partial \nu} \right|_- = \dfrac{\partial x^i}{\partial \nu} \quad \text{on } \partial B \, .$$

This identity together with the second equation in (12.12) yields

$$\dfrac{\mu}{\mu_0} \left. \dfrac{\partial(\mathcal{S}_B^0 \widehat{\psi}_i)}{\partial \nu} \right|_+ - \left. \dfrac{\partial(\mathcal{S}_B^0 \widehat{\psi}_i)}{\partial \nu} \right|_- = \left(1 - \dfrac{\mu}{\mu_0} \right) \dfrac{\partial x^i}{\partial \nu} \, .$$

In view of the relation (11.4), we have

$$\dfrac{\mu}{\mu_0} \left(\dfrac{1}{2} I + \mathcal{K}_B^* \right) \widehat{\psi}_i - \left(-\dfrac{1}{2} I + \mathcal{K}_B^* \right) \widehat{\psi}_i = \left(1 - \dfrac{\mu}{\mu_0} \right) \dfrac{\partial x^i}{\partial \nu} \, ,$$

where $\mathcal{K}_B^*$ is the operator defined in (11.6) when $k = 0$. Therefore, we have

$$\widehat{\psi}_i = (\lambda I - \mathcal{K}_B^*)^{-1}\left(\frac{\partial x^i}{\partial \nu}\bigg|_{\partial B}\right), \tag{12.13}$$

where

$$\lambda := \frac{\frac{\mu}{\mu_0} + 1}{2(1 - \frac{\mu}{\mu_0})} = \frac{\frac{\mu_0}{\mu} + 1}{2(\frac{\mu_0}{\mu} - 1)}. \tag{12.14}$$

Here we have used the fact from Theorem 2.8 that the operator $\lambda I - \mathcal{K}_B^*$ on $L^2(\partial B)$ is invertible. Observe that if $|i| = 0$, then $\widehat{\psi}_i = 0$ and $\mathcal{S}_B^0 \widehat{\varphi}_i = 1$. Hence we obtain $\psi_i = O(\delta)$ and $\mathcal{S}_B^{k\delta}\varphi_i = 1 + O(\delta)$. Moreover, since $\mathcal{S}_B^{k\delta}\varphi_i$ depends on δ analytically and $(\Delta + k^2\delta^2)\mathcal{S}_B^{k\delta}\varphi_i = 0$ in B, we conclude that

$$\psi_i = O(\delta) \quad \text{and} \quad \mathcal{S}_B^{k\delta}\varphi_i = 1 + O(\delta^2), \quad |i| = 0. \tag{12.15}$$

It also follows from (12.13) that if $|i| = |j| = 1$, then

$$W_{ij} = \int_{\partial B} x^j (\lambda I - \mathcal{K}_B^*)^{-1}\left(\frac{\partial y^i}{\partial \nu}\bigg|_{\partial B}\right)(x)\, d\sigma(x) + O(\delta). \tag{12.16}$$

The first quantity in the right-hand side of (12.16) is the polarization tensor M_{ij} as defined in (3.1). In summary, we obtained that

$$W_{ij} = M_{ij} + O(\delta), \quad |i| = |j| = 1. \tag{12.17}$$

Suppose that either $i = 0$ or $j = 0$. By (11.4) and (12.6), we have

$$\psi_i = \frac{\partial(\mathcal{S}_B^{k_0\delta}\psi_i)}{\partial\nu}\bigg|_+ - \frac{\partial(\mathcal{S}_B^{k_0\delta}\psi_i)}{\partial\nu}\bigg|_-$$

$$= \frac{\mu_0}{\mu}\frac{\partial(\mathcal{S}_B^{k\delta}\varphi_i)}{\partial\nu}\bigg|_- - \frac{\partial x^i}{\partial\nu} - \frac{\partial(\mathcal{S}_B^{k_0\delta}\psi_i)}{\partial\nu}\bigg|_-. \tag{12.18}$$

It then follows from the divergence theorem that

$$\int_{\partial B} x^j \psi_i\, d\sigma = -k^2\delta^2\frac{\mu_0}{\mu}\int_B x^j \mathcal{S}_B^{k\delta}\varphi_i\, dx + k_0^2\delta^2\int_B x^j \mathcal{S}_B^{k_0\delta}\psi_i\, dx \tag{12.19}$$

$$+ \frac{\mu_0}{\mu}\int_{\partial B}\frac{\partial x^j}{\partial\nu}\mathcal{S}_B^{k\delta}\varphi_i\, d\sigma - \int_{\partial B}\frac{\partial x^j}{\partial\nu}\mathcal{S}_B^{k_0\delta}\psi_i\, d\sigma.$$

From (12.19), we can observe the following.

$$W_{ij} = -k^2\delta^2\frac{\mu_0}{\mu}|B| + O(\delta^3) = -\delta^2\omega^2\varepsilon\mu_0|B| + O(\delta^3), \quad |i| = |j| = 0, \tag{12.20}$$

$$W_{ij} = O(\delta^2), \quad |i| = 1, |j| = 0, \tag{12.21}$$

$$W_{ij} = O(\delta^2), \quad |i| = 0, |j| = 1. \tag{12.22}$$

In fact, (12.20) and (12.22) follow from (12.15) and (12.19), and (12.21) immediately follows from (12.19). As a consequence of (12.20), (12.21), (12.22), and (12.10), we obtain

$$\frac{\partial(G_D^{k_0}\psi)}{\partial\nu}(x) = O(\delta^d) , \quad \text{uniformly on } x \in \partial\Omega .$$

Since the center z is apart from $\partial\Omega$, it follows from (11.24) that

$$|H(z) - U(z)| + |\nabla H(z) - \nabla U(z)| = O(\delta^d) .$$

We now consider the case $|i| = 2$ and $|j| = 0$. In this case, one can show using (12.18) that

$$\int_{\partial B} \psi_i \, d\sigma = - \int_B \Delta x^i \, dx + O(\delta^2) .$$

Therefore, if $|j| = 0$, then

$$\sum_{|i|=2} \frac{1}{i!j!} \partial^i H(z) W_{ij} = -\Delta H(z)|B| + O(\delta^2) = k_0^2 H(z)|B| + O(\delta^2) . \quad (12.23)$$

So (12.9) together with (12.17)-(12.23) yields the following expansion formula of Vogelius-Volkov [259]. In fact, in [259], the formula is expressed in terms of the free space Green's function Γ_k instead of the Green's function G_{k_0}. However, these two formulae are the same, as we can see using the relation (11.20).

Theorem 12.3 *For any $x \in \partial\Omega$,*

$$\frac{\partial u}{\partial\nu}(x) = \frac{\partial U}{\partial\nu}(x)$$

$$+ \delta^d \left(\nabla U(z) M \frac{\partial \nabla_z G_{k_0}(x,z)}{\partial\nu_x} + \omega^2 \mu_0 (\varepsilon - \varepsilon_0)|B|U(z)\frac{\partial G_{k_0}(x,z)}{\partial\nu_x} \right) \quad (12.24)$$

$$+ O(\delta^{d+1}) ,$$

where M is the polarization tensor defined in (3.1) with λ given by (12.14).

Before returning to (12.9) let us make the following important remark. The tensors W_{ij} play the same role as the generalized polarization tensors. As defined in Chap. 3 the GPT's are given for $i, j \in \mathbb{N}^d$ by

$$M_{ij} := \int_{\partial B} w^j \widehat{\psi}_i(w) \, d\sigma(w) ,$$

where $\widehat{\psi}_i$ is defined by (12.12). The following result makes the connection between W_{ij} and M_{ij}. Its proof is immediate.

Lemma 12.4 *Suppose that a_i are constants such that $\sum_i a_i w^i$ is a harmonic polynomial. Then*

$$\sum_i a_i W_{ij} \to \sum_i a_i M_{ij} \quad as \ \delta \to 0 \ .$$

Observing now that the formula (12.9) still contains $\partial^i H$ factors, the remaining task is to convert (12.9) to a formula given solely by U and its derivatives. Substitution of (12.10) into (11.24) yields that, for any $x \in \Omega$,

$$
\begin{aligned}
H(x) = U(x) \\
-\delta^{d-2} \sum_{|j|=0}^{n+1} \sum_{|i|=0}^{n+1-|j|} \frac{\delta^{|i|+|j|}}{i!j!} \partial^i H(z) S_\Omega^{k_0} \left(\frac{\partial \partial_z^j G_{k_0}(x,z)}{\partial \nu_x} \right) W_{ij} \qquad (12.25) \\
+O(\delta^{n+d}) \ .
\end{aligned}
$$

In (12.25) the remainder $O(\delta^{n+d})$ is uniform in the $\mathcal{C}^n$-norm on any compact subset of Ω for any $n \in \mathbb{N}$ and therefore

$$(\partial^\gamma H)(z) + \delta^{d-2} \sum_{|j|=0}^{n+1} \sum_{|i|=0}^{n+1-|j|} \delta^{|i|+|j|} \partial^i H(z) P_{ij\gamma} = (\partial^\gamma U)(z) + O(\delta^{d+n}) \ ,$$

for all $\gamma \in \mathbb{N}^d$ with $|\gamma| \le n+1$ where

$$P_{ij\gamma} = \frac{1}{i!j!} W_{ij} \partial^\gamma S_\Omega^{k_0} \left(\frac{\partial \partial_z^j G_{k_0}(\cdot, z)}{\partial \nu_x} \right) \Big|_{x=z} \ .$$

Define the operator $\mathcal{P}_\delta$ by

$$\mathcal{P}_\delta : (w_\gamma)_{\gamma \in \mathbb{N}^d, |\gamma| \le n} \mapsto \left(w_\gamma + \delta^{d-2} \sum_{|j|=0}^{n+1} \sum_{|i|=0}^{n+1-|j|} \delta^{|i|+|j|} w_i P_{ij\gamma} \right)_{\gamma \in \mathbb{N}^d, |\gamma| \le n} \ .$$

Observe from (12.11) that $\mathcal{P}_\delta$ can be written as

$$\mathcal{P}_\delta = I + \delta^d \mathcal{P}_1 + \ldots + \delta^{n+d-1} \mathcal{P}_{n-1} + O(\delta^{n+d}) \ .$$

Defining as in (4.17) $\mathcal{Q}_p$, $p = 1, \ldots, n-1$, by

$$(I + \delta^d \mathcal{P}_1 + \ldots + \delta^{n+d-1} \mathcal{P}_{n-1})^{-1} = I + \delta^d \mathcal{Q}_1 + \ldots + \delta^{n+d-1} \mathcal{Q}_{n-1} + O(\delta^{n+d}) \ ,$$

we finally obtain that

$$((\partial^i H)(z))_{i \in \mathbb{N}^d, |i| \le n+1} = (I + \sum_{p=1}^n \delta^{d+p-1} \mathcal{Q}_p)((\partial^i U)(z))_{i \in \mathbb{N}^d, |i| \le n+1} + O(\delta^{d+n}) \ ,$$

which yields the main result of this chapter.

Theorem 12.5 *The following pointwise asymptotic expansion on $\partial\Omega$ holds:*

$$\frac{\partial u}{\partial \nu}(x) = \frac{\partial U}{\partial \nu}(x) + \delta^{d-2} \sum_{|j|=0}^{n+1} \sum_{|i|=0}^{n+1-|j|} \frac{\delta^{|i|+|j|}}{i!j!} \times$$

$$\left[\left((I + \sum_{p=1}^{n+2-|i|-|j|-d} \delta^{d+p-1}\mathcal{Q}_p)(\partial^\gamma U(z)) \right)_i \frac{\partial \partial_z^j G_{k_0}(x,z)}{\partial \nu_x} W_{ij} \right]$$

$$+ O(\delta^{n+d}) ,$$

where the remainder $O(\delta^{d+n})$ is dominated by $C\delta^{d+n}\|f\|_{W_{\frac{1}{2}}^2(\partial\Omega)}$ for some C independent of $x \in \partial\Omega$.

When $n = d$, we have a simpler formula

$$\frac{\partial u}{\partial \nu}(x) = \frac{\partial U}{\partial \nu}(x) + \delta^{d-2} \sum_{|j|=0}^{d+1} \sum_{|i|=0}^{d+1-|j|} \frac{\delta^{|i|+|j|}}{i!j!} \partial^i U(z) \frac{\partial \partial_z^j G_{k_0}(x,z)}{\partial \nu_x} W_{ij} \quad (12.26)$$

$$+ O(\delta^{2d}) .$$

Let us now consider the case when there are several well separated inclusions. The inclusion D takes the form $\cup_{s=1}^m (\delta B_s + z_s)$. The magnetic permeability and electric permittivity of the inclusion $\delta B_s + z_s$ are μ_s and ε_s, $s = 1, \ldots, m$. By iterating the formula (12.26) we can derive the following theorem.

Theorem 12.6 *The following pointwise asymptotic expansion on $\partial\Omega$ holds:*

$$\frac{\partial u}{\partial \nu}(x) = \frac{\partial U}{\partial \nu}(x)$$

$$+ \delta^{d-2} \sum_{s=1}^m \sum_{|j|=0}^{d+1} \sum_{|i|=0}^{d+1-|j|} \frac{\delta^{|i|+|j|}}{i!j!} \partial^i U(z) \frac{\partial \partial_z^j G_{k_0}(x,z)}{\partial \nu_x} W_{ij}^s + O(\delta^{2d}) . \quad (12.27)$$

Here W_{ij}^s is defined by (12.8) with B, μ, ε replaced by $B_s, \mu_s, \varepsilon_s$.

We conclude this chapter by making one final remark. In this chapter, we only derive the asymptotic formula for the solution to the Dirichlet problem. However, by the same method, we can derive an asymptotic formula for the Neumann problem as well.

13

Reconstruction Algorithms

Our goal in this chapter is to use the expansion (12.27) for efficiently determining the locations and/or shapes of the small electromagnetic inclusions from boundary measurements at a fixed frequency.

Assume that $d = 3$ only for the sake of simplicity. We develop two algorithms that use plane wave sources for identifying the small electromagnetic inclusions. We suppose that the boundary condition is given by

$$f = e^{ik_0\theta \cdot x},$$

where θ is a vector on the unit sphere S^2 in $\mathbb{R}^3$, $\theta \cdot \theta = 1$. We propose in this chapter two efficient and robust non-iterative algorithms for reconstructing the electromagnetic inclusions $\{D_s\}_{s=1}^m$ from limited voltage-to-current pairs

$$\left(u = e^{ik_0\theta \cdot x}|_{\partial\Omega}, \frac{\partial u}{\partial \nu}|_{\partial\Omega} \right).$$

The first algorithm, like the variational method in Sect. 5.4, reduces the reconstruction problem of the small inclusions to the calculation of an inverse Fourier transform. The second one is the MUSIC (standing for MUltiple-Signal-Classification) algorithm. We explain how it applies to imaging of small electromagnetic inclusions.

Another algorithm based on projections on three planes was proposed and successfully tested by Volkov in [261].

Note that algorithms similar to those proposed in this chapter can be designed in the context of the full time-harmonic Maxwell equations.

13.1 Asymptotic Expansion of a Weighted Combination of Voltage-to-Current Pairs

According to (12.27) the following asymptotic formula holds uniformly on $\partial\Omega$:

$$\frac{\partial u}{\partial \nu}(x) = \frac{\partial U}{\partial \nu}(x) + \delta^3 \sum_{s=1}^{m} \left[\nabla_y \frac{\partial G_{k_0}(x, z_s)}{\partial \nu_x} \cdot M^s \nabla U(z_s) \right.$$

$$\left. + k_0^2 (\frac{\varepsilon_s}{\varepsilon_0} - 1) \frac{\partial G_{k_0}(x, z_s)}{\partial \nu_x} |B_s| U(z_s) \right] + O(\delta^4) , \quad (13.1)$$

where the remainder $O(\delta^4)$ is independent of the set of points $\{z_s\}_{s=1}^{m}$ provided that the inclusions are well-separated from each other and from the boundary, and M^s is a the polarization tensor of Pólya–Szegö, associated with the s-th inclusion B_s and the conductivity μ_s/μ_0.

Let $\mathcal{H}(x/|x|, \theta, k_0)$ be defined as the function satisfying

$$-\mathcal{S}_\Omega(\frac{\partial u}{\partial \nu}|_{\partial \Omega}(y))(x) + \mathcal{D}_\Omega(e^{ik_0\theta \cdot y}|_{\partial \Omega})(x) = \mathcal{H}(\frac{x}{|x|}, \theta, k_0)\frac{e^{ik|x|}}{4\pi|x|} + O\left(\frac{1}{|x|^2}\right)$$

as $|x| \to \infty$. Note that $\mathcal{H}(x/|x|, \theta, k_0)$ is directly computed from the current-to-voltage pairs $(e^{ik_0\theta \cdot y}|_{\partial \Omega}, \partial u/\partial \nu|_{\partial \Omega})$. The following asymptotic formula holds uniformly on $\hat{x} = x/|x|$ and θ.

Theorem 13.1 *We have*

$$\mathcal{H}(\hat{x}, \theta, k_0)$$

$$= \delta^3 k_0^2 \sum_{s=1}^{m} \left[\hat{x} \cdot M^s \cdot \theta + (\frac{\varepsilon_s}{\varepsilon_0} - 1)|B_s| \right] e^{ik_0(\theta - \hat{x}) \cdot z_s} + O(\delta^4) , \quad (13.2)$$

for any $\hat{x}$ and $\theta \in S^2$, where $O(\delta^4)$ is independent of the set of points $\{z_s\}_{s=1}^{m}$.

Proof. It immediately follows from (13.1) that

$$- \mathcal{S}_\Omega(\frac{\partial u}{\partial \nu}|_{\partial \Omega}) + \mathcal{D}_\Omega(u|_{\partial \Omega})$$

$$= -\mathcal{S}_\Omega(\frac{\partial U}{\partial \nu}|_{\partial \Omega}) + \mathcal{D}_\Omega(U|_{\partial \Omega}) - \delta^3 \sum_{s=1}^{m} \left[\nabla_y \mathcal{S}_\Omega(\frac{\partial G_{k_0}(\cdot, z_s)}{\partial \nu}) \cdot M^s \nabla U(z_s) \right.$$

$$\left. + k_0^2 (\frac{\varepsilon_s}{\varepsilon_0} - 1)\mathcal{S}_\Omega(\frac{\partial G_{k_0}(\cdot, z_s)}{\partial \nu})|B_s| U(z_s) \right] + O(\frac{\delta^4}{|x|}) .$$

By (11.19) we have

$$\mathcal{S}_\Omega(\frac{\partial G_{k_0}(\cdot, y)}{\partial \nu})(x) = \Gamma_{k_0}(x - y) , \quad \forall \, x \in \mathbb{R}^3 \setminus \overline{\Omega} , \forall \, y \in \Omega .$$

Combining this relation with the following easy-to-check fact

$$-\mathcal{S}_\Omega(\frac{\partial U}{\partial \nu}|_{\partial \Omega})(x) + \mathcal{D}_\Omega(U|_{\partial \Omega})(x) = 0, \quad \forall \, x \in \mathbb{R}^3 \setminus \overline{\Omega} ,$$

we readily get that for $x \in \mathbb{R}^3 \setminus \overline{\Omega}$

$$-\mathcal{S}_\Omega(\frac{\partial u}{\partial \nu}|_{\partial\Omega})(x) + \mathcal{D}_\Omega(u|_{\partial\Omega})(x) = -\delta^3 \sum_{s=1}^{m} \left[\nabla_y \Gamma_{k_0}(x - z_s) \cdot M^s \nabla U(z_s) \right.$$

$$\left. + k_0^2(\frac{\varepsilon_s}{\varepsilon_0} - 1)\Gamma_{k_0}(x - z_s)|B_s|U(z_s)\right] + O(\frac{\delta^4}{|x|}) .$$

Since

$$\Gamma_{k_0}(x - z_s) = -\frac{e^{ik_0|x|}}{|x|}\frac{e^{-ik_0\frac{x}{|x|}\cdot z_s}}{4\pi} + O(\frac{1}{|x|^2})$$

and

$$\nabla_y \Gamma_{k_0}(x - z_s) = \frac{e^{ik_0|x|}}{|x|}\frac{ik_0 x}{4\pi|x|}e^{-ik_0\frac{x}{|x|}\cdot z_s} + O(\frac{1}{|x|^2}) ,$$

as $|x| \to \infty$, we obtain the desired asymptotic formula (13.2) which holds uniformly on $\hat{x} = x/|x|$ and θ in S^2. $\square$

In the following section we develop an algorithm for reconstructing the locations and order of magnitude of the inclusions from the function $\mathcal{H}(\hat{x}, \theta, k_0)$ at fixed k_0.

13.2 Reconstruction of Multiple Inclusions

13.2.1 The Fourier Transform Algorithm

In this subsection we present a linear method to determine the locations and the polarization tensors of several small inclusions from limited voltage-to-current pairs. Based on the asymptotic expansion (13.2) we reduce the reconstruction of the small electromagnetic inclusions from limited voltage-to-current pairs to the calculation of an inverse Fourier transform. This method follows the lines of the method proposed in [16] for reconstructing a collection of small inclusions from their scattering amplitude at a fixed frequency.

For convenience we are going to assume that B_s, for $s = 1, \ldots, m$, are balls. In this case, according to (3.22) the polarization tensors M^s have the following explicit forms:

$$M^s = m_s I_3 ,$$

where I_3 is the 3×3 identity matrix and the scalars m_s are given by

$$m_s = 3|B_s|\frac{\mu_s - \mu_0}{\mu_s + 2\mu_0} .$$

We are in possession of $\mathcal{H}(\hat{x}_l, \theta_{l'}, k_0)$ for a collection of pairs $(\hat{x}_l, \theta_{l'})$, where $l = 1, \ldots, L$ and $l' = 1, \ldots, L'$. Let, for $\hat{x}, \theta \in S^2$,

$$g(\hat{x}, \theta) := \delta^3 k_0^2 \sum_{s=1}^{m} e^{ik_0(\theta - \hat{x})\cdot z_s} \left[m_s\hat{x} \cdot \theta + (\frac{\varepsilon_s}{\varepsilon_0} - 1)|B_s| \right] .$$

We first observe that

$$g(\hat{x}, \theta) = g(-\theta, -\hat{x}) , \quad \forall \, \hat{x}, \theta \in S^2 .$$

Define, for $l = 1, \ldots, L$ and $l' = 1, \ldots, L'$, the coefficients $a_{l,l'}$ by

$$a_{l,l'} = \mathcal{H}(\hat{x}_l, \theta_{l'}, k_0) .$$

The reconstruction procedure is divided into three steps.

Step 1: Given that

$$g(\hat{x}_l, \theta_{l'}) \approx a_{l,l'} ,$$

we can compute using the Fast Fourier Transform (FFT) an accurate approximation of $g(\hat{x}, \theta)$ on $S^2 \times S^2$.

Step 2: Let $\mathcal{M}$ denote the following complex variety

$$\mathcal{M} = \left\{ \xi \in \mathbb{C}^3, \xi \cdot \xi = 1 \right\} .$$

It is easy to see that $g(\hat{x}, \theta)$ has an analytic continuation to $\mathcal{M} \times \mathcal{M}$. Let $(Y_{p,q})_{-p \leq q \leq p, p=0,1,\ldots}$ denote the normalized (in $L^2(S^2)$) spherical harmonics. Denote by $g_{p,q}$ the Fourier coefficients of g

$$g(\hat{x}, \theta) = \sum_{p,q} g_{p,q}(\hat{x}) Y_{p,q}(\theta) , \quad \forall \, \hat{x}, \theta \in S^2 . \tag{13.3}$$

Recall that from Step 1 we are in fact in possession of an accurate approximation of $g_{p,q}(\hat{x})$ on S^2 for $-p \leq q \leq p$ and $p \leq P$ for some P. In view of (13.3), the analytic continuation of the truncated Fourier series

$$\sum_{p,q;p \leq P} g_{p,q}(\hat{x}) Y_{p,q}(\theta)$$

of $g(\hat{x}, \theta)$ on $\mathcal{M} \times \mathcal{M}$ can be obtained by using the standard analytic continuation of the spherical harmonics $(Y_{p,q}(\theta))_{p,q}$ on the complex variety $\mathcal{M}$ followed by other analytic continuation of the Fourier expansion in $\hat{x}$. We know that the analytic continuation of g defined from $S^2 \times S^2$ to $\mathcal{M} \times \mathcal{M}$ is unique. In fact, $\mathcal{M}$ is a two dimensional complex variety and S^2 is a totally real two dimensional real sub-manifold of $\mathcal{M}$. Thus an analytic function which vanishes on S^2 must be 0, see for example [43].

Step 3: Recalling that given $a_{l,l'}$ for $l = 1, \ldots, L$ and $l' = 1, \ldots, L'$ we have constructed by Step 1 and Step 2 an accurate approximation of the function $g(\hat{x}, \theta)$ that is analytic on $\mathcal{M} \times \mathcal{M}$ and is such that

$$g(\hat{x}_l, \theta_{l'}) \approx a_{l,l'} , \quad \forall \, l = 1, \ldots L \text{ and } l' = 1, \ldots, L' .$$

But for any $\xi \in \mathbb{R}^3$ we know that there exists ξ_1 and ξ_2 in $\mathcal{M}$ such that $\xi = (\xi_1 - \xi_2)/k_0$. It suffices to choose

$$\xi_1 = \frac{\xi}{2k_0} + r\zeta + i\eta, \xi_2 = -\frac{\xi}{2k_0} + r\zeta + i\eta$$

with $r \in \mathbb{R}$ and $\zeta, \eta \in \mathbb{R}^3$ such that

$$\xi \cdot \zeta = \xi \cdot \eta = \zeta \cdot \eta = 0, |\zeta| = 1$$

and

$$|\eta|^2 = \frac{|\xi|^2}{4k_0^2} + r^2 - 1 .$$

Let us now view $(a_{l,l'})$ as a function of $\xi \in \mathbb{R}^3$. We have

$$g(\xi_1, \xi_2) = \delta^3 k_0^2 \sum_{s=1}^{m} e^{-i\xi \cdot z_s} \left[m_s \xi_1 \cdot \xi_2 + (\frac{\varepsilon_s}{\varepsilon_0} - 1)|B_s| \right] ,$$

and since

$$\xi_1 \cdot \xi_2 = 1 - \frac{1}{2k_0^2}|\xi|^2 ,$$

we can rewrite g as follows

$$g(\xi_1, \xi_2) = \delta^3 k_0^2 |B_s| \sum_{s=1}^{m} e^{-i\xi \cdot z_s} \left[\frac{3(\mu_s - \mu_0)}{\mu_s + 2\mu_0}(1 - \frac{1}{2k_0^2}|\xi|^2) + (\frac{\varepsilon_s}{\varepsilon_0} - 1) \right] .$$

$$(13.4)$$

Define

$$\mathcal{E}(\xi) = g(\xi_1, \xi_2) ,$$

and note that we are now in possession of an approximation to $\mathcal{E}(\xi)$ for any $\xi \in \mathbb{R}^3$. Here we rely on the fact that the analytic continuation is unique.

Recall that $e^{-i\xi \cdot z_s}$ (up to a multiplicative constant) is exactly the Fourier transform of the Dirac function δ_{z_s} (a point mass located at z_s). Multiplication by powers of ξ in Fourier space corresponds to differentiation of the Dirac function. Therefore, using the inverse Fourier transform we obtain

$$\check{\mathcal{E}} = \sum_{s=1}^{m} \delta^3 L_s \delta_{z_s} ,$$

where L_s are, in view of (13.4), second-order constant coefficient differential operators.

Hence $\mathcal{E}(\xi)$ is the Fourier transform of a distribution with its support at the locations of the centers of inclusions z_s. Therefore, we think that a discrete inverse Fourier transform of a sample of $\mathcal{E}(\xi)$ will efficiently pin down the z_s's. The method of locating the points z_s is then similar to that proposed for the conductivity problem in Sect. 5.4. Recall that the number of data (sampling) points needed for an accurate discrete Fourier inversion of $\mathcal{E}(\xi)$ follows from the Shannon's sampling theorem . We need (conservatively) of order $(h/\delta)^3$ sampled values of ξ to reconstruct, with resolution of order δ, a collection of inclusions that lie inside a square of side h. Note, however, that real measurements are only taken in Step 1. It remains to be seen how many such measurements are needed. Once the locations $\{z_s\}_{s=1}^m$ are known we may calculate $|B_s|$ by solving the appropriate linear system arising from (13.4).

If B_s are general domains our calculations become more complex and eventually we have to deal with pseudo-differential operators (independent of the space variable x) applied to the same Dirac functions. The feasibility of this approach is illustrated in the following numerical examples from [146].

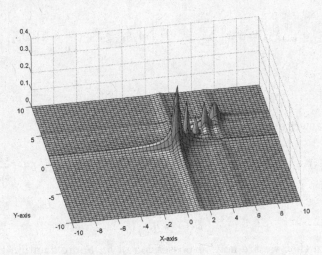

Fig. 13.1. Reconstruction of five electromagnetic inclusions of the shape of balls in $[-10, 10]^3$.

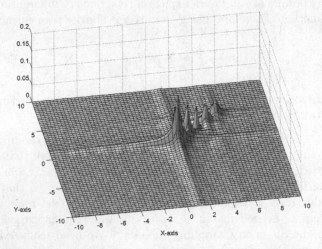

Fig. 13.2. Reconstruction of five electromagnetic inclusions with 10% noise.

13.2.2 The MUSIC Algorithm

Following Cheney [77] we briefly present the MUSIC algorithm. This is essentially a method of characterizing the range of a self-adjoint operator. Suppose A is a self-adjoint operator with eigenvalues $\lambda_1 \geq \lambda_2 \geq \ldots$ and corresponding eigenvectors $v_1, v_2, \ldots$. Suppose the eigenvalues $\lambda_{n+1}, \lambda_{n+2}, \ldots$ are all zero, so that the vectors $v_{n+1}, v_{n+2}, \ldots$ span the null space of A. Alternatively, $\lambda_{n+1}, \lambda_{n+2}, \ldots$ could merely be very small, below the noise level of the system represented by A; in this case we say that the vectors $v_{n+1}, v_{n+2}, \ldots$ span the noise subspace of A. We can form the projection onto the noise subspace; this projection is given explicitly by $P_{\text{noise}} = \sum_{p>n} v_p \overline{v_p}^T$, where the subscript T denotes the transpose and the bar denotes the complex conjugate. The (essential) range of A, meanwhile, is spanned by the vectors $v_1, v_2, \ldots, v_n$.

The key idea of MUSIC is this: because A is self-adjoint, we know that the noise subspace is orthogonal to the (essential) range. Therefore, a vector f is in the range of A if and only if its projection onto the noise subspace is zero, *i.e.*, if $||P_{\text{noise}} f|| = 0$, or equivalently,

$$\frac{1}{||P_{\text{noise}} f||} = +\infty . \tag{13.5}$$

Equation (13.5) is the MUSIC characterization of the range of A. If A is not self-adjoint, MUSIC can be used with the singular-value decomposition (SVD) instead of the eigenvalue decomposition.

MUSIC is generally used in signal processing problems [251] as a method for estimating the individual frequencies of multiple time-harmonic signals. Devaney [99] has recently applied the MUSIC algorithm to the problem of estimating the locations of a number of point-like scatterers. See also [61] and [74].

In this subsection we apply the MUSIC algorithm to determine the locations of several small inclusions from limited voltage-to-current pairs.

Let $(\theta_1, \ldots, \theta_n)$ and $(\hat{x}_1, \ldots, \hat{x}_n) \in (S^2)^n$ be n directions of incidence and observation, respectively. Our inverse problem is to determine the locations $z_1, \ldots, z_m$ from $\mathcal{H}(\hat{x}_l, \theta_{l'}, k_0)$.

Defining the matrix $A = (A_{ll'})_{l,l'=1}^n \in \mathbb{C}^{n \times n}$ by

$$A_{ll'} = \sum_{s=1}^m \left(-\theta_l M^s \theta_{l'} + (\frac{\varepsilon_s}{\varepsilon_0} - 1)|B_s| \right) e^{ik_0(\theta_l + \theta_{l'}) \cdot z_s}, \quad l, l' = 1, \ldots, n ,$$

we observe that

$$\delta^{-3} k_0^{-2} \mathcal{H}(\hat{x}_l, \theta_{l'}, k_0) \Big|_{\hat{x}_l = -\theta_l} \simeq A_{ll'} .$$

Introduce the notation

$$v_s = \left((1, \theta_1)^T e^{ik_0 \theta_1 \cdot z_s}, \ldots, (1, \theta_n)^T e^{ik_0 \theta_n \cdot z_s} \right)^T$$

to rewrite the matrix A as a sum of outer products:

$$A = \sum_{s=1}^{m} v_s \begin{pmatrix} (\frac{\varepsilon_s}{\varepsilon_0} - 1)|B_s| & 0 \\ 0 & M^s \end{pmatrix} v_s^T .$$

Our matrix A, called the multi-static response matrix, is symmetric, but it is not Hermitian. We form a Hermitian matrix $\widetilde{A} = \overline{A}A$. We note that $\overline{A}$ is the frequency-domain version of a time-reversed multi-static response matrix; thus $\widetilde{A}$ corresponds to performing an experiment, time-reversing the received signals and using them as input for a second experiment [201, 233, 234, 61]. The matrix $\widetilde{A}$ can be written as follows

$$\widetilde{A} = \sum_{s=1}^{m} \overline{v_s} \begin{pmatrix} (\frac{\varepsilon_s}{\varepsilon_0} - 1)|B_s| & 0 \\ 0 & M^s \end{pmatrix} \overline{v_s}^T \sum_{s=1}^{m} v_s \begin{pmatrix} (\frac{\varepsilon_s}{\varepsilon_0} - 1)|B_s| & 0 \\ 0 & M^s \end{pmatrix} v_s^T .$$

For simplicity we consider only the case $n > 3m$. For any point $z \in \Omega$ we define g_z by

$$g_z = \left((1, \theta_1)^T e^{ik_0\theta_1 \cdot z}, \ldots, (1, \theta_n)^T e^{ik_0\theta_n \cdot z} \right)^T .$$

It can be shown that there exists $n_0 \in \mathbb{N}$ such that for any $n \geq n_0$ the following statement holds [178, 146, 15]

$$g_z \in \text{Range}(\widetilde{A}) \text{ if and only if } z \in \{z_1, \ldots, z_m\} .$$

The MUSIC algorithm can now be used as follows to determine the location of the inclusions. Let $P_{\text{noise}} = I - P$, where P is the orthogonal projection onto the range of $\widetilde{A}$. Given any point $z \in \Omega$, form the vector g_z. The point z coincides with the location of an inclusion if and only if $P_{\text{noise}}g_z = 0$. Thus we can form an image of the inclusions by plotting, at each point z, the quantity $1/\|P_{\text{noise}}g_z\|$. The resulting plot will have large peaks at the locations of the inclusions.

As pointed out the eigenvectors of the Hermitian matrix $\widetilde{A}$ can be computed by the SVD of the response matrix A. The eigenvalues of $\widetilde{A}$ are the squares of the singular values of A. An immediate application of the SVD of A is the determination of the number of inclusions. If, for example, $\mu_s \neq \mu_0$ and $\varepsilon_s \neq \varepsilon_0$ for all $s = 1, \ldots, m$, then there are exactly $3m$ significant singular values of A and the rest are zero or close to zero. If therefore the SVD of A has no significant singular values, then there are no detectable inclusions in the medium. Now, when there are detectable inclusions in the medium, we can use the singular vectors of A to locate them since these vectors span the range of $\widetilde{A}$. We have, in fact, a one-to-one correspondence between the singular vectors and the inclusions [233, 235].

A

Appendices

A.1 Theorem of Coifman, McIntosh, and Meyer

The proof of Theorem 2.4 is based on the following celebrated theorem of Coifman, McIntosh, and Meyer [87].

Theorem A.1 *Let A, φ be Lipschitz functions on $\mathbb{R}^{d-1}$. The singular integral operator with the integral kernel*

$$\frac{A(x') - A(y')}{(|x' - y'|^2 + (\varphi(x') - \varphi(y'))^2)^{\frac{d}{2}}}$$

is bounded on $L^2(\mathbb{R}^{d-1})$.

Theorem A.1 was proved by reducing the matter to the one dimension using the method of rotation of Calderón, and then by using the following general theorem obtained in the same paper.

Theorem A.2 *Let K be a compact convex subset in the complex plane, U be an open set containing K, and $F : U \to \mathbb{C}$ be a holomorphic function. Let A and B be Lipschitz functions on $\mathbb{R}$ such that*

$$\frac{A(x) - A(y)}{x - y} \in K .$$

Then the principal value operator defined by the kernel

$$\frac{B(x) - B(y)}{(x - y)^2} F\left(\frac{A(x) - A(y)}{x - y}\right)$$

is bounded on $L^2(\mathbb{R})$.

The L^2-boundedness of the operators $\mathcal{K}_D$ and $\mathcal{K}_D^*$ in Theorem 2.4 follows immediately from Theorem A.1. In order to keep the technicalities to the minimum, we suppose that $d \geq 3$, and the domain D is given by a Lipschitz graph, namely, $D = \{(x', x_d) : x_d = \varphi(x')\}$, where $\varphi : \mathbb{R}^{d-1} \rightarrow \mathbb{R}$ is a Lipschitz function. If $x = (x', x_d), y = (y', y_d)$, then $\mathcal{K}_D$ is the principle value operator with the kernel

$$\frac{1}{\omega_d} \frac{\varphi(y') - \varphi(x') - \langle y' - x', \nabla\varphi(y')\rangle}{(|x' - y'|^2 + (\varphi(x') - \varphi(y'))^2)^{\frac{d}{2}}},$$

and $\mathcal{K}_D^*$ is the principle value operator with the kernel

$$\frac{1}{\omega_d} \frac{(\varphi(x') - \varphi(y') - \langle x' - y', \nabla\varphi(x')\rangle)\sqrt{1 + |\nabla\varphi(y')|^2}}{(|x' - y'|^2 + (\varphi(x') - \varphi(y'))^2)^{\frac{d}{2}}\sqrt{1 + |\nabla\varphi(x')|^2}}.$$

From Theorem A.1 (with first $A(x') = x'$, then $A(x') = \varphi(x')$) and the boundedness of $\nabla\varphi(x')$ we conclude that $\mathcal{K}_D$ is a bounded operator on $L^2(\partial D)$.

The integral kernel for the same operator $\mathcal{K}_D$ for the Lamé system involves terms defined by

$$\frac{(x_j' - y_j')^2(x_k' - y_k')}{|x' - y'|^{d+2}}.$$

The L^2-boundedness of such operators can be proved in a similar way using the method of rotation and Theorem A.2.

A.2 Continuity Method

Theorem A.3 *For $0 \leq t \leq 1$ suppose that the family of operators $A_t : L^2(\mathbb{R}^{d-1}) \rightarrow L^2(\mathbb{R}^{d-1})$ satisfy*

(i) $\|A_t\phi\|_{L^2(\mathbb{R}^{d-1})} \geq C\|\phi\|_{L^2(\mathbb{R}^{d-1})}$, *where C is independent of t,*
(ii) $t \mapsto A_t$ *is continuous in norm,*
(iii) $A_0 : L^2(\mathbb{R}^{d-1}) \rightarrow L^2(\mathbb{R}^{d-1})$ *is invertible.*

Then, $A_1 : L^2(\mathbb{R}^{d-1}) \rightarrow L^2(\mathbb{R}^{d-1})$ is invertible.

We give a brief proof for the readers' sake. Let

$$T := \left\{ t \in [0, 1] : A_t \text{ is invertible on } L^2(\mathbb{R}^{d-1}) \right\}.$$

Then T is nonempty by (iii). We can infer from (ii) that T is an open subset of $[0, 1]$. To prove that T is closed, choose a sequence $t_j, j = 1, 2, \ldots$, from T and assume that t_j converges to t_0 as $j \rightarrow \infty$. For a given $g \in L^2(\mathbb{R}^{d-1})$ let f_j be such that $A_{t_j} f_j = g$. Then by (i) there is a subsequence of f_j, which is still denoted by f_j, converging weakly to, say, f_0. We claim that $A_{t_0} f_0 = g$. In fact, if $h \in L^2(\mathbb{R}^{d-1})$, then

$$\langle A_{t_0} f_0 - g, h \rangle = \langle A_{t_0}(f_0 - f_j)g, h \rangle + \langle (A_{t_0} - A_{t_j})f_j, h \rangle$$
$$= \langle (f_0 - f_j)g, A_{t_0}^* h \rangle + \langle (A_{t_0} - A_{t_j})f_j, h \rangle \rightarrow 0 \quad \text{as } j \rightarrow \infty.$$

A.3 Collectively Compact Operators

Let $\{K_n\}_{n=1}^{+\infty}$ be a sequence of bounded, linear operators of a Banach space B (into itself). We say that the family of operators $\{K_n\}_{n=1}^{+\infty}$ is collectively compact iff the set $\{K_n(x) : n \geq 1, ||x|| \leq 1\}$ is relatively compact (its closure is compact) in B. The following result is the first assertion in Theorem 4.3 in [35].

Theorem A.4 *Let K and $K_n, n \geq 1$, be bounded, linear operators of a Banach space B. Assume that $K_n \to K$, pointwise, and that $\{K_n - K\}_{n=1}^{+\infty}$ is collectively compact. For any scalar, λ, the following two statements are equivalent*

(i) $\lambda I - K$ is an isomorphism.
(ii) There exists N such that $\lambda I - K_n$ is an isomorphism for $n \geq N$, and the set $\{(\lambda I - K_n)^{-1} : n \geq N\}$ is norm bounded.

A.4 Uniqueness for the Inverse Conductivity Problem

Let Ω be a simply connected Lipschitz domain in $\mathbb{R}^d, d \geq 2$, and let D be a compact subdomain of Ω. Let $g \in L_0^2(\partial\Omega)$. Fix $0 < k \neq 1 < +\infty$ and let u and U be the solutions of (2.37) and (2.38).

The inverse conductivity problem is to find D (and k) given $f = u|_{\partial\Omega}$ for one g (one boundary measurement) or for all g (many boundary measurements). In many applied situations it is f that is prescribed on $\partial\Omega$ and g that is measured on $\partial\Omega$. This makes some difference (not significant theoretically and computationally) in the case of single boundary measurements but makes almost no difference in the case of many boundary measurements, since actually it is the set of Cauchy data $\{f, g\}$ that is given.

A.4.1 Uniqueness With Many Measurements

Our purpose here is to state and prove a special case of the general uniqueness result due to Isakov [156]. We will need the following lemma which was first obtained in [172].

Lemma A.5 *Let u and U be as in Lemma 2.21. Then there are positive constants C_1 and C_2 depending only on k such that*

$$C_1 \left| \int_{\partial\Omega} (U - u)g \, d\sigma \right| \leq \int_D |\nabla U|^2 \, dx \leq C_2 \left| \int_{\partial\Omega} (U - u)g \, d\sigma \right|. \tag{A.1}$$

Proof. This lemma is a direct consequence of Lemma 2.21. Suppose first that $k > 1$. It follows from (2.58) that

$$\int_{\partial\Omega} (U - u)g \, d\sigma > 0,$$

and hence by (2.59)

$$\int_{\partial\Omega} (U - u)g \, d\sigma \leq (k - 1) \int_D |\nabla U|^2 \, dx \, .$$

On the other hand, using (2.58), we arrive at

$$\int_D |\nabla U|^2 \, dx \leq \int_D |\nabla u|^2 \, dx + \int_D |\nabla(u - U)|^2 \, dx \leq \frac{k}{k-1} \int_{\partial\Omega} (U - u)g \, d\sigma \, .$$

If $k < 1$, then

$$\int_{\partial\Omega} (U - u)g \, d\sigma < 0,$$

and we can proceed in the same way to prove the claim. □

We define the set of Cauchy data

$$\mathcal{C}_{D,k} = \left\{ (u|_{\partial\Omega}, \frac{\partial u}{\partial \nu}|_{\partial\Omega}) \, : \, u \in W^{1,2}(\Omega), \, \Delta u = 0 \text{ in } (\Omega\backslash\overline{D})\cup D, \, \frac{\partial u}{\partial \nu}|_+ = k\frac{\partial u}{\partial \nu}|_- \right\} \, .$$

In fact, $\mathcal{C}_{D,k}$ is a graph, namely

$$\mathcal{C}_{D,k} = \left\{ (f, \Lambda(f)) \in W^2_{\frac{1}{2}}(\partial\Omega) \times W^2_{-\frac{1}{2}}(\partial\Omega) \right\},$$

where $\Lambda(f) = \partial u/\partial\nu|_{\partial\Omega}$ with $u \in W^{1,2}(\Omega)$ the solution of

$$\begin{cases} \nabla \cdot \left(1 + (k - 1)\chi(D) \right)\nabla u = 0 & \text{in } \Omega \, , \\ u|_{\partial\Omega} = f \, . \end{cases}$$

The operator Λ is the Dirichlet-to-Neumann map in this case.

The following theorem is a special case of the general uniqueness theorem due to Isakov [156].

Theorem A.6 *Let Ω be a Lipschitz bounded domain in $\mathbb{R}^d, d \geq 2$. Suppose that D_1 and D_2 are bounded Lipschitz domains such that, for $p = 1, 2$, $\overline{D_p} \subset \Omega$ and $\Omega \setminus \overline{D_p}$ are connected. Suppose that the conductivity of D_p is $0 < k_p \neq 1 < +\infty$, $p = 1, 2$. If $\mathcal{C}_{D_1,k_1} = \mathcal{C}_{D_2,k_2}$, then $D_1 = D_2$ and $k_1 = k_2$.*

Proof.   For a fixed but arbitrary $g \in L^2_0(\partial\Omega)$, let $u_p, p = 1, 2$, be the solution to

$$\begin{cases} \nabla \cdot \left(1 + (k_p - 1)\chi(D_p) \right)\nabla u_p = 0 & \text{in } \Omega \, , \\ \dfrac{\partial u_p}{\partial \nu}\bigg|_{\partial\Omega} = g \in L^2_0(\partial\Omega), \quad \int_{\partial\Omega} u_p = 0 \, , \end{cases}$$

and U be the solution to (2.38). If $\mathcal{C}_{D_1,k_1} = \mathcal{C}_{D_2,k_2}$, then $u_1 = u_2$ on $\partial\Omega$, and hence

$$\int_{\partial\Omega} (U - u_1)g\,d\sigma = \int_{\partial\Omega} (U - u_2)g\,d\sigma .$$

It then follows from Lemma A.5 that

$$\int_{D_1} |\nabla U|^2\,dx \approx \int_{D_2} |\nabla U|^2\,dx . \tag{A.2}$$

Observe that (A.2) holds for all $U \in W^{1,2}(\Omega)$ harmonic in Ω.

Suppose that $D_1 \neq D_2$ and assume that D_1 is not a subset of D_2 without loss of generality. Then there is $z_0 \in \partial D_1$ such that z_0 is away from $\overline{D_2}$. For $z \notin \overline{D_1 \cup D_2}$ let $\Gamma_z(x) := \Gamma(x - z)$ where Γ is the fundamental solution for Δ. Then, Γ_z is harmonic in a neighborhood of $\overline{D_1 \cup D_2}$. Therefore, by the Runge approximation, there is a sequence of entire harmonic functions which converges uniformly on $\overline{D_1 \cup D_2}$ to $\Gamma_z(x)$. It then follows from (A.2) that

$$\int_{D_1} |\nabla \Gamma_z|^2\,dx \approx \int_{D_2} |\nabla \Gamma_z|^2\,dx , \tag{A.3}$$

regardless of z. As $z \to z_0$, the left-hand side of (A.3) goes to ∞ while the right-hand side stays bounded since z_0 is away from $\overline{D_2}$. This contradiction forces us to conclude that $D_1 = D_2$.

Let $g \in L_0^2(\partial\Omega)$ be nontrivial. Let $f = u_1 = u_2$ on $\partial\Omega$ and $\lambda_p = (k_p + 1)/(2(k_p - 1)), p = 1, 2$. Let $H = -\mathcal{S}_\Omega f + \mathcal{D}_\Omega g$ in $\mathbb{R}^d \setminus \partial\Omega$. By the representation formula (2.39), it follows that for $p = 1, 2$,

$$u_p = H + \mathcal{S}_{D_p}(\lambda_p I - \mathcal{K}_{D_p}^*)^{-1}\left(\frac{\partial H}{\partial\nu}\Big|_{\partial D_p}\right) \quad \text{in } \Omega .$$

Then

$$\mathcal{S}_{D_1}(\lambda_1 I - \mathcal{K}_{D_1}^*)^{-1}\left(\frac{\partial H}{\partial\nu}\Big|_{\partial D_1}\right) = \mathcal{S}_{D_1}(\lambda_2 I - \mathcal{K}_{D_1}^*)^{-1}\left(\frac{\partial H}{\partial\nu}\Big|_{\partial D_1}\right) \quad \text{in } \Omega .$$

Consequently, we can see that $\partial u_1/\partial\nu|_{\pm} = \partial u_2/\partial\nu|_{\pm}$ on $\partial D_1(= \partial D_2)$ and thus

$$(k_1 - k_2)\frac{\partial u_1}{\partial\nu}\Big|_{-} = 0 \quad \text{on } \partial D_1 .$$

It then suffices to prove that $\partial u_1/\partial\nu$ is not identically zero on ∂D_1. Suppose that $\partial u_1/\partial\nu \equiv 0$ on ∂D_1. From the uniqueness of a solution to the Neumann problem, it follows that u_1 is constant in D and hence

$$\partial u_1/\partial\nu|_{+} = \partial u_1/\partial\nu|_{-} = 0 \text{ on } \partial D_1 .$$

Therefore

$$(\lambda_1 I - \mathcal{K}_{D_1}^*)^{-1}\left(\frac{\partial H}{\partial\nu}\Big|_{\partial D_1}\right) = 0 ,$$

which implies that $\partial H/\partial\nu = 0$ on ∂D_1 and so the harmonic function H is constant everywhere in Ω. This leads us to a contradiction. $\square$

A.4.2 Uniqueness of Disks With One Measurement

Let Ω be a simply connected Lipschitz domain in $\mathbb{R}^d$ and let $D_p, p = 1, 2$, be compact subdomains of Ω. Fix $0 < k \neq 1 < +\infty$ and let $u_p, p = 1, 2$ be the solutions of

$$\begin{cases} \nabla \cdot \left(1 + (k-1)\chi(D_p)\right)\nabla u_p = 0 & \text{in } \Omega, \\ \dfrac{\partial u_p}{\partial \nu}\bigg|_{\partial\Omega} = g \in L_0^2(\partial\Omega), \quad \displaystyle\int_{\partial\Omega} u_p = 0. \end{cases} \tag{A.4}$$

The uniqueness question here is whether from $u_1 = u_2$ on $\partial\Omega$ for a certain g, it follows that $D_1 = D_2$. This question has been studied extensively recently. However, it is still wide open. The global uniqueness results are only obtained when D is restricted to convex polyhedrons and balls in three-dimensional space and polygons and disks in the plane (see [48, 46, 122, 159, 242, 169, 170]). Even the uniqueness within the classes of ellipses and ellipsoids is not known. We give here a proof due to Kang and Seo [168] for the unique determination of disks with one measurement.

Theorem A.7 *Let Ω be a simply connected Lipschitz domain in $\mathbb{R}^d$ and let D_1 and D_2 be two disks compactly contained in Ω. For any nonzero $g \in L_0^2(\partial\Omega)$ if $u_1 = u_2$ on $\partial\Omega$ then $D_1 = D_2$.*

Proof.   Let $f = u_1 = u_2$ on $\partial\Omega$ and $\lambda = (k+1)/(2(k-1))$. Let $H = -\mathcal{S}_\Omega f + \mathcal{D}_\Omega g$ in $\mathbb{R}^2 \setminus \partial\Omega$. By the representation formula (2.39), it follows that for $p = 1, 2$,

$$u_p = H + \frac{1}{\lambda}\mathcal{S}_{D_p}\left(\frac{\partial H}{\partial \nu}\bigg|_{\partial D_p}\right) \quad \text{in } \Omega.$$

(i) The monotone case: Assume that $D_1 \subset D_2$. Then $u_1 = u_2$ on $\partial\Omega$ implies

$$\int_\Omega (1 + (k-1)\chi(D_1))\nabla u_1 \cdot \nabla \eta = \int_\Omega (1 + (k-1)\chi(D_2))\nabla u_2 \cdot \nabla \eta,$$

and hence for all $\eta \in W^{1,2}(\Omega)$,

$$\int_\Omega (1 + (k-1)\chi(D_1))\nabla(u_1 - u_2) \cdot \nabla\eta = (k-1)\int_{D_2\setminus D_1} \nabla u_2 \cdot \nabla\eta. \tag{A.5}$$

Consequently, substituting $\eta = u_1$ and $\eta = u_1 - u_2$ in (A.5), we obtain

$$\int_\Omega (1 + (k-1)\chi(D_1))|\nabla(u_1 - u_2)|^2 + (k-1)\int_{D_2\setminus D_1} |\nabla u_2|^2 = 0.$$

Here we have used the fact that

$$\int_\Omega (1 + (k-1)\chi(D_1))\nabla(u_1 - u_2) \cdot \nabla U = 0.$$

So, if $k > 1$ then $u_1 = u_2$ in Ω and therefore by the transmission condition we conclude that $D_1 = D_2$ since otherwise $u_1 = u_2 \equiv 0$ in Ω. If $0 < k < 1$ we interchange the roles of D_1 and D_2 to arrive at the same conclusion.

(ii) The disjoint case: If D_1 and D_2 are disjoint, then

$$\mathcal{S}_{D_1}\left(\frac{\partial H}{\partial \nu}\bigg|_{\partial D_1}\right) = \mathcal{S}_{D_2}\left(\frac{\partial H}{\partial \nu}\bigg|_{\partial D_2}\right) \quad \text{in } \mathbb{R}^2 \setminus D_1 \cup D_2$$

implies that $\mathcal{S}_{D_1}(\frac{\partial H}{\partial \nu}|_{\partial D_1})$ is harmonic on $\mathbb{R}^2$ and hence $\partial H/\partial \nu = 0$ on ∂D_1 and H is a constant function which is a contradiction.

(iii) The non-monotone case: Recall that if D_p is a disk then $\mathcal{K}_{D_p}^* \equiv 0$ on $L_0^2(\partial D_p)$. From $u_1 = u_2$ on $\partial \Omega$ it follows by using the representation formula (2.39) that

$$\mathcal{S}_{D_1}\left(\frac{\partial H}{\partial \nu}\bigg|_{\partial D_1}\right) = \mathcal{S}_{D_2}\left(\frac{\partial H}{\partial \nu}\bigg|_{\partial D_2}\right) \quad \text{in } \mathbb{R}^2 \setminus D_1 \cup D_2 \, .$$

Assume that none of D_1 and D_2 contains the other. Assume that D_1 and D_2 are not disjoint. Since

$$\frac{\partial}{\partial \nu} \mathcal{S}_{D_p}\left(\frac{\partial H}{\partial \nu}\bigg|_{D_p}\right)\bigg|_{-} = -\frac{1}{2}\frac{\partial H}{\partial \nu}\bigg|_{D_p} \quad \text{on } \partial D_p \, , \ p = 1, 2 \, ,$$

it follows from the uniqueness of a solution to the Neumann boundary value problem for the Laplacian that

$$\mathcal{S}_{D_p}\left(\frac{\partial H}{\partial \nu}\bigg|_{D_p}\right) = -\frac{1}{2}H + c_p \quad \text{in } D_p$$

for some constant c_p. Hence

$$\mathcal{S}_{D_1}\left(\frac{\partial H}{\partial \nu}\bigg|_{D_1}\right) = \mathcal{S}_{D_2}\left(\frac{\partial H}{\partial \nu}\bigg|_{D_2}\right) + \text{ constant } \quad \text{in } D_1 \cap D_2 \, .$$

Since

$$\mathcal{S}_{D_1}\left(\frac{\partial H}{\partial \nu}\bigg|_{D_1}\right) = \mathcal{S}_{D_2}\left(\frac{\partial H}{\partial \nu}\bigg|_{D_2}\right) \quad \text{in } \mathbb{R}^2 \setminus D_1 \cup D_2 \, ,$$

we get by the continuity of the single layer potential that

$$\mathcal{S}_{D_1}\left(\frac{\partial H}{\partial \nu}\bigg|_{D_1}\right) = \mathcal{S}_{D_2}\left(\frac{\partial H}{\partial \nu}\bigg|_{D_2}\right) \quad \text{in } D_1 \cap D_2 \, .$$

Hence

$$\mathcal{S}_{D_1}\left(\frac{\partial H}{\partial \nu}\bigg|_{D_1}\right) = \mathcal{S}_{D_2}\left(\frac{\partial H}{\partial \nu}\bigg|_{D_2}\right) \quad \text{on } \partial(D_1 \setminus D_2) \cup \partial(D_2 \setminus D_1) \, ,$$

and by the maximum principle

$$\mathcal{S}_{D_1}\left(\frac{\partial H}{\partial \nu}\bigg|_{D_1}\right) = \mathcal{S}_{D_2}\left(\frac{\partial H}{\partial \nu}\bigg|_{D_2}\right) \quad \text{in } \mathbb{R}^2 \, .$$

Therefore $\mathcal{S}_{D_1}(\partial H/\partial \nu|_{D_1})$ is a harmonic function in the entire domain $\mathbb{R}^2$, and so $\partial H/\partial \nu = 0$ on ∂D_1 and H is a constant function which is a contradiction. $\square$

Note that in the monotone case global uniqueness holds for general domains [47, 2].

References

1. G. Alessandrini, Stable determination of conductivity by boundary measurements, Applicable Anal., 27 (1988), 153–172.
2. G. Alessandrini, Remark on a paper of Bellout and Friedman, Boll. Unione. Mat. Ita., 7 (1989), 243–250.
3. G. Alessandrini, Singular solutions of elliptic equations and the determination of conductivity by boundary measurements, J. Diff. Equat., 84 (1990), 252–272.
4. G. Alessandrini, Examples of instability in inverse boundary value problems, Inverse Problems, 13 (1997), 887–897.
5. G. Alessandrini, V. Isakov, and J. Powell, Local uniqueness in the inverse conductivity problem with one measurement, Trans. Amer. Math. Soc., 347 (1995), 3031–3041.
6. G. Alessandrini, A. Morassi, and E. Rosset, Detecting an inclusion in an elastic body by boundary measurements, SIAM J. Math. Anal., 33 (2002), 1247–1268.
7. G. Alessandrini, A. Morassi, and E. Rosset, Detecting cavities by electrostatic boundary measurements, Inverse Problems, 18 (2002), 1333–1353.
8. G. Alessandrini, A. Morassi, and E. Rosset, Size estimates in *Inverse Problems: Theory and Applications*, Contemp. Math., 333, Amer. Math. Soc., Providence, RI, 2003.
9. G. Alessandrini and E. Rosset, The inverse conductivity problem with one measurement: bounds on the size of the unknown object, SIAM J. Appl. Math., 58 (1998), 1060–1071.
10. G. Alessandrini, E. Rosset, and J.K. Seo, Optimal size estimates for the inverse conductivity problem with one measurement, Proc. Amer. Math. Soc., 128 (2000), 53–64.
11. D.L. Alumbaugh and H.F. Morrison, Monitoring subsurface changes over time with cross-well electromagnetic tomography, Geophys. Prospect., 43 (1995), 873–902.
12. C. Alves and H. Ammari, Boundary integral formulae for the reconstruction of imperfections of small diameter in an elastic medium, SIAM J. Appl. Math., 62 (2002), 94–106.
13. H. Ammari, M. Asch, and H. Kang, Boundary voltage perturbations caused by small conductivity inhomogeneities nearly touching the boundary, preprint, 2003.

14. H. Ammari, E. Beretta, and E. Francini, Reconstruction of thin conductivity imperfections, Applicable Anal., 83 (2004), 63–78.

15. H. Ammari, E. Iakovleva, and D. Lesselier, A MUSIC algorithm for locating small inclusions buried in a half-space from the scattering amplitude at a fixed frequency, preprint, 2004.

16. H. Ammari, E. Iakovleva, and S. Moskow, Recovery of small inhomogeneities from the scattering amplitude at a fixed frequency, SIAM J. Math. Anal., 34 (2003), 882–900.

17. H. Ammari and H. Kang, High-order terms in the asymptotic expansions of the steady-state voltage potentials in the presence of conductivity inhomogeneities of small diameter, SIAM J. Math. Anal., 34 (2003), 1152–1166.

18. ——————, Properties of generalized polarization tensors, Multiscale Modeling and Simulation: A SIAM Interdisciplinary Journal, 1 (2003), 335–348.

19. ——————, A new method for reconstructing electromagnetic inhomogeneities of small volume, Inverse Problems, 19 (2003), 63–71.

20. ——————, Boundary layer techniques for solving the Helmholtz equation in the presence of small inhomogeneities, to appear in J. Math. Anal. Appl. (2004).

21. H. Ammari, H. Kang, E. Kim, and M. Lim, Reconstruction of closely spaced small inclusions, to appear in SIAM J. Numer. Anal. (2004).

22. H. Ammari, H. Kang, and M. Lim, Polarization tensors and their applications, preprint, 2004.

23. H. Ammari, H. Kang, G. Nakamura, and K. Tanuma, Complete asymptotic expansions of solutions of the system of elastostatics in the presence of an inclusion of small diameter and detection of an inclusion, J. Elasticity, 67 (2002), 97–129.

24. H. Ammari, H. Kang, and K. Touibi, Boundary layer techniques for deriving the effective properties of composite materials, to appear in Asymp. Anal. (2004).

25. H. Ammari and A. Khelifi, Electromagnetic scattering by small dielectric inhomogeneities, J. Math. Pures Appl., 82 (2003), 749–842.

26. H. Ammari, O. Kwon, J.K. Seo, and E.J. Woo, Anomaly detection in T-scan trans-admittance imaging system, to appear in SIAM J. Appl. Math. (2004).

27. H. Ammari and S. Moskow, Asymptotic expansions for eigenvalues in the presence of small inhomogeneities, Math. Meth. Appl. Sci., 26 (2003), 67–75.

28. H. Ammari, S. Moskow, and M.S. Vogelius, Boundary integral formulas for the reconstruction of electromagnetic imperfections of small diameter, ESAIM: Cont. Opt. Calc. Var., 9 (2003), 49–66.

29. H. Ammari and J.K. Seo, An accurate formula for the reconstruction of conductivity inhomogeneities, Adv. Appl. Math., 30 (2003), 679–705.

30. H. Ammari and G. Uhlmann, Reconstruction of the Potential from Partial Cauchy Data for the Schrödinger Equation, Indiana Univ. Math. J., 53 (2004), 169–184.

31. H. Ammari, M.S. Vogelius, and D. Volkov, Asymptotic formulas for perturbations in the electromagnetic fields due to the presence of imperfections of small diameter II. The full Maxwell equations, J. Math. Pures Appl. 80 (2001), 769–814.

32. H. Ammari and D. Volkov, Correction of order three for the expansion of two dimensional electromagnetic fields perturbed by the presence of inhomogeneities of small diameter, J. Comput. Phys., 189 (2003), 371–389.

33. S. Andrieux, Fonctionnelles d'écart à la réciprocité généralisée et identification de fissure par les mesures surabondantes de surface, C. R. Acad. Sci., Paris I, 320 (1995), 1553–1559.

34. S. Andrieux and A. Ben Abda, Identification of planar cracks by complete overdetermined data: inversion formulae, Inverse Problems, 12 (1996), 553–564.

35. P.M. Anselone, *Collectively Compact Operator Approximation Theory and Applications to Integral Equations,* Prentice-Hall, Englewood Cliffs, NJ, 1971.

36. S.R. Arridge, Optical tomography in medical imaging, Inverse Problems, 15 (1999), R41–R93.

37. M. Assenheimer, O. Laver-Moskovitz, D. Malonek, D. Manor, U. Nahliel, R. Nitzan, and A. Saad, The T-scan technology: Electrical impedance as a diagnostic tool for breast cancer detection, Physiol. Meas., 22 (2001), 1–8.

38. A. El Badia and T. Ha-Duong, An inverse source problem in potential analysis, Inverse Problems, 16 (2000), 651–663.

39. G. Bal, Optical tomography of small volume absorbing inclusions, Inverse Problems, 19 (2003), 371–386.

40. C. Bandle, *Isoperimetric Inequalities and Applications*, Monogr. Stud. Math. 7, Pitman, Boston, MA, 1980.

41. G. Bao and P. Li, Inverse medium scattering for three-dimensional time harmonic Maxwell equations, Inverse Problems, 20 (2004), L1-L7.

42. G. Bao, F. Ma, and Y. Chen, An error estimate for recursive linearization of the inverse scattering problems, J. Math. Anal. Appl., 247 (2000), 255–271.

43. M.S. Baouendi, P. Ebenfelt, and L.P. Rothschild, *Real Submanifolds in Complex Space and Their Mappings*, Princeton University Press, Princeton, NJ, 1999.

44. L. Baratchart, J. Leblond, F. Mandréa, and E.B. Saff, How can the meromorphic approximation help to solve some 2D inverse problems for the Laplacian ?, Inverse Problems, 15 (1999), 79–90.

45. D.C. Barber and B.H. Brown, Applied potential tomography, J. Phys. Sci. Instrum., 17 (1984), 723–733.

46. B. Barcelo, E. Fabes, and J.K. Seo, The inverse conductivity problem with one measurement, uniqueness for convex polyhedra, Proc. Amer. Math. Soc., 122 (1994), 183–189.

47. H. Bellout and A. Friedman, Identification problems in potential theory, Arch. Rational Mech. Anal., 101 (1988), 143–160.

48. H. Bellout, A. Friedman, and V. Isakov, Stability for an inverse problem in potential theory, Trans. Amer. Math. Soc., 332 (1992), 271–296.

49. F. Ben Hassen and E. Bonnetier, Asymptotic formulas for the voltage potential in a composite medium containing close or touching disks of small diameter, preprint, 2003.

50. J. Bercoff, S. Chaffai, M. Tanter, L. Sandrin, S. Catheline, M. Fink, J.L. Gennisson, and M. Meunier, In vivo breast tumor detection using transient elastography, Ultrasound in Med. Bio., 29 (2003), 1387–1396.

51. E. Beretta and E. Francini, Asymptotic formulas for perturbations in the electromagnetic fields due to the presence of thin inhomogeneities in *Inverse Problems: Theory and Applications*, 49–63, Contemp. Math., 333, Amer. Math. Soc., Providence, RI, 2003.

52. E. Beretta, E. Francini, and M.S. Vogelius, Asymptotic formulas for steady state voltage potentials in the presence of thin inhomogeneities. A rigorous error analysis, J. Math. Pures Appl., 82 (2003), 1277–1301.

53. E. Beretta, A. Mukherjee, and M.S. Vogelius, Asymptotic formuli for steady state voltage potentials in the presence of conductivity imperfections of small area, Z. Angew. Math. Phys., 52 (2001), 543–572.

54. P.M. van den Berg and R.E. Kleinman, A total variation enhanced modified gradient algorithm for profile reconstruction, Inverse Problems, 11 (1995), L5–L10.

55. J. Bergh and J. Löfström, *Interpolation Spaces. An Introduction,* Grundlehren der Mathematischen Wissenschaften, 223, Springer-Verlag, Berlin-New York, 1976.

56. J. Blitz, *Electrical and Magnetic Methods of Nondestructive Testing,* IOP Publishing, Adam Hilger, 1991.

57. M. Bonnet, T. Burczynski, and M. Nowakowski, Sensitivity analysis for shape perturbation of cavity or internal crack using BIE and adjoint variable approach, Internat. J. Solids Structures, 39 (2002), 2365–2385.

58. L. Borcea, Electrical impedance tomography, Inverse Problems, 18 (2002), 99–136.

59. L. Borcea, J.G. Berrymann, and G.C. Papanicolaou, High contrast impedance tomography, Inverse Problems, 12 (1996), 1–24.

60. L. Borcea, J.G. Berrymann, and G.C. Papanicolaou, Matching pusruit for imaging high-contrast conductivity, Inverse Problems, 15 (1999), 811–849.

61. L. Borcea, G.C. Papanicolaou, C. Tsogka, and J.G. Berrymann, Imaging and time reversal in random media, Inverse Problems, 18 (2002), 1247–1279.

62. M. Brühl, Explicit characterization of inclusions in electrical impedance tomography, SIAM J. Math. Anal., 32 (2001), 1327–1341.

63. M. Brühl and M. Hanke, Numerical implementation of two noniterative methods for locating inclusions by impedance tomography, Inverse Problems, 16 (2000), 1029–1042.

64. M. Brühl, M. Hanke, and M.S. Vogelius, A direct impedance tomography algorithm for locating small inhomogeneities, Numer. Math., 93 (2003), 635–654.

65. K. Bryan, Numerical recovery of certain discontinuous electrical conductivities, Inverse Problems, 7 (1991), 827–840.

66. K. Bryan and M.S. Vogelius, A computational algorithm to determine crack locations from electrostatic boundary measurements. The case of multiple cracks, Int. J. Engng. Sci, 32 (1994), 579–603.

67. H.D. Bui, *Introduction aux Problèmes Inverses en Mécanique des Matériaux,* Eyrolles, Paris, 1992.

68. A.P. Calderón, Cauchy integrals on Lipschitz curves and related operators, Proc. Nat. Acad. Sci. U.S.A., 74 (1977), 1324–1327.

69. A.P. Calderón, On an inverse boundary value problem, Seminar on Numerical Analysis and its Applications to Continuum Physics, Soc. Brasileira de Matemática, Rio de Janeiro, 1980, 65–73.

70. Y. Capdeboscq and M.S. Vogelius, A general representation formula for the boundary voltage perturbations caused by internal conductivity inhomogeneities of low volume fraction, Math. Modelling Num. Anal., 37 (2003), 159–173.

71. ——————, Optimal asymptotic estimates for the volume of internal inhomogeneities in terms of multiple boundary measurements, Math. Modelling Num. Anal., 37 (2003), 227–240.

72. ——————, A review of some recent work on impedance imaging for inhomogeneities of low volume fraction, to appear in Proceedings of the Pan-American Advanced Studies Institute on PDEs, Inverse Problems and Nonlinear Analysis, January 2003. Contemporary Mathematics (2004).

73. D.J. Cedio-Fengya, S. Moskow, and M.S. Vogelius, Identification of conductivity imperfections of small diameter by boundary measurements: Continuous dependence and computational reconstruction, Inverse Problems, 14 (1998), 553–595.

74. D.H. Chambers and J.G. Berryman, Time-reversal analysis for scatterer characterization, Phys. Rev. Lett., 92 (2004), 023902-1–023902-4.

75. Y. Chen and V. Rokhlin, On the inverse scattering problem for the Helmholtz equation in one dimension, Inverse Problems, 8 (1992), 365–391.

76. Y. Chen, Inverse scattering via Heisenberg's Uncertainty Principle, Inverse Problems, 13 (1997), 253–282.

77. M. Cheney, The linear sampling method and the MUSIC algorithm, Inverse Problems, 17 (2001), 591–595.

78. M. Cheney, D. Isaacson, and J.C. Newell, Electrical impedance tomography, SIAM Rev., 41 (1999), 85–101.

79. M. Cheney and D. Isaacson, Distinguishability in impedance imaging, IEEE Trans. Biomed. Engr., 39 (1992), 852–860.

80. M. Cheney, D. Isaacson, J.C. Newell, S. Simske, and J. Goble, NOSER: an algorithm for solving the inverse conductivity problem, Int. J. Imag. Syst. Technol., 22 (1990), 66–75.

81. H. Cheng and L. Greengard, A method of images for the evaluation of electrostatic fields in systems of closely spaced conducting cylinders, SIAM J. Appl. Math., 58 (1998), 122–141.

82. V.A. Cherepenin, A. Karpov, A. Korjenevsky, V. Kornienko, A. Mazaletskaya, D. Mazourov, and D. Meister, A 3D electrical impedance tomography (EIT) system for breast cancer detection, Physiol. Meas., 22 (2001), 9–18.

83. V.A. Cherepenin, A. Y. Karpov, A. V. Korjenevsky, V. N. Kornienko, Y. S. Kultiasov, M. B. Ochapkin, O. V. Trochanova, and J. D. Meister, Three-dimensional EIT imaging of breast tissues: system design and clinical testing, IEEE Trans. Med. Imag., 21 (2002), 662–667.

84. T.C. Choy, Effective Medium Theory. Principles and Applications, International Series of Monographs on Physics, 102, Oxford Science Publications, New York, 1999.

85. P.G. Ciarlet, Mathematical Elasticity, Vol. I, Norh-Holland, Amsterdam (1988).

86. C. Cohen-Bacrie and R. Guardo, Regularized reconstruction in electrical impedance tomography using a variance uniformization constraint, IEEE Trans. Med. Imag., 16 (1997), 562–571.

87. R.R. Coifman, A. McIntosh, and Y. Meyer, L'intégrale de Cauchy définit un opérateur bourné sur L^2 pour les courbes lipschitziennes, Ann. Math., 116 (1982), 361–387.

88. R.R. Coifman, M. Goldberg, T. Hrycak, M. Israel, and V. Rokhlin, An improved operator expansion algorithm for direct and inverse scattering computations, Waves Random Media, 9 (1999), 441–457.

89. D. Colton and A. Kirsch, A simple method for solving inverse scattering problems in the resonance region, Inverse Problems, 12 (1996), 383–393.

90. D. Colton and R. Kress, *Integral Equation Methods in Scattering Theory*, John Wiley, New York, 1983.

91. D. Colton and R. Kress, *Inverse Acoustic and Electromagnetic Scattering Theory*, Applied Math. Sciences 93, Springer-Verlag, New York, 1992.

92. R.D. Cook, G.J. Saulnier, D.G. Gisser, J.C. Goble, J.C. Newell, and D. Isaacson, ACT3: a high-speed, high-precision electrical impedance tomography, IEEE Trans. Biomed. Engr., 41 (1994), 713–722.

93. B.E. Dahlberg, C.E. Kenig, and G. Verchota, Boundary value problem for the systems of elastostatics in Lipschitz domains, Duke Math. Jour., 57 (1988), 795–818.

94. G. Dassios, Low-frequency moments in inverse scattering theory, J. Math. Phys., 31 (1990), 1691–1692.

95. G. Dassios and R.E. Kleinman, On Kelvin inversion and low-frequency scattering, SIAM Rev., 31 (1989), 565–585.

96. I. Daubechies, *Ten Lectures on Wavelets*, SIAM, Philadelphia, 1992.

97. G. David and J.-L. Journé, A boundedness criterion for generalized Calderón-Zygmund operators, Ann. of Math., 120 (1984), 371–397.

98. L. Desbat and A.G. Ramm, Finding small objects from tomographic data, Inverse Problems, 13 (1997), 1239–1246.

99. A.J. Devaney, Super-resolution processing of multi-static data using reversal and MUSIC, to appear in J. Acoust. Soc. Am. (2003).

100. D.C. Dobson and F. Santosa, An image-enhancement technique for electrical impedance tomography, Inverse Problems, 10 (1994), 317–334.

101. D.C. Dobson and F. Santosa, Resolution and stability analysis of an inverse problem in electrical impedance tomography: dependence of the input current patterns, SIAM J. Appl. Math., 54 (1994), 1542–1560.

102. O. Dorn, A transport-backtransport method for optical tomography, Inverse Problems, 14 (1998), 1107–1130.

103. O. Dorn, H. Bertete-Aguirre, J.G. Berryman, and G.C. Papanicolaou, A nonlinear inversion method for 3D electromagnetic imaging using adjoint fields, Inverse Problems, 15 (1999), 1523–1558.

104. J.F. Douglas and A. Friedman, Coping with complex boundaries, IMA Series on Mathematics and its Applications Vol. 67, 166–185, Springer, New York, 1995.

105. J.F. Douglas and E.J. Garboczi, Intrinsic viscosity and polarizability of particles having a wide range of shapes, Adv. Chem. Phys., 91 (1995), 85–153.

106. V. Druskin, The unique solution of the inverse problem of electrical surveying and electrical well-logging for piecewise-continuous conductivity, Izvestiya, Earth Physics, 18 (1982), 51–53.

107. B. Duchêne, M. Lambert, and D. Lesselier, On the characterization of objects in shallow water using rigorous inversion methods, in *Inverse Problems in Underwater Acoustics*, 127–147, edited by M.I. Taroudakis and G.N. Makrakis, Springer-Verlag, New York, 2001.

108. P. Edic, D. Isaacson, G. Saulnier, H. Jain, and J.C. Newell, An iterative Newton-Raphson method to solve the inverse conductivity problem, IEEE Trans. Biomed. Engr., 45 (1998), 899–908.

109. M.R. Eggleston, R.J. Schwabe, D. Isaacson, and L.F. Coffin, The application of electric current computed tomography to defect imaging in metals, in *Review of Progress in Quantitative NDE*, D.O. Thompson and D.E. Chimenti, eds., Plenum, New York, 1989.

110. L. Escauriaza, E.B. Fabes, and G. Verchota, On a regularity theorem for weak solutions to transmission problems with internal Lipschitz boundaries, Proc. Amer. Math. Soc., 115 (1992), 1069–1076.

111. L. Escauriaza and J.K. Seo, Regularity properties of solutions to transmission problems, Trans. Amer. Math. Soc., 338 (1) (1993), 405–430.

112. G. Eskin and J. Ralston, The inverse backscattering in three dimensions, Comm. Math. Phys., 124 (1989), 169–215.

113. G. Eskin and J. Ralston, Inverse backscattering in two dimensions, Comm. Math. Phys., 138 (1991), 451–486.

114. G. Eskin and J. Ralston, Inverse backscattering, J. Anal. Math., 58 (1992), 177–190.

115. E.B. Fabes, M. Jodeit, and N.M. Riviére, Potential techniques for boundary value problems on $\mathcal{C}^1$ domains, Acta Math., 141 (1978), 165–186.

116. E. Fabes, H. Kang, and J.K. Seo, Inverse conductivity problem with one measurement: Error estimates and approximate identification for perturbed disks, SIAM J. Math. Anal., 30 (1999), 699–720.

117. E. Fabes, M. Sand, and J.K. Seo, The spectral radius of the classical layer potentials on convex domains, The IMA volumes in Mathematics and its Applications, 42 (1992), 129–137.

118. S.N. Fata, B.B. Guzina, and M. Bonnet, Computational framework for the BIE solution to inverse scattering problems in elastodynamics, Comput. Mechanic., 32 (2003), 370–380.

119. G.B. Folland, *Introduction to Partial Differential Equations*, Princeton University Press, Princeton, NJ, 1976.

120. A. Friedman, Detection of mines by electric measurements, SIAM J. Appl. Math., 47 (1987), 201–212.

121. A. Friedman and B. Gustafsson, Identification of the conductivity coefficient in an elliptic equation, SIAM J. Math. Anal., 18 (1987), 777–787.

122. A. Friedman and V. Isakov, On the uniqueness in the inverse conductivity problem with one measurement, Indiana Univ. Math. J., 38 (1989), 553–580.

123. A. Friedman and M.S. Vogelius, Identification of small inhomogeneities of extreme conductivity by boundary measurements: a theorem on continuous dependence, Arch. Rat. Mech. Anal., 105 (1989), 299–326.

124. L.F. Fuks, M. Cheney, D. Isaacson, D.G. Gisser, and J.C. Newell, Detection and imaging of electric conductivity and permittivity at low frequencies, IEEE Trans. Biomed. Engr., 3 (1991), 1106–1110.

125. N. Garofalo and F. Lin, Monotonicity properties of variational integrals, A_p weights and unique continuation, Indiana Univ. Math. J., 35 (1986), 245–268.

126. S. Garreau, Ph. Guillaume, and M. Masmoudi, The topological asymptotic for PDE systems: the elasticity case, SIAM J. Control Optim., 39 (2001), 1756–1778.

127. F. Gesztesy and A.G. Ramm, An inverse problem for point inhomogeneities, Meth. Funct. Anal. Topology, 6 (2000), 1–12.

128. D. Gilbarg and N.S. Trudinger, *Elliptic Partial Differential Equations of Second Order*, Grundlehren der Mathematischen Wissenschaften, 224, Springer-Verlag, Berlin-New York, 1977.

129. D. Gisser, D. Isaacson, and J.C. Newell, Electric current tomography and eigenvalues, SIAM J. Appl. Math., 50 (1990), 1623–1634.

130. Yu.A. Gryazin, M.V. Klibanov, and T.R. Lucas, Two numerical methods for an inverse problem for the 2-D Helmholtz equation, J. Comput. Phys., 184 (2003), 122-148.

131. Yu.A. Gryazin, M.V. Klibanov, and T.R. Lucas, Numerical solution of a subsurface imaging inverse problem, SIAM J. Appl. Math., 62 (2001), 664–683.

132. Ph. Guillaume and K. Sid Idris, The topological asymptotic expansion for the Dirichlet problem, SIAM J. Control Optim., 41 (2003), 1042–1072.

133. S. Gutman and M.V. Klibanov, Three-dimensional inhomogeneous media imaging, Inverse Problems, 10 (1994), L39–L46.

134. S. Gutman and M.V. Klibanov, Iterative method for multi-dimensional inverse scattering problems at fixed frequencies, Inverse Problems, 10 (1994), 573–599.

135. T.M. Habashy, R.W. Groom, and B.R. Spies, Beyond the Born and Rytov approximations: a nonlinear approach to electromagnetic scattering, J. Geophys. Res., 98 (1993), 1759–1775.

136. Q. Han and F. Lin, *Elliptic Partial Differential Equations*, Courant Lecture Notes in Mathematics, 1, New York University, Courant Institute of Mathematical Sciences, New York, Amer. Math. Soc., Providence, RI, 1997.

137. L. Han, J.A. Noble, and M. Burcher, A novel ultrasound identification system for measuring biomechanical properties of in vivo soft issue, Ultrasound in Med. Biol., 29 (2003), 813–823.

138. P. Hähner, An inverse problem in electrostatics, Inverse Problems, 15 (1999), 961–975.

139. Z. Hashin and S. Shtrickman, A variational approach to the theory of effective magnetic permeability of multiphase materials, J. Appl. Phys., 33 (1962), 3125–3131.

140. S. He and V.G. Romanov, Identification of small flaws in conductors using magnetostatic measurements, Math. Comput. Simul., 50 (1999), 457–471.

141. G.C. Herman, Transmission of elastic waves through solids containing smallscale heterogeneities, Geophys. J. Int., 145 (2001), 436–446.

142. B. Hofmann, Approximation of the inverse electrical impedance tomography by an inverse transmission problem, Inverse Problems, 14 (1998), 1171–1187.

143. D. Holder, *Clinical and Physiological Applications of Electrical Impedance Tomography*, UCL Press, London, 1993.

144. S.C. Hsieh and T. Mura, Nondestructive cavity identification in structures, Internat. J. Solids Structures, 30 (1993), 1579–1587.

145. A.L. Hyaric and M.K. Pidcock, An image reconstruction algorithm for three-dimensional electrical impedance tomography, IEEE Trans. Biomed. Engr., 48 (2001), 230–235.

146. E. Iakovleva, Inverse Scattering from Small Inhomogeneities, Ph.D. Thesis, Ecole Polytechnique, 2004.

147. M. Ikehata, Enclosing a polygonal cavity in a two-dimensional bounded domain from Cauchy data, Inverse Problems, 15 (1999), 1231–1241.

148. —————, Reconstruction of the support function for inclusion from boundary measurements, J. Inverse Ill-Posed Probl., 8 (2000), 367–378.

149. —————, On reconstruction in the inverse conductivity problem with one measurement, Inverse Problems, 16 (2000), 785–793.

150. —————, Reconstruction of inclusion from boundary measurements, J. Inverse Ill-Posed Probl., 10 (2002), 37–65.

151. M. Ikehata and T. Ohe, A numerical method for finding the convex hull of polygonal cavities using the enclosure method, Inverse Problems, 18 (2002), 111–124.

152. M. Ikehata and S. Siltanen, Numerical method for finding the convex hull of an inclusion in conductivity from boundary measurements, Inverse Problems, 16 (2000), 1043–1052.

153. D. Isaacson, Distinguishability of conductivities by electric current computed tomography, IEEE Trans. Medical Imag., 5 (1986), 91–95.

154. D. Isaacson and M. Cheney, Effects of measurements precision and finite numbers of electrodes on linear impedance imaging algorithms, SIAM J. Appl. Math., 51 (1991), 1705–1731.

155. D. Isaacson and E.L. Isaacson, Comments on Calderón's paper: "On an inverse boundary value problem", Math. Compt., 52 (1989), 553–559.

156. V. Isakov, On uniqueness of recovery of a discontinuous conductivity coefficient, Comm. Pure Appl. Math., 41 (1988), 865–877.

157. ——————, Inverse Source Problems, Math. Surveys and Monograph Series Vol. 34, AMS, Providence, RI, 1990.

158. ——————, Inverse Problems for Partial Differential Equations, Springer-Verlag, New York, 1998.

159. V. Isakov and J. Powell, On the inverse conductivity problem with one measurement, Inverse Problems, 6 (1990), 311–318.

160. V. Isakov and A. Sever, Numerical implementation of an integral equation method for the inverse conductivity problem, Inverse Problems, 12 (1996), 939–953.

161. V. Isakov and S.F. Wu, On theory and application of the Helmholtz equation least squares method in inverse acoustics, Inverse Problems, 18 (2002), 1147–1159.

162. D.S. Jerison and C. Kenig, The Neumann problem in Lipschitz domains, Bull. Amer. Math. Soc., 4 (1981), 203–207.

163. J. Jossinet, E. Marry, and A. Montalibet, Electrical impedance endotomography: imaging tissue from inside, IEEE Trans. Medical Imag., 21 (2002), 560–565.

164. J. Jossinet, E. Marry, and A. Matias, Electrical impedance endo-tomography, Phys. Med. Biol., 47 (2002), 2189–2202.

165. J. Jossinet and M. Schmitt, A review of parameters for the bioelectrical characterization of breast tissue, Ann. New York Academy of Sci., 873 (1999), 30–41.

166. H. Kang, E. Kim, and J. Lee, Identification of Elastic Inclusions and Elastic Moment Tensors by Boundary Measurements, Inverse Problems, 19 (2003), 703–724.

167. H. Kang, E. Kim, and K. Kim, Anisotropic polarization tensors and determination of an anisotropic inclusion, SIAM J. Appl. Math., 65 (2003), 1276–1291.

168. H. Kang and J.K. Seo, Layer potential technique for the inverse conductivity problem, Inverse Problems, 12 (1996), 267–278.

169. ——————, Identification of domains with near-extreme conductivity: Global stability and error estimates, Inverse Problems, 15 (1999), 851–867.

170. ——————, Inverse conductivity problem with one measurement: Uniqueness of balls in R^3, SIAM J. Appl. Math., 59 (1999), 1533–1539.

232   References

171. —————, Recent progress in the inverse conductivity problem with single measurement, in Inverse Problems and Related Fields, CRC Press, Boca Raton, FL, 2000, 69–80.

172. H. Kang, J.K. Seo, and D. Sheen, The inverse conductivity problem with one measurement: stability and estimation of size, SIAM J. Math. Anal., 28 (1997), 1389–1405.

173. T. Kato, *Perturbation Theory for Linear Operators*, Springer-Verlag, 1980.

174. J.B. Keller, Removing small features from computational domains, J. Comput. Phys., 113 (1994), 148–150.

175. O.D. Kellogg, *Foundations of Potential Theory*, Dover, New York, 1953.

176. A. Kirsch, *An Introduction to the Mathematical Theory of Inverse Problems*, Applied Mathematical Sciences 120, Springer-Verlag, New York, 1996.

177. A. Kirsch, Characterization of the shape of the scattering obstacle using the spectral data of the far field operator, Inverse Problems, 14 (1998), 1489–1512.

178. A. Kirsch, The MUSIC algorithm and the factorization method in inverse scattering theory for inhomogeneous media, Inverse Problems, 18 (2002), 1025–1040.

179. R.E. Kleinman and P.M. van den Berg, A modified gradient method for two-dimensional problems in tomography, J. Comput. Appl. Math., 42 (1992), 17–35.

180. R.E. Kleinman and T.B.A. Senior, Rayleigh scattering in *Low and High Frequency Asymptotics*, 1–70, edited by V.K. Varadan and V.V. Varadan, North-Holland, 1986.

181. R.V. Kohn and A. McKenny, Numerical implementation of a variational method for electrical impedance tomography, Inverse Problems, 6 (1990), 389–414.

182. R.V. Kohn and M.S. Vogelius, Determining conductivity by boundary measurements, Comm. Pure Appl. Math., 37 (1984), 289–298.

183. —————, Determining conductivity by boundary measurements, interior results, II, Comm. Pure Appl. Math., 38 (1985), 643–667.

184. —————, Relaxation of a variational method for impedance computed tomography, omm. Pure Appl. Math., 40 (1987), 745–777.

185. S.M. Kozlov, Geometric aspects of averaging, Usp. Mat. Nauk., 44 (1989), 79–120.

186. S.M. Kozlov, On the domain of variations of added masses, polarization and effective characteristics of composites, J. Appl. Math. Mech., 56 (1992), 102–107.

187. R. Kress, On the low wave number asymptotics for the two-dimensional exterior problem for the reduced wave equation, Math. Meth. Appl. Sci., 9 (1987), 335–341.

188. —————, *Linear Integral Equations*. Second edition. Applied Mathematical Sciences, 82. Springer-Verlag, New York, 1999.

189. V.D. Kupradze, *Potential Methods in the Theory of Elasticity*, Jerusalem, 1965.

190. O. Kwon and J.K. Seo, Total size estimation and identification of multiple anomalies in the inverse electrical impedance tomography, Inverse Problems, 17 (2001), 59–75.

191. O. Kwon, J.K. Seo, and J.R. Yoon, A real-time algorithm for the location search of discontinuous conductivities with one measurement, Comm. Pure Appl. Math., 55 (2002), 1–29.

192. O. Kwon, J.R. Yoon, J.K. Seo, E.J. Woo, and Y.G. Cho, Estimation of anomaly location and size using impedance tomography, IEEE Trans. Biomed. Engr., 50 (2003), 89–96.

193. S.K. Lehman and A.J. Devaney, Transmission mode time-reversal super-resolution imaging, J. Acoust. Soc. Am., 113 (2003), 2742–2753.

194. D. Lesnic, A numerical investigation of the inverse potential conductivity problem in a circular inclusion, Inverse Probl. Engr., 9 (2001), 1–17.

195. D. Lesselier and B. Duchêne, Buried two-dimensional penetrable objects illuminated by line sources: FFT-based iterative computations of the anomalous fields, in *Application of Conjugate Gradient Methods to Electromagnetics and Signal Analysis*, 400–438, edited by T.K. Sarkar, Elsevier, New York, 1991.

196. T. Lewiński and Sokolowski, Energy change due to the appearance of cavities in elastic solids, International J. Solids Structures, 40 (2003), 1765–1803.

197. M. Lim, Reconstruction of Inhomogeneities via Boundary Measurements, Ph.D. thesis, Seoul National University, 2003.

198. R. Lipton, Inequalities for electric and elastic polarization tensors with applications to random composites, J. Mech. Phys. Solids, 41 (1993), 809–833.

199. K.A. Lurie and A.V. Cherkayev, Exact estimates of conductivity of composites formed by two isotropically conducting media taken in prescribed proportion, Proc. Roy. Soc. Edinburgh, 99 A (1984), 71–87.

200. M.L. Mansfield, J.F. Douglas, and E.J. Garboczi, Intrinsic viscosity and electrical polarizability of arbitrary shaped objects, Physical Review E, 64 (2001), 061401.

201. T.D. Mast, A. Nachman, and R.C. Waag, Focusing and imagining using eigenfunctions of the scattering operator, J. Acoust. Soc. Am., 102 (1997), 715–725.

202. V.G. Maz'ya and S.A. Nazarow, The asymptotic behavior of energy integrals under small perturbations of the boundary near corner points and conical points (in Russian). Trudy Moskovsk. Matem. Obshch. Vol. 50, English Translation: Trans. Moscow Math. Soc. (1988), 77–127.

203. R.C. McPhedran and A.B. Movchan, The Rayleigh multipole method for linear elasticity, J. Mech. Phys. Solids, 42 (1994), 711–727.

204. K. Miller, Stabilized numerical analytic prolongation with poles, SIAM J. Appl. Math., 18 (1970), 346–363.

205. G.W. Milton, On characterizing the set of possible effective tensors of composites: the variational methods and the translation methods, Commun. Pure Appl. Math., 43 (1990), 63–125.

206. G.W. Milton, *The Theory of Composites*, Cambridge Monographs on Applied and Computational Mathematics, Cambridge University Press, 2001.

207. D. Mitrea and M. Mitrea, Uniqueness for inverse conductivity and transmission problems in the class of Lipschitz domains, Commun. Part. Diff. Eqns, 23 (1998), 1419–1448.

208. C.B. Morrey, *Multiple Integrals in the Calculus of Variations*, Springer-Verlag, New York, 1966.

209. A.B. Movchan, Integral characteristics of elastic inclusions and cavities in the two-dimensional theory of elasticity, European J. Appl. Math., 3 (1992), 21–30.

210. A.B. Movchan and N.V. Movchan, *Mathematical Modelling of Solids with Nonregular Boundaries*, CRC Press, Boca Raton, 1995.

211. A.B. Movchan, N.V. Movchan, and C.G. Poulton, *Asymptotic Models of Fields in Dilute and Densely Packed Composites*, Imperial College Press, London, 2002.

212. A.B. Movchan and S.K. Serkov, The Pólya-Szegö matrices in asymptotic models of dilute composite, Euro. J. Appl. Math., 8 (1997), 595–621.

213. J. Mueller, D. Isaacson, and J. Newell, A reconstruction algorithm for electrical impedance tomography data collected on rectangular electrode arrays, IEEE Trans. Biomed. Engr., 46 (1999), 1379–1386.

214. T. Mura and T. Koya, *Variational Methods in Mechanics*, The Clarendon Press, Oxford University Press, New York, 1992.

215. F. Murat and L. Tartar, Optimality conditions and homogenization, Research Notes in Mathematics, 127, 1–8, Pitman, London, 1985.

216. N.I. Muskhelishvili, *Some Basic Problems of the Mathematical Theory of Elasticity*, English translation, Noordhoff International Publishing, Leyden, 1977.

217. A. Nachmann, Reconstructions from boundary measurements, Ann. of Math., 128 (1988), 531–587.

218. —————, Global uniqueness for a two-dimensional inverse boundary value problem, Ann. Math., 142 (1996), 71–96.

219. G. Nakamura and G. Uhlmann, Identification of Lamé parameters by boundary observations, American J. Math., 115 (1993), 1161–1187.

220. F. Natterer and F. Wübbeling, A propagation backpropagation method for ultrasound tomography, Inverse Problems, 11 (1995), 1225–1232.

221. F. Natterer and F. Wübbeling, *Mathematical Methods in Image Reconstruction*, SIAM Monographs on Mathematical Modeling and Computation, SIAM, Philadelphia, 2001.

222. S.A. Nazarov and J. Sokolowski, Asymptotic analysis of shape functionals, J. Math. Pures Appl., 82 (2003), 125–196.

223. J. Nečas, *Les Méthodes Directes en Théorie des Équations Elliptiques*, Academia, Prague, 1967.

224. J.C. Nédélec, *Acoustic and Electromagnetic Equations. Integral Representations for Harmonic Problems,* Springer-Verlag, New-York, 2001.

225. T. Ohe and K. Ohnaka, A precise estimation method for locations in an inverse logarithmic potential for point mass models, Appl. Math. Modelling, 18 (1994), 446–452.

226. —————, Determination of locations of point-like masses in an inverse source problem of the Poisson equation, J. Comput. Appl. Math., 54 (1994), 251–261.

227. S. Ozawa, Singular variation of domains and eigenvalues of the Laplacian, Duke Math. J., 48 (1981), 767–778.

228. —————, Spectra of domains with small spherical Neumann boundary, J. Fac. Sci. Univ. Tokyo, Sect IA, 30 (1983), 259–277.

229. L. Payne, Isoperimetric inequalities and their applications, SIAM Rev., 9 (1967), 453–488.

230. L. Payne and G.A. Philippin, Isoperimetric inequalities for polarization and virtual mass, J. Anal. Math., 47 (1986), 255–267.

231. L. Payne and H. Weinberger, New bounds in harmonic and biharmonic problems, J. Math. Phys., 33 (1954), 291–307.

232. G. Pólya and G. Szegö, *Isoperimetric Inequalities in Mathematical Physics*, Annals of Mathematical Studies Number 27, Princeton University Press, Princeton, NJ, 1951.

233. C. Prada and M. Fink, Eigenmodes of the time reversal operator: a solution to selective focusing in multiple-target media, Wave Motion, 20 (1994), 151–163.

234. C. Prada, J.L. Thomas, and M. Fink, The iterative time reversal process: analysis of the convergence, J. Acoust. Soc. Am., 97 (1995), 62–71.

235. C. Prada, S. Manneville, D. Spolianski, and M. Fink, Decomposition of the time reversal operator: detection and selective focusing on two scatterers, J. Acoust. Soc. Am., 99 (1996), 2067–2076.

236. A.G. Ramm, Finding small inhomogeneities from surface scattering data, J. Inverse Ill-Posed Problems, 8 (2000), 205–210.

237. F. Rellich, Darstelling der eigenwerte von $\Delta u = \lambda u$ durch ein randintegral, Math Z., 46 (1940), 635–646.

238. P.C. Sabatier, Past and future of inverse problems, J. Math. Phys., 41 (2000), 4082–4124.

239. B. Samet, S. Amstutz, and M. Masmoudi, The topological asymptotic for the Helmholtz equation, SIAM J. Control Optim., 42 (2004), 1523–1544.

240. F. Santosa and M.S. Vogelius, A backprojection algorithm for electrical impedance imaging, SIAM J. Appl. Math., 50 (1990), 216–243.

241. M. Schiffer and G. Szegö, Virtual mass and polarization, Trans. Amer. Math. Soc., 67 (1949), 130–205.

242. J.K. Seo, A uniqueness result on inverse conductivity problem with two measurements, J. Fourier Anal. Appl., 2 (1996), 227–235.

243. J.K. Seo, O. Kwon, H. Ammari, and E.J. Woo, Mathematical framework and anomaly estimation algorithm for breast cancer detection using TS2000 configuration, to appear in IEEE Trans. Biomedical Engineering (2004).

244. S. Siltanen, J. Mueller, and D. Isaacson, An implementation of the reconstruction algorithm of A. Nachman for the 2D inverse conductivity problem, Inverse Problems, 16 (2000), 681–699.

245. J.E. Silva, J.P. Marques, and J. Jossinet, Classification of breast tissue by electrical impedance spectroscopy, Med. Biol. Eng. Comput., 38 (2000), 26–30.

246. E. Somersalo, M. Cheney, and D. Isaacson, Existence and uniqueness for electrode models for electric current computed tomography, SIAM J. Appl. Math., 52 (1992), 1023–1040.

247. E. Somersalo, M. Cheney, D. Isaacson, and E. Isaacson, Layer-stripping: a direct numerical method for impedance imaging, Inverse Problems, 7 (1991), 899–926.

248. J. Sylvester and G. Uhlmann, A global uniqueness theorem for an inverse boundary value problem, Ann. Math., 125 (1987), 153–169.

249. ——————, The Dirichlet to Neumann map and applications, Inverse Problems in Partial Differential Equations, SIAM, Philadelphia (1990), 197–221.

250. M.I. Taroudakis and G.N. Makrakis, editors, *Inverse Problems in Underwater Acoustics*, Springer-Verlag, New York, 2001.

251. C.W. Therrien, *Discrete Random Signals and Statistical Signal Processing*, Englewood Cliffs, NJ, Prentice-Hall, 1992.

252. A. Timonov and M.V. Klibanov, An efficient algorithm for solving the inverse problem of locating the interfaces using the frequency sounding data, J. Comput. Phys., 183 (2002), 422–437.

253. C.F. Tolmasky and A. Wiegmann, Recovery of small perturbations of an interface for an elliptic inverse problem via linearization, Inverse Problems, 15 (1999), 465–487.

254. R. Torres and G. Welland, The Helmholtz equation and transmission problems with Lipschitz interfaces, Indiana Univ. Math. J., 42 (1993), 1457–1485.

255. G. Uhlmann, Inverse boundary value problems for partial differential equations, Proceedings of the International Congress of Mathematicians, Berlin (1998), Documenta Mathematica Vol. III, 77–86.

256. ——————, Developments in inverse problems since Calderón's foundational paper, Chapter 19 in *Harmonic Analysis and Partial Differential Equations*, 295–345, edited by M. Christ, C. Kenig, and C. Sadosky, University of Chicago Press, 1999.

257. M. Vauhkonen, D. Vadasz, P.A. Karjalainen, E. Somersalo, and J.P. Kaipio, Tikhonov regularization and prior information in electrical impedance tomography, IEEE Trans. Med. Imag., 17 (1998), 285–293.

258. G.C. Verchota, Layer potentials and boundary value problems for Laplace's equation in Lipschitz domains, J. Funct. Anal., 59 (1984), 572–611.

259. M.S. Vogelius and D. Volkov, Asymptotic formulas for perturbations in the electromagnetic fields due to the presence of inhomogeneities, Math. Model. Numer. Anal., 34 (2000), 723–748.

260. D. Volkov, An Inverse Problem for the Time Harmonic Maxwell Equations, Ph.D. thesis, Rutgers University, New Brunswick, NJ, 2001.

261. ——————, Numerical methods for locating small dielectric inhomogeneities, Wave Motion, 38 (2003), 189–206.

262. S.H. Ward and G.W. Hohmann, Electromagnetic theory for geophysical applications, in *Electromagnetic Methods in Applied Geophysics–Theory*, vol. 1, 131–311, edited by M.N. Nabighian, Tulsa, 1987.

263. E.J. Woo, P. Hua, J.G. Webster, and W.J. Tompkins, Measuring lung resistivity using electrical impedance tomography, IEEE Trans. Biomed. Engr., 39 (1992), 756–760.

264. E.J. Woo, J.G. Webster, and W.J. Tompkins, A robust image reconstruction algorithm and its parallel implementation in electrical impedance tomography, IEEE Trans. Med. Imag., 12 (1993), 137–146.

265. K. Yamatani, T. Ohe and K. Ohnaka, An identification method of electric current dipoles in spherically symmetric conductor, J. Comput. Appl. Math., 143 (2002), 189–200.

266. T. Yorkey, J. Webster, and W. Tompkins, Comparing reconstruction algorithms for electrical impedance tomography, IEEE Trans. Biomed. Engr., 34 (1987), 843–852.

Index

Printing and Binding: Strauss GmbH, Mörlenbach